COMPSTAT

Proceedings in
Computational Statistics

9th Symposium held
at Dubrovnik,
Yugoslavia, 1990

Edited by
K. Momirović and V. Mildner

With 29 Figures

Physica-Verlag Heidelberg

Professor Dr. Konstantin Momirović
Vesna Mildner, MA
University of Zagreb
Engelsova bb
41000 Zagreb/Yugoslavia

ISBN 978-3-7908-0475-1 ISBN 978-3-642-50096-1 (eBook)
DOI 10.1007/978-3-642-50096-1

Cover: Erich Kirchner, Heidelberg

Preface

Although no-one is, probably, too enthused about the idea, it is a fact that the development of most empirical sciences to a great extent depends on the development of data analysis methods and techniques, which, due to the necessity of application of computers for that purpose, actually means that it practically depends on the advancement and orientation of computer statistics.

Every other year the International Association for Statistical Computing sponsors the organizition of meetings of individuals professionally involved in computational statistics. Since these meetings attract professionals from all over the world, they are a good sample for the estimation of trends in this area which some believe is a statistics proper while others claim it is computer science. It seems, though, that an increasing number of colleagues treat it as an independent scientific or at least technical discipline.

This volume contains six invited papers, 41 contributed papers and, finally, two papers which are, formally, software descriptions, but it was agreed by the Program Committee that they should be included in a separate section entitled "Notes about new developments in statistical software", due to their special significance for current trends in computational statistics. All contributions are grouped into ten topics:

- *Classification*
- *Algorithms and statistical software*
- *Expert systems in statistics*
- *Multivariate data analysis and model building*
- *Optimization techniques and nonlinear models*
- *Computing for robust statistics*
- *Statistics and database management*
- *Time dependent models*
- *Analysis of spatial data*
- *Computational inference*

The structure and content of these papers, together with the 72 similarly grouped short communications issued separately by the Organizer, are a good indicator of the current interests in computational statistics. As in the past years, this, 9th, COMPSTAT is dominated by Multivariate data analysis, Expert systems in statistics and Algorithms and statistical software.

The Program Committee, which consisted of J. Antoch, Y. Dodge, D. Edwards, D. Hand, I.W. Molenaar, K. Momirović (Chairman) and P. Naeve had a difficult task of selecting among many very good papers those which could be accepted and published for their quality and topicality, as well as because of their relevance for computational statistics. To them the editors wish to express their deepest gratitude for their efforts and the job well done.

Konstantin Momirović
Vesna Mildner

Zagreb
June 1990

Contents

Contributors

Addresses of authors will be found at the end of this volume

Classification

Stochastic Algorithms for Clustering

G. Celeux and G. Govaert, Le Chesnay

1. Introduction

It is well known that, generally, the solution provided by a partitioning algorithm depends upon its initial position. In this paper, we consider two algorithms which incorporate random perturbations to reduce the initial-position dependence. Both appear to be variations of a general Classification EM algorithm (CEM), conceived to optimizing Classification Maximum Likelihood (CML) criteria in the mixture context. In Section 2, we present the CEM algorithm and we give its main characteristics. In Section 3, we present a Stochastic version (SEM) of the CEM algorithm and a simulated annealing algorithm for clustering (CAEM) conceived in the same framework. Both algorithms can be performed to optimize most of clustering criteria, but here we focus on the variance criterion. In Section 4, we summarize the conclusions of numerical experiments performed to analyze the practical behaviour of both algorithms to optimizing the variance criterion.

2. The CEM algorithm

A probabilistic clustering model (see, for instance, Bock 1989) is to suppose that the objects to be classified, characterized by $\mathbf{R}^d$-valued vectors $(\mathbf{x}_1, ..., \mathbf{x}_n)$, are a sample from a mixture of densities

$$f(\mathbf{x}) = \sum_{k=1}^{K} p_k \, f(\mathbf{x},\mathbf{a}_k) \tag{2.1}$$

where the p_k's are the mixing weights and the $f(\mathbf{x},\mathbf{a}_k)$'s are densities from the same parametric family. The CML criterion associated to this model is

$$C_1(P,\mathbf{a}) = \sum_{k=1}^{K} \log L(P_k,\mathbf{a}_k) \tag{2.2}$$

where $P = (P_1, ..., P_K)$ is a partition of $(\mathbf{x}_1, ..., \mathbf{x}_n)$, $\mathbf{a} = (\mathbf{a}_1,\ldots,\mathbf{a}_k)$ and where $L(P_k,\mathbf{a}_k)$ is the likelihood of the cluster P_k for the density $f(\mathbf{x},\mathbf{a}_k)$. Since C_1 tends to produce equal size clusters, Symons (1981) has proposed a modified CML criterion

$$C_2(P,\mathbf{p},\mathbf{a}) = \sum_{k=1}^{K} \sum_{\mathbf{x}_i \in P_k} \log \{p_k \, f(\mathbf{x},\mathbf{a}_k)\} \tag{2.3}$$

which can be written

Compstat 1990

$$C_2(\mathbf{P},\mathbf{p},\mathbf{a}) = C_1(\mathbf{P},\mathbf{a}) + \sum_{k=1}^{K} n_k \log p_k \qquad (2.4)$$

where $\mathbf{p} = (p_1, ..., p_K)$ and where $n_k = \# \mathrm{P}_k$ $(1 \le k \le K)$.

To optimizing C_2 criterion, we consider the following Classification version of the EM algorithm (Dempster, Laird and Rubin 1977), the so-called CEM algorithm, which incorporates a Classification step (C-step) between the classical E-step and M-step of the EM algorithm. Starting from P^0, a CEM iteration $(m > 0)$ is defined as follows:

E-step: Compute for $i = 1, ..., n$ and $k = 1, ..., K$ the current posterior probabilities that $\mathbf{x}_i$ belongs to P_k: $t_k^m(\mathbf{x}_i) = p_k^m \mathrm{f}(\mathbf{x}_i,\mathbf{a}_k^m) / \Sigma_j p_j^m \mathrm{f}(\mathbf{x}_i,\mathbf{a}_j^m)$ for the current parameter estimate $\mathbf{a}^m$.

C-step: Assign each $\mathbf{x}_i$ to the cluster which provides the maximum posterior probability $t_k^m(\mathbf{x}_i)$, $1 \le k \le K$, (if the maximum posterior probability is not unique, we choose the cluster with the smallest index). Let P^m denotes the resulting partition.

M-step: For $k = 1, ..., K$ compute the maximum likelihood (m.l.) estimates $(p_k^{m+1}, \mathbf{a}_k^{m+1})$ using the sub-samples P_k^m.

We have the following proposition.

Proposition 1: *Any sequence* $(\mathrm{P}^m, \mathbf{p}^m, \mathbf{a}^m)$ *of the CEM algorithm increases the CML criterion* C_2 *and converges to a stationary value. Moreover, if the m.l. estimates of the parameters are well-defined, the sequence* $(\mathrm{P}^m, \mathbf{p}^m, \mathbf{a}^m)$ *converges to a stationary position.*

Proof: We first show that the criterion C_2 is increasing.

Since $(p_k^{m+1}, \mathbf{a}_k^{m+1})$ is maximizing $\sum_{\mathbf{x}_i \in \mathrm{P}_k^m} \log \{p_k \mathrm{f}(\mathbf{x}_i,\mathbf{a}_k)\}$, we have from (2.3)

$$C_2(\mathrm{P}^m, \mathbf{p}^{m+1}, \mathbf{a}^{m+1}) \ge C_2(\mathrm{P}^m, \mathbf{p}^m, \mathbf{a}^m),$$

and since $\mathbf{x}_i \in \mathrm{P}_k^{m+1}$ is equivalent to $t_k^{m+1}(\mathbf{x}_i) \ge t_h^{m+1}(\mathbf{x}_i)$ for $h \neq k$ which implies $p_k^{m+1}\mathrm{f}(\mathbf{x}_i,\mathbf{a}_k^{m+1}) \ge p_h^{m+1}\mathrm{f}(\mathbf{x}_i,\mathbf{a}_h^{m+1})$, we have

$$C_2(\mathrm{P}^{m+1}, \mathbf{p}^{m+1}, \mathbf{a}^{m+1}) \ge C_2(\mathrm{P}^m, \mathbf{p}^{m+1}, \mathbf{a}^{m+1}).$$

Since there is a finite number of partitions of the sample into K clusters, the increasing sequence $C_2(\mathrm{P}^m, \mathbf{p}^m, \mathbf{a}^m)$ takes a finite number of values, and, thus, converges to a stationary value. Hence $C_2(\mathrm{P}^m, \mathbf{p}^m, \mathbf{a}^m) = C_2(\mathrm{P}^m, \mathbf{p}^{m+1}, \mathbf{a}^{m+1}) = C_2(\mathrm{P}^{m+1}, \mathbf{p}^{m+1}, \mathbf{a}^{m+1})$ for m large enough; from the first equality and from the assumption that the m.l. estimate $\mathbf{p}^m$ and $\mathbf{a}^m$ are well-defined, we deduce that $\mathbf{p}^m = \mathbf{p}^{m+1}$ and $\mathbf{a}^m = \mathbf{a}^{m+1}$. From the second equality and the very definition of the C-step, it follows that $\mathrm{P}^m = \mathrm{P}^{m+1}$.

Remark: the assumption that the m.l. estimator of the parameter **a** of the density f(.,**a**) is well-defined appears to be mild, since it is true for a large class of densities (for instance, densities from an exponential family).

We focus, in this paper, on the widely-used variance criterion. This criterion, to be minimized, is

$$W(P) = \sum_{k=1}^{K} \sum_{\mathbf{x}_i \in P_k} ||\mathbf{x}_i - \mathbf{g}_k||^2 \tag{2.5}$$

where $\mathbf{g}_k$ is the center of the cluster P_k ($1 \le k \le K$). This criterion is often optimized by using a k-means-type algorithm. Starting from a partition P^0, an iteration m ($m \ge 1$) of the k-means algorithm can be summarized as follows:

1. *Representation step*: Compute the centers $\mathbf{g}_k^m$ of the clusters P_k^m ($1 \le k \le K$).

2. *Assignment step*: Define $P^{m+1} = (P_1^{m+1}, ..., P_K^{m+1})$ where

$$P_k^{m+1} = \{\mathbf{x}_i / ||\mathbf{x}_i - \mathbf{g}_k^m||^2 \le ||\mathbf{x}_i - \mathbf{g}_h^m||^2, h \ne k\}.$$

This algorithm does not necessarily provide the global optimum but only a local sub-optimal solution. Thus, it is usually repeated several times with different starting positions and the best of results is selected. We have the following proposition.

Proposition 2: *Maximizing the* C_2 *criterion for a Gaussian mixture with equal mixing weights and equal variance matrix of the form* $\sigma^2 I$ *(*σ^2 *unknown) is equivalent to minimizing the variance criterion* W.

Proof: In this situation, we have $\mathbf{a}_k = (\mathbf{m}_k, \sigma)$ and $p_k = 1/K$. For a fixed partition $P = (P_1, ..., P_K)$, it can easily be proved that the m.l. estimate of m_k is the center of cluster P_k. In these conditions, C_2 can be written

$$C_2(P, \mathbf{p}, \mathbf{a}) = \frac{-1}{\sigma^2} W(P) - np \log(\sigma^2) + A \tag{2.6}$$

where A denotes a constant. The Proposition 2 follows immediately from (2.6) and it can be seen that the estimate s^2 of σ^2 is $W(P)/np$.

From Proposition 2, we deduce that the k-means algorithm described above can be viewed as a specific version of the CEM algorithm. In fact, as most of the classical clustering criteria can be viewed as CML criteria (see, for instance, Celeux 1988), CEM can be considered as a quite general clustering algorithm.

3. Two stochastic versions of CEM

3.1 The SEM algorithm

The SEM algorithm (Celeux and Diebolt 1985), which is a stochastic version of the EM algorithm (Dempster, Laird and Rubin 1977), conceived for the m.l. estimation of the mixture parameters (p_k, $\mathbf{a}_k$, $k = 1, ..., K$), works as follows from an initial position p_k^0, $\mathbf{a}_k^0$, $1 \leq k \leq K$.

E-step: Compute for $i = 1, ..., n$ and $k = 1, ..., K$ the posterior probabilities that $\mathbf{x}_i$ belongs to P_k: $t_k^m(\mathbf{x}_i) = p_k^m f(\mathbf{x}_i, \mathbf{a}_k^m) / \Sigma_j p_j^m f(\mathbf{x}_i, \mathbf{a}_j^m)$.

S-step: Assign at random each x_i to one of the clusters $P_1, ..., P_K$ with probabilities $(t_k^m(\mathbf{x}_i)$, $k = 1, ..., K)$. Denote P^m the resulting partition.

M-step: For $k = 1, ..., K$ compute the m.l. estimates $(p_k^m, \mathbf{a}_k^m)$ using the sub-samples P_k^m.

It is clear that SEM can be thought of as a stochastic version of CEM as well as EM: the S-step appears to be simply a stochastic version of the C-step. Thus, from Proposition 2, it is straightforward to define a version of SEM optimizing the variance criterion which can be viewed as a stochastic k-means algorithm. For the estimation of mixture parameters, SEM has been proved to have nice properties (Celeux and Diebolt 1985) and, especially, it avoids convergence to local sub-optima. As defined above, an original feature of SEM is that, due to the S-step, the sequence $(\mathbf{p}^m, \mathbf{a}^m)$ does not converge pointwise: $(\mathbf{p}^m, \mathbf{a}^m)$ is a random vector, and so, P^m is a random partition. It has been proved (Celeux and Diebolt 1985) that $(\mathbf{p}^m, \mathbf{a}^m)$ converges in distribution to an unique stationary probability ψ. Therefore, in order to to obtain an unique solution, we proceed as follows: We run SEM a few dozen iterations, so that we are ensured that the sequence $(P^m, \mathbf{p}^m, \mathbf{a}^m)$ has reached its stationary position, and we ended the algorithm with a few CEM iterations. In Section 4, we discuss the practical ability of the "k-means version" of SEM to produce sensible local optima.

3.2. Simulated annealing clustering

Simulated annealing (SA) is a stochastic relaxation algorithm which has been used successfully in many optimization problems (see, for instance, van Laarhoven and Aarts 1987). Experiments of simulated annealing in clustering have been performed by Klein and Dubes (1989). These authors used a classical SA scheme that we sketch briefly hereafter. Starting from an initial partition, a clustering criterion, say W(P), is minimized in the following way: Perturb the partition P to partition P' (using, for instance, an exchange algorithm). If $\Delta W = W(P') - W(P) \leq 0$, accept partition P', else accept partition P' with probability $\exp(-\Delta W/\tau)$ where τ is control parameter called temperature which decreases along the iterations. Hence, as SEM, SA introduces random perturbations to produce sensible local optima. But, notice that SEM is more closely related to the underlying clustering model than classical SA. So, we have defined a "simulated annealing" version of the CEM algorithm, the so-called CAEM (Classification Annealing EM) algorithm. As SEM, CAEM is allowing random perturbations, but following the SA scheme, the magnitude of the random perturbations is decreasing along the iterations by using a sequence ($\tau_m, m \geq 0$) of temperatures decreasing to 0 as m tends to infinity from $\tau_0 = 1$. An CAEM iteration is the following.

E-step: Compute for $i = 1, ..., n$ and $k = 1, ..., K$ scores associated to the posterior probabilities that $\mathbf{x}_i$ belongs to P_k: $t_k^m(\mathbf{x}_i) = \{p_k^m f(\mathbf{x}_i, \mathbf{a}_k^m)\}^{1/\tau_m} / \Sigma_j \{p_j^m f(\mathbf{x}_i, \mathbf{a}_j^m)\}^{1/\tau_m}$ where τ_m is the temperature at the iteration m.

CA-step: For $i = 1, ..., N$ assign at random each x_i to one of the clusters $P_1, ..., P_K$ with probabilities $(t_k^m(\mathbf{x}_i), k = 1, ..., K)$. Denote P^m the resulting partition.

M-step: For $k = 1, ..., K$ compute the m.l. estimates $(p_k^m, \mathbf{a}_k^m)$ using the sub-samples P_k^m.

For $\tau = 1$, the CAEM iteration is exactly the SEM iteration, whereas for $\tau = 0$, it is exactly the CEM iteration. The crucial part of this algorithm is the cooling schedule (i.e. the convergence rate to 0 of the sequence (τ_m)). The choice of good cooling schedule is very difficult. For simplicity, we have chosen the cooling schedule defined by $\tau_{m+1} = a\ \tau_m$ with $0.9 \leq a \leq 1$ since it is well known that to give good performances SA needs a slow convergence rate of (τ_m) to 0 (van Laarhoven and Aarts 1987).

Notice that CAEM algorithm, conceived in a clustering context, has a median position between SEM and CEM, as the simulated annealing type EM algorithm (Celeux and Diebolt 1990), conceived in a mixture context, has a median position between SEM and EM.

On an other hand, CAEM has actually a simulated annealing flavor: the clustering criterion is not monotonically decreasing, but as τ_m goes to 0, the criterion tends to be strictly decreasing. In contrast, since CAEM is closely related with SEM, the choice of the starting temperature ($\tau_0 = 1$) is natural which is not the case for standard SA.

4. Comparative experiments

We have performed numerical experiments to compare the three algorithms (CEM, SEM, and, CAEM) in differents situations (simulated and real data, small and large data, ...) for the variance criterion. We have no place to report significant tables from these experiments. These tables can be found in Celeux and Govaert (1990). Here, we can summarize these numerical experiments as follows:

CEM algorithm produces multiple solutions except with strong clustering structures. Generally, the best solution from 20 runs is a good local optimum. But, this is not the case for large data with no evident clustering structure.

For a very slow cooling rate, simulated annealing (CAEM) produces most often sensible local optima, but since it needs slow cooling it spends a lot of time. As standard SA, the CAEM results depend greatly on the cooling schedule. We recommend $\tau_{m+1} = 0.99\ \tau_m$ as a good rule for decreasing. On an other hand, it is noteworthy that CAEM initiated with a sub-optimal solution of CEM algorithm does not generally provide an improved solution.

SEM algorithm generally produces sensible local optima; moreover, initiated with a sub-optimal solution of CEM algorithm, it generally converges to a better optimum. In this sense, SEM outperforms CEM and CAEM. However, this behaviour of SEM is not so general for small samples. The only real drawback of SEM is that it needs a large number of iterations to ensure the best results: according to the data and to its initial position, it needs a few or many iterations to converge. But we cannot anticipate this point and, thus, its computational cost remains high.

The conclusion of our experiments is that stochastic algorithms are an interesting approach to clustering despite their computational costs. Research on the choice of optimum cooling schedule for CAEM and on stationarity tests for SEM may make stochastic algorithms very powerful in clustering. It is possible that an efficient implementation of these stochastic algorithms could be as follows: Initiate the stochastic algorithm (SEM or CAEM) with the partition derived from a standard clustering algorithm, as CEM, and run it a few dozen iterations. If the initial position is a local sub-optimum, the stochastic algorithm is expected to propose a more steady local optimum in many cases.

References

BOCK H.H. (1989), Probabilistic aspects in cluster analysis, *Proceedings 13th Conf. of GFKI.* Springer-Verlag, Berlin-Heidelberg.

CELEUX G. (1988), Classification et modèles, *Revue de Statistique Appliquée* **36** n° 4, 43-58.

CELEUX G. and DIEBOLT J. (1985), The SEM algorithm: a probabilistic teacher algorithm derived from the EM algorithm for the mixture problem, *Computational Statistics Quarterly.* **2**, 73-82.

CELEUX G. and DIEBOLT J. (1990), Une version de type recuit simulé de l'algorithme EM, *C. R. Acad. Sci.* Série 1, **310**, 119-124.

CELEUX G. and GOVAERT G. (1990), *Stochastic algorithms for clustering,* rapport de recherche INRIA (to appear).

DEMPSTER A. LAIRD N. and RUBIN D. (1977), Maximum likelihood estimation from incomplete data via the EM algorithm (with discussion), *JRSS*, B. **39**, 1-38.

KLEIN R.W. and DUBES R.C. (1989), Experiments in projection and clustering by simulated annealing, *Pattern Recognition* **22**, 213-220.

VAN LAARHOVEN P.J.M. and AARTS E.H.L. (1987), *Simulated Annealing: Theory and Applications*, D. Reidel, Hingham, MA.

SYMONS M.J. (1981), Clustering criteria and multivariate normal mixture, *Biometrics* **37**, 35-43.

Keywords: Clustering, k-means, Classification maximum likelihood, Stochastic EM, Simulated annealing.

A Classification Algorithm for Binary Observations Based on Euclidean Representations of Hypergraphs

M. Bolla, Budapest

Table of Contents

1. Introduction

We are given a sample for n binary random variables. The objects form the edges of the hypergraph $H = (V, E)$ on n vertices. The aim of the present paper is to classify the vertices of this hypergraph in the way that "similar" vertices (those having many incident edges in common) be of the same cluster. The problem is formulated as follows: given a connected hypergraph with n vertices and fixing the integer k $(1 \leq k \leq n)$, we are looking for k-partition of the set of vertices such that the edges of the corresponding cut-set be as few as possible. Some combinatorial measures — the minimal k-sector, $\theta_k(H)$ and the minimal weighted cut, $\nu_k(H)$ — characterizing this structural property are introduced and the following relations between them and the eigenvalues $0 = \lambda_1 \leq \lambda_2 \leq \cdots \leq \lambda_n$ of H are proved:

$$c_n \theta_k(H) \leq \sum_{j=1}^{k} \lambda_j \leq \nu_k(H),$$

where the constant c_n depends only on n. The notion of *spectra of hypergraphs* — which is the generalization of C-spectra of graphs (see Fiedler [4]) — is also introduced together with *k-dimensional Euclidean representations*. We shall show that the existence of k "small" eigenvalues is a necessary but not sufficient condition for the existence of a good classification. In addition, the representatives of the vertices in an optimal k-dimensional Euclidean represententation of the hypergraph should be well separated by means of their Euclidean distances. In this case the k-partition giving the optimal clustering is also obtained by this classification method and the estimation $\nu_k(H) \leq q^2 \sum_{j=1}^{k} \lambda_j$ holds true, where the constant q depends on n and the diameters of the clusters.

Compstat 1990

2. Spectra and Optimal Euclidean Representations of Hypergraphs

The spectrum of the hypergraph H is defined by the eigenvalues of its *Laplacian*

$$\mathbf{C}(H) := \mathbf{D}_v(H) - \mathbf{A}(H)\,\mathbf{D}_e^{-1}(H)\,\mathbf{A}^T(H),$$

where $\mathbf{A}(H)$ is the vertex-edge incidence matrix of H, while $\mathbf{D}_v(H)$ and $\mathbf{D}_e(H)$ are diagonal matrices with the vertex- and edge-valencies of H in their main diagonals, respectively. In [2], the following *Representation Theorem* is proved: Let the $k \times n$ matrix $\mathbf{X}_k(H)$ contain *k-dimensional representatives* of the vertices, as column vectors. The representative of the vertex v_j is denoted by $\mathbf{x}_j$, $(j = 1, \ldots, n)$. The minimum of the quadratic form $L(\mathbf{X}_k(H)) = \sum_{e \in E} L(e, \mathbf{X}_k(H))$ conditioned on $\mathbf{X}_k(H)\mathbf{X}_k(H)^T = \mathbf{I}_k$ — where $L(e, \mathbf{X}_k(H)) = \frac{1}{2|e|}\sum_{v_i \in e}\sum_{v_j \in e}\|\mathbf{x}_i - \mathbf{x}_j\|^2$ — is the sum of the k smallest eigenvalues of the Laplacian $\mathbf{C}(H)$, and the rows of an optimal $\mathbf{X}_k^*(H)$ consist of an orthonormal set of the corresponding eigenvectors.

3. Classification Theorems Based on Spectra

Let $H = (V, E)$, $|V| = n$, $|E| = m$ be a hypergraph without loops and multiple edges, its eigenvalues being $0 = \lambda_1 \leq \lambda_2 \cdots \leq \lambda_n$ in increasing order. Now we shall give upper and lower bounds for combinatorial measures characterizing k-partitions of the vertex-set of H by means of the k smallest eigenvalues, where k is any natural number between 2 and n. First of all let us introduce the following notions:

DEFINITION 1. *A k-tuple $(V_1, \ldots, V_k)$ of non-empty subsets of V is called a k-partition of the set of vertices, if $V_i \cap V_j = \emptyset$ for $i \neq j$ and $\cup_{i=1}^k V_i = V$. Sometimes a k-partition is denoted by P_k, while the set of all k-partitions by $\mathcal{P}_k$. The volume $v(P_k)$ of the k-partition $P_k = (V_1, \ldots, V_k)$ is defined by*

$$v(P_k) := \sum_{e \in E} \frac{1}{|e|} \sum_{1 \leq i < j \leq k} a_i(e)\, a_j(e)$$

and its weighted volume $u(P_k)$ by

$$u(P_k) := \sum_{e \in E} \frac{1}{|e|} \sum_{1 \leq i < j \leq k} \left(\frac{1}{n_i} + \frac{1}{n_j}\right) a_i(e)\, a_j(e),$$

where $a_i(e) = |e \cap V_i|$ and $n_i = |V_i|$.
The minimal k-cut of H is defined by

$$\mu_k(H) = \min_{P_k \in \mathcal{P}_k} v(P_k), \tag{3.1}$$

while the minimal weighted k-cut by

$$\nu_k(H) = \min_{P_k \in \mathcal{P}_k} u(P_k). \tag{3.2}$$

DEFINITION 2. *The cut-set of the k-partition $P_k = (V_1, \ldots, V_k)$ consists of those edges e for which $|e \cap V_i| \neq \emptyset$ holds for at least two different parts of P_k, and it is denoted by $H(P_k)$. The k-partition P_k defines a colouring c of the vertices in the following way: $c(v) := i$, if $v \in V_i$. An edge e is said to be multi-coloured in this colouring, if it contains two different vertices v, v' such that $c(v) \neq c(v')$. Thus the cut-set $H(P_k)$ consists of the multi-coloured edges. $H(P_k^*)$ is called a minimal k-sector of H, if*

$$|H(P_k^*)| = \min_{P_k \in \mathcal{P}_k} |H(P_k)|,$$

and its cardinality is denoted by $\theta_k(H)$.

THEOREM 1. *For the sum of the k smallest eigenvalues of the hypergraph H the upper and lower estimations*

$$c_n \theta_k(H) \leq \sum_{j=1}^{k} \lambda_j \leq \nu_k(H) \tag{3.3}$$

hold, where $c_n = \frac{6}{n(n^2-1)}$.

The upper bound shows that the existence of k relatively small eigenvalues is a necessary condition for the existence of a good colouring (with a small minimal weighted cut). Thus the spectrum can give us some idea about the choice of the number k of the different colours for which good colouring may exist. But the spectrum itself does not say anything about the optimal k-partition, moreover it does not give a sufficient condition for the existence of a good colouring. The lower bound in (3.3) depends on the constant c_n, and there are graphs for which the lower bound is attained in order of magnitude.

Now we want to recognize optimal k-partitions by means of classification of k-dimensional representatives of the vertices in an optimal k-dimensional Euclidean representation of the hypergraph. The classification is performed by the k-means method (introduced by Mac Queen). We shall be confined to the case, when a "very" well-separated k-partition of the above k-dimensional points exists.

DEFINITION 3. *A k-partition $P_k = (V_1, \ldots, V_k)$ is called a well-separated k-partition of the vertex-set V in the k-dimensional Euclidean representation $\mathbf{X} = (\mathbf{x}_1, \ldots, \mathbf{x}_n)$ of the vertices, if for the colouring c belonging to P_k the relation $\alpha(P_k) > 1$ holds, where*

$$\alpha(P_k) := \frac{\min\limits_{c(v_i) \neq c(v_j)} \|\mathbf{x}_i - \mathbf{x}_j\|}{\max\limits_{c(v_i) = c(v_j)} \|\mathbf{x}_i - \mathbf{x}_j\|} \tag{3.4}$$

THEOREM 2. *Assume that for some $k < n$ there exists a well-separated k-partition of the vertex set V, for the clusters of which the diameters are at most ε, where $\varepsilon < \frac{1}{2\sqrt{n}}$ is a small positive number. Then*

$$\nu_k(H) \leq q^2 \sum_{j=1}^{k} \lambda_j, \tag{3.5}$$

where $q = 1 + \frac{2\varepsilon}{1-\varepsilon\sqrt{n}}$.

Comparing the results of Theorem 1 and 2, under the constraints of Theorem 2 we obtain that

$$\sum_{j=1}^{k} \lambda_j \leq \nu_k(H) \leq q^2 \sum_{j=1}^{k} \lambda_j, \quad \text{where} \quad 1 < q < 2.$$

This means, that provided ε is less than $\frac{1}{2\sqrt{n}}$, then q is at most 2, and the combinatorial and analytical measures of H, $\nu_k(H)$ and $\sum_{j=1}^{k} \lambda_j$ differ at most by a factor of 4.

4. A Classification Algorithm Based on Euclidean Representations

On the basis of these spectral techniques the following classification algorithm is introduced: Let k be a fixed integer, $(1 \leq k \leq n)$, $(E_1, \ldots, E_k)$ be a k-partition of the edge-set of H and $H_i = (V, E_i)$ be the hypergraph of the subsample E_i $(i = 1, \ldots, k)$. The following cost function is constructed: $Q = \sum_{i=1}^{k} Q_{d_i}(H_i)$, where

$$Q_{d_i}(H_i) := c2^{n-d_i} + L(\mathbf{X}^*_{d_i}(H_i)), \quad (i = 1, \ldots, k),$$

c being a constant chosen previously according to the size of problem.

To minimize the cost function Q — with respect to k-partitions of the edges and dimensions of the edge-clusters — the following iteration is introduced. First let us choose k disjoint clusters $E_1, \ldots, E_k$ of the objects (e.g. by the k-means method).

i. Fixing the clusters $E_1, \ldots, E_k$: the spectra and optimal Euclidean representations of the sub-hypergraps of the edge-clusters are calculated.

ii. The function $Q_{d_i}(H_i)$ is minimized with respect to the dimension d_i, $(1 \leq d_i \leq n)$ for each i separately. The i-th optimum is denoted by d_i^*.

iii. Fixing the d_i^*-dimensional optimal Euclidean representations $\mathbf{X}^*_{d_i^*}(H_i)$-s: an object e is replaced into the cluster E_i, for which $L(e, \mathbf{X}^*_{d_i^*}(H_i))$ is minimal. In this way a new disjoint classification $E_1^*, \ldots, E_k^*$ of the objects is obtained. From now on we go back to step i. with starting classification $E_1^*, \ldots, E_k^*$, etc.

As the cost function Q is in each step decreased and in steps ii. and iii. discrete minimizations are performed the algorithm must converge to a local minimum of Q in finite steps. As a new step of the iteration, a minimization with respect to k could be introduced, but it would be very time-demanding.

During the iteration some edge-clusters may become empty. Also the hypergraph $H_i = (V, E_i)$ may contain isolated vertices. Let us denote by V_i the set of the non-isolated vertices of H_i. In this way not necessarily disjoint subsets of the vertices are obtained, that define so-called ***characteristic associations*** of the binary variables in question.

The algorithm is included in the DISTAN (DIscrete STatistical ANalysis) program package as procedure BinClus (clustering of binary data).

References

[1] Alon, N., Eigenvalues and Expanders. Combinatorica 6 (2) (1986), 83-96.

[2] Bolla,M., Spectra, Euclidean Representations and Vertex-Colourings of Hypergraphs. Preprint of the Mathematical Institute of the Hungarian Academy of Sciences, No.78, 1989.

[3] Cvetković, D.M., Doob, M., Sachs, H., Spectra of Graphs. Academic Press, New York-San Francisco-London, 1979.

[4] Fiedler, M., Algebraic Connectivity of Graphs. Czechoslovak. Math. J. 23 (1973), 298-305.

Agglomerative Hierarchical Multicriteria Clustering Using Decision Rules

V. Batagelj and A. Ferligoj, Ljubljana

Abstract

In a multicriteria clustering problem optimization over more than one criterion is required. The problem can be treated in different ways. In this paper the agglomerative hierarchical method based on decision rules for making decisions under uncertainty is proposed for solving multicriteria clustering problem. An application of proposed approach to Rosenberg and Kim kinship data is presented.

Key words: multicriteria optimization, agglomerative hierarchical clustering methods, decision rules for making decisions under uncertainty.

1 Multicriteria clustering problem

There are clustering problems which can not be appropriately solved with classical clustering methods because they require optimization over more then one criterion. In general, optimal solutions according to each particular criterion are not the same. Because of this the problem arises of how to find the best solution so as to satisfy as much as possible all of the considered criteria.

The usual *clustering problem* (Φ, P) can be formulated as an optimization problem: determine the clustering C^* for which

$$P(C^*) = \min_{C \in \Phi} P(C)$$

where Φ is the set of *feasible clusterings*, C is a clustering and $P : \Phi \to \mathbb{R}$ the (single) *criterion function*.

In a *multicriteria clustering problem* $(\Phi, P_1, P_2, \ldots, P_k)$ we have several criterion functions $P_t, t = 1, \ldots, k$ over the same set of feasible clusterings Φ and our aim is to determine the clustering $C \in \Phi$ in such a way that

$$P_t(C) \to \min, \qquad t = 1, \ldots, k$$

The problem arises of how to find a solution to the problem as good as is possible according to each of the given criteria.

[1]This work was supported in part by the Research Council of Slovenia, Yugoslavia.

The multicriteria clustering problem can be treated in different ways: by reduction to a clustering problem with the single criterion obtained as a combination of the given criteria; by clustering with constraints methods where a selected criterion is considered as the clustering criterion and all others determine the constraints; or by direct methods. In [5] two types of direct methods for solving multicriteria clustering problem were proposed: the adapted relocation method, and the adapted agglomerative method. Different elaborations of these two types of methods based on the notion of Pareto-clustering were discussed and compared.

In this paper we elaborate further the agglomerative hierarchical approach based on the criteria compositon.

2 Agglomerative hierarchical approach

We assume that we have k dissimilarity matrices $D^t, t = 1, \ldots, k$, summarizing relevant information on the relationships between the n units, obtained, for example, in k different situations. The problem is to find the best hierarchical solution in a way which satisfies as much as possible all k dissimilarity matrices.

In [5] we proposed two types of adapted agglomerative hierarchical methods for solving the multicriteria clustering problem.

The first approach is to combine the given dissimilarity matrices (at each step) into a composed matrix on which the selection of the nearest pair of clusters is made.

The second approach is to perform the selection step by searching for the *Pareto nearest* pair of clusters. The deficiency of this approach is that the proposed procedure gives several (Pareto) hierarchical solutions. If a smaller set of solutions is desired, additional decision rules or user decisions have to be built into the procedure. Also, there is no single fusion level at each step – there is no simple graphical representation of solution by a dendrogram.

In this paper we follow the first approach. We propose to use at the composition and selection step one of *decision rules* (pessimistic, optimistic, Hurwicz, Laplace, Savage) *for making decisions under uncertainty* [3, 6]. We obtain the following scheme of the adapted agglomerative method:

> Each unit is a cluster: $C_i = \{X_i\}$, $X_i \in E$, $i = 1, 2, \ldots, n$;
> normalize each dissimilarity matrix $D^t, t = 1, \ldots, k$;
> **repeat** while there exist at least two clusters:
>   determine the nearest pair of clusters C_p in C_q, $d_{p,q} = d(C_p, C_q)$
>   according to a given decision rule;
>   fuse clusters C_p and C_q into a new cluster $C_r = C_p \cup C_q$;
>   replace the clusters C_p and C_q by the cluster C_r;
>   **for each** dissimilarity matrix $D^t, t = 1, \ldots, k$:
>   determine the dissimilarities d^t between the cluster C_r
>   and other clusters according to a given fusion strategy.

The normalization step is not always necessary, especially when dissimilarities are obtained according to the same variables and the same dissimilarity measure but for example on different occasions.

In the pair selection step of the algorithm the decision rules get the form:

Wald's (pessimistic) rule:

$$d_{p,q} = \min_{i,j} \max_{t} d_{i,j}^t$$

Optimistic rule:

$$d_{p,q} = \min_{i,j} \min_{t} d_{i,j}^t$$

Hurwicz's rule with pessimism index α, $0 \leq \alpha \leq 1$:

$$d_{p,q} = \min_{i,j}(\alpha \max_{t} d_{i,j}^t + (1-\alpha) \min_{t} d_{i,j}^t)$$

Laplace's principle of insufficient reason:

$$d_{p,q} = \frac{1}{k} \min_{i,j} \sum_{t=1}^{k} d_{i,j}^t$$

The Savage's minmax regret decision rule

$$d_{p,q} = \min_{i,j} \max_{t}(d_{i,j}^t - \min_{t} d_{i,j}^t)$$

can not be directly applied in this context.

The obtained hierarchical solution can be graphically represented by the dendrogram. The dendrogram levels are the dissimilarities $d_{p,q}$ from the pair selection step.

3 Application: The Rosenberg and Kim data

To illustrate the proposed approach for multicriteria clustering we used the well-known data set on kinship terms [8]. The data are based upon subjective sortings of fifteen kinship terms. Different groups of subjects sorted kinship terms under different sorting instructions with which were obtained six data matrices [1, 62-63]. Two structuring principles could be hypothesized: the first based on kinship and the second based on gender. The data were analyzed by means of hierarchical clustering techniques [8], INDSCAL and INDCLUS models [1], multicriteria relocation methods [4], with algorithm for fitting general graphs to proximity data [7] and other techniques.

The dendrograms obtained with the proposed agglomerative multicriteria clustering method for Wald's, optimistic, Laplace's and Hurwicz's ($\alpha = 0.25, 0.50$ and 0.75) rules and Ward's criterion function are presented in figures.

All the dendrograms determine the same clustering into three clusters: the first cluster forms the *nuclear family* (mother, father, daughter, son, sister, brother), the second cluster consists of *collaterals* (aunt, uncle, cousin, niece, nephew), and the third cluster are the *grand*parents and *grand*children.

When analyzing each of the six data matrices separately with hierarchical and relocation clustering methods four times the same (previuosly mentioned) clustering was obtained [4].

The dendrograms obtained with four decision rules differ at lower clustering levels: the optimistic and Laplace's dendrograms are very similar – the terms are clustered

CLUSE – plot [0.00, 260.00] May- 1-1990
Wald

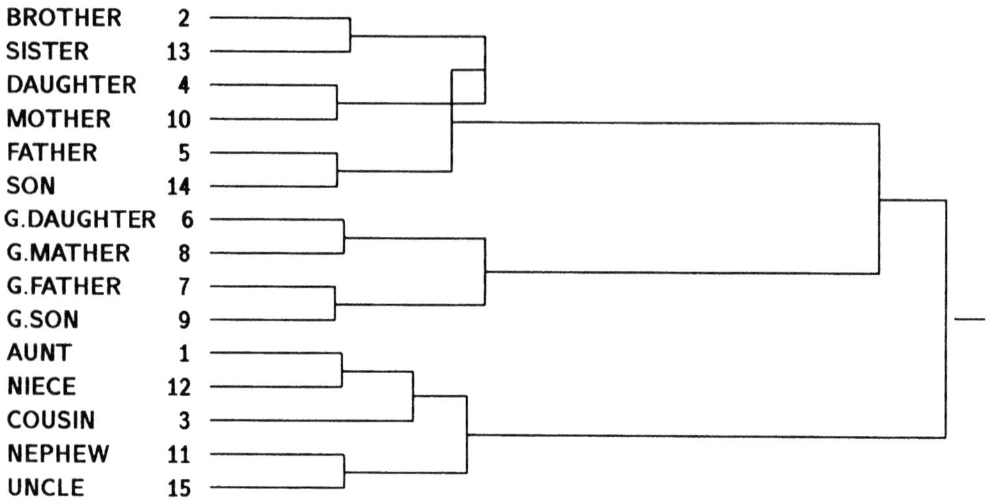

CLUSE – plot [0.00, 130.00] May- 1-1990
Optimist

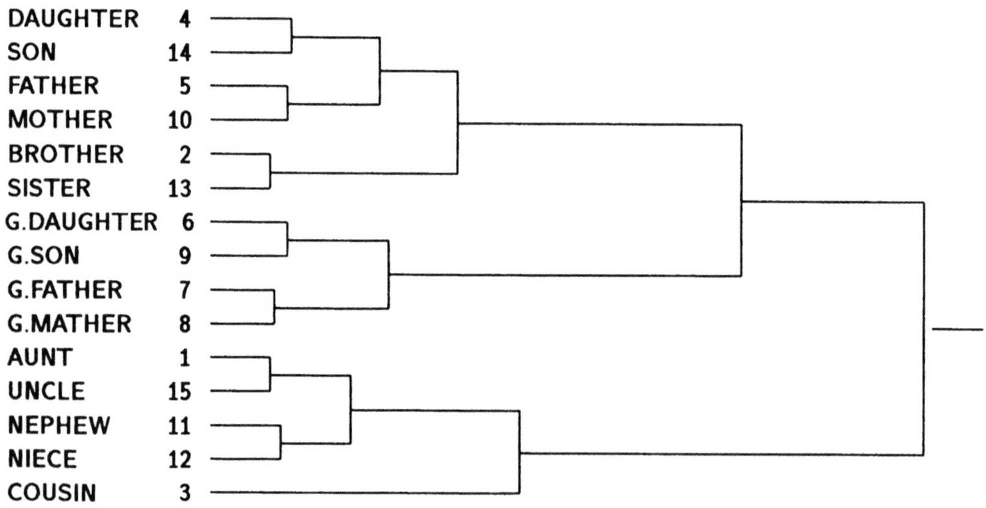

CLUSE – plot [0.00, 200.00] May- 1-1990
Hurwicz 0.5

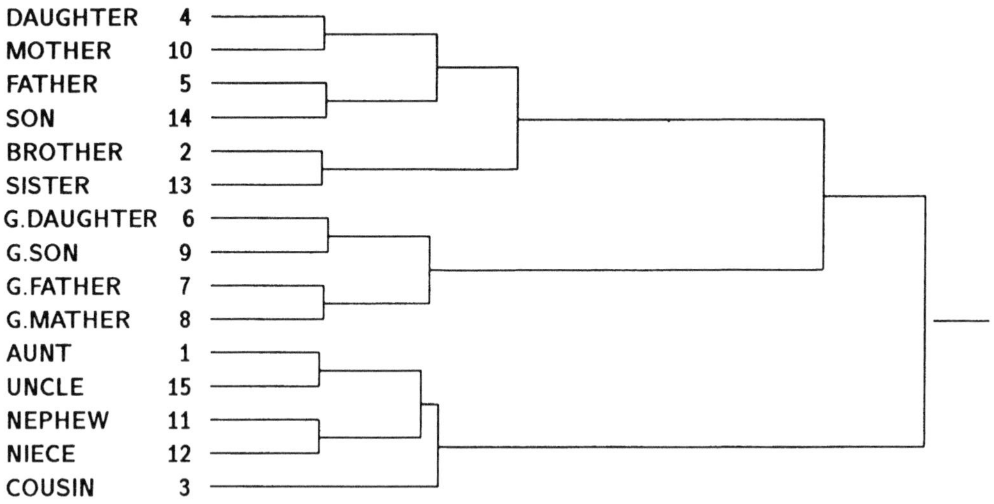

CLUSE – plot [0.00, 190.00] May- 1-1990
Laplace

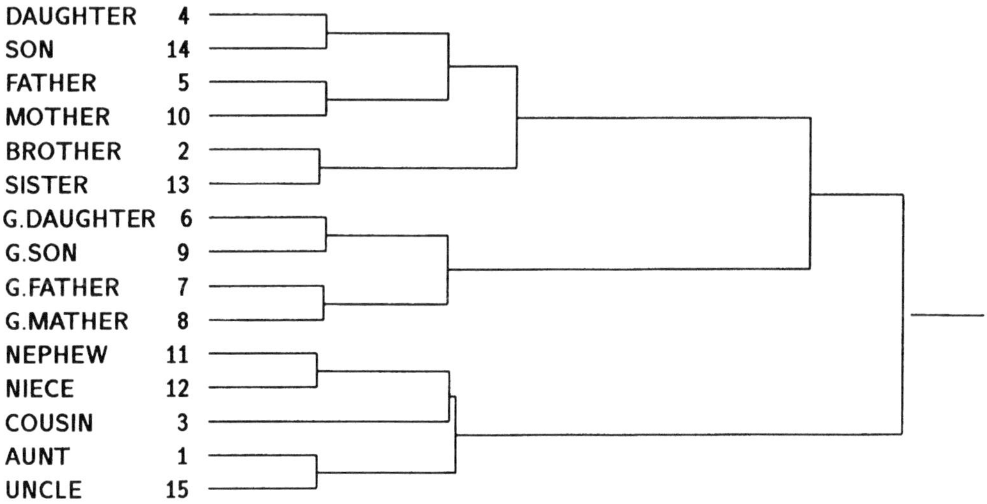

on the kinship basis (mother-father, aunt-uncle, brother-sister, ...). Similar is also the clustering produced by the Hurwicz's decision rule with exception of mother and father which are in different clusters. The most different solution was obtained with Wald's rule where the clusters on the lower levels are based on gendre with exception of one cluster (sister, brother).

In summary, the obtained findings are congruent with previous analyses. It is evident that the kinship level has stronger impact to the clustering of kinship terms than gender.

Additional experiments with this approach on different data sets are needed to study the impact of particular decision rule to the corresponding clusterings.

All computations were carried out with the system of clustering programs CLUSE [2].

References

[1] Arabie P., Carroll J.D., and DeSarbo W.S (1987): Three-way Scaling and Clustering. Newbury Park, CA: Sage.

[2] Batagelj V. (1989): CLUSE for IBM PC - Manual. Ljubljana.

[3] Chankong V., and Haimes Y.Y. (1983): Multiobjective Decision Making. New York: North-Holland.

[4] Ferligoj A., and Batagelj V. (1988): Metoda prestavljanj za večkriterijsko razvrščanje v skupine. Statistička revija, 38, 140-154.

[5] Ferligoj A., and Batagelj V. (1989): Direct multicriteria clustering methods. To appear in Journal of Classification.

[6] French S. (1986): Decision Theory, an introduction to mathematics of rationality. New York: John-Wiley.

[7] Klauer K.C., and Carroll J.D. (1989): A mathematical programming approach to fitting general graphs. Journal of Classification, 6, 247-270.

[8] Rosenberg S., and Kim M.P. (1975): The method of sorting as a data-gathering procedure in multivariate research. Multivariate Behaviour, 10, 489-502.

GLIMTREE: RECPAM Trees with the Generalized Linear Model

A. Ciampi, Q. Lin and G. Yousif, Montreal

1. Introduction

Many problems of data analysis can be reduced to the construction of a linear prediction rule. Typically, one studies the relationship of a variable of special importance, y, which we shall call *outcome variable*, to other variables, here referred to as *predictors*. A linear prediction rule is a linear combination of the predictors whose values represent the expected value of the response variable, given the predictors. In the classical paradigm, the outcome variable is assumed to be continuous and normally distributed, and the coefficients of the linear predictor are estimated by least squares.

The Generalized Linear Model (McCullagh & Nelder, 1989) (GLIM, from the initials of the well-known software package which has made the model accessible to applied statisticians) has greatly extended the scope of classical *linear* data analysis by removing the restriction of normality and continuity for the outcome variable, and by weakening the assumption of linearity: in GLIM linearity is assumed, but only up to a variable transformation. Continuous, discrete and count outcome variables from a large class of distributions can be easily treated by GLIM. The linear predictor links the covariate vector **z** to the expected value of the outcome μ = E(y) by the so-called *link function*:

$$g(\mu) = \beta_0 + \beta_1 z_1 + \ldots + \beta_p z_p \qquad (1)$$

The coefficients of the linear predictor for η are estimated by minimizing the *deviance*, a measure of the lack-of-fit of the predicted values to the data, based on the log-likelihood of the model assumed for y (see below).

GLIM remains anchored to the (modified) assumption of linearity. To handle non-linearities, several methods have been proposed in the context of computationally intensive Statistics (Härdle, 1989). Recently, some authors have proposed approaches that combine the advantages of GLIM with those of non-linear techniques. For example, Hastie & Tibshirani (1986),further weaken linearity and work with *generalized additive models*.

One of the most powerful approaches to non-linear prediction is tree-structured regression (Breiman *et al.*, 1984; Ciampi *et al.*, 1987, 1989). In this work we propose *generalized tree-structured prediction*, or GLIMTREE, an approach that combines the advantages of GLIM with those of trees and that further extends a tree-growing method known as RECPAM (Ciampi *et al.*, 1987, 1989). Just as GLIM can build a linear prediction rule for an oucome variable distributed according to an exponential family, GLIMTREE can build a tree-structured predictor for outcome variables of this family

2. The RECPAM method

Suppose we have a data matrix D obtained from measuring two classes of variables on a population *P*: *criterion variables*, **u**, and *predictors*, **z**. The

Compstat 1990

columns of **D** are correspondingly partitioned as [**U**|**Z**]. The criterion variables are measured in order to gather information on a parameter, called *criterion* and denoted by γ, and on auxiliary parameters α and β. It is assumed that the predictors contain information on γ and on α, but not on β, whence the term *local* for α and *global* for β.

The primary aim of a RECPAM analysis is to construct from **D**: a) a tree structure on the predictors; b) a partition $\mathcal{P}$ of P , related to the tree structure, into subpopulations which are *homogeneous* and *distinct* with respect to γ; c) a prediction of γ for each subpopulation. An auxiliary aim is to construct, at the same time: d) a tree-structured partition $\mathcal{L}$ not coarser than $\mathcal{P}$ with distinct estimates for α on each set of $\mathcal{L}$; e) an estimate for the local parameter β.

Essential to the RECPAM method is a definition of a measure of the *information content pertaining to* γ of a row-partition $\mathcal{P}$ of a data matrix, given the relevant portion of the data **U**. We assume that such measure has been chosen and will denote it by $IC(\gamma|\mathcal{P}, \mathbf{U})$. The information content measure is used for constructing the classification by RECursive Partitioning or tree-growing, followed by AMalgamation, whence the acronym RECPAM.

In most of the applications of RECPAM, the distribution of **u** is known to belong to a family of distributions, $\{f(\mathbf{u};\ \gamma\ ,\ \beta)\}$ indexed by the criterion γ and by the nuisance parameter β. A natural definition for $IC(\gamma|\mathcal{P}, \mathbf{U})$ is the likelihood ratio statistic (LRS) of the hypothesis H_1 that γ varies across $\mathcal{P}$ to the hypothesis H_0 that it does not. As is well known, this LRS is an estimate of the discriminant information of the distribution for **u** under H_1 against the distribution for **u** under H_0.

In spite of its appeal, the definition of the information content as a LRS has a major drawback. By definition, if $\mathcal{P}'$ is a partition finer than $\mathcal{P}$, then it has a higher information content than $\mathcal{P}$. This implies that complex predictions are *always* better (have more information) than simpler ones. The approach taken in RECPAM is, as we shall see, to prefer a more complex partition to a simpler one only if the increase of information content exceeds a certain threshold.

Let Q denote a binary question concerning **z**, *i.e.* a question with a yes or no answer, and let Q denote the associated binary partition, or *split*. With slight abuse of language, we shall refer to $I(\gamma|Q,\mathbf{U})$ as the *information content of the question* Q.

A (binary) prediction tree T consists of sets of hierarchically structured (binary) questions, with a prediction for γ attached to any complete set of answers. (In what follows we will consider only binary questions and trees and therefore the term binary will be omitted). Each question defines a split of an (internal) node, of the tree, and each complete set of answers defines a terminal node or *leaf* of the tree. The set of the leaves of T constitutes a partition, $\mathcal{L}_T$. With the same abuse of language as above, we shall refer to the information content of $\mathcal{L}_T$ as the *information content of the tree* T.

The RECPAM tree construction proceeds in three steps:

Step 1: Construction of a *large tree*.

RECPAM starts from the original population, which is identified as the *root node*, and searches among all questions of a specified class, known as the SDQ family (family of split-defining questions). RECPAM chooses as the first question, the one with the largest information content and as first branching, the corresponding split. The same search is performed recursively on the descendents of the root node, thus building a binary tree with nodes defined by populations and branching defined by questions of the SDQ family. The process is stopped when there are no more SDQ's: this defines the terminal nodes or *leaves* of the *large tree*, T_{Max}.

Step 2: *Pruning* of the *large tree* and selection of the *honest tree.*
RECPAM builds a sequence of rooted subtrees of T_{Max}, beginning with T_{Max}, and ending with the *trivial* tree, *i.e.* the one which contains only the root. Specifically, let us define the *information loss* (IL) of a subtree as the difference of the IC's of the large tree and of the subtree. Then, each subtree of the sequence is obtained from the preceding one by removing the pair of leaves with the same parent node, resulting in the smallest IL. As already remarked, the sequence thus built is of decreasing IC. The process is known as *pruning*. By *honest tree* we mean the smallest tree whose information loss with respect to T_{Max} is smaller than a chosen threshold. Several methods have been proposed for chosing the threshold when the information content of a tree is an LRS (Ciampi *et al.*, 1987, 1989): these are based on the idea of *cross-validation*. Without going into the details, we mention here the *AIC approach*, according to which the threshold, different for each tree of the pruning sequence, is simply twice the difference of the number of parameters associated to T_{Max} and the one associated to the subtree under consideration.

Step 3: Construction of the *amalgamation tree* and selection of the RECPAM classification.
RECPAM amalgamates the leaves of the honest tree successively, by joining, at each step, the two subpopulations such that the resulting information loss is minimal. A tree is constructed, going from the leaves towards the root. The partition corresponding to each step of this construction, is coarser than that of the preceeding one, and thus the information loss increases. Again, the chosen partition is the simplest (coarsest) one such that the information loss with respect to the leaves of T_{Max}, is smaller than a certain threshold. The AIC approach for the case of LRS as information content, can be invoked here too. The tree is cut when this threshold is exceeded, and the resulting partition gives the RECPAM classes.

To recapitulate a RECPAM algorithm is defined by a number of choices:

Choice 1: Specification of the criterion variables, the predictors and the criterion. This often implies a specification of a statistical model for **u**.

Choice 2: Specification of the Information Content of a partitition, $IC(c|\mathcal{P}, U)$.

Choice 3: Specification of the SDQ in step 1

Choice 4: Specification of thresholds in step 2.

Choice 5: Specification of thresholds in step 3.

In what follows we will specify these choices for GLIMTREE.

3. GLIMTREE

Choice 1:
We shall assume that **U** is further partitioned as [Y|**W**]. This partition corresponds to the classification of the criterion variables as *outcome*, y (one-dimensional) and *covariates*, **w**.

In GLIMTREE we assume for y, given **w** and z, a distribution from the exponential family:

$$f(y;\ \theta,\ \psi,\ \phi) = \exp\{\phi[y\theta - b(\theta)]/\psi + c(y,\phi)\} \qquad (2)$$

where θ, and ψ are individual-specific parameters, φ is a "scale factor", the same for all individuals, and b(θ) and c(y,φ) are known functions.

We chose for the link function, g(μ), the model of the form:

$$g(\mu) = C(\mathbf{w};\gamma)\cdot I_{\mathcal{P}}(z) + \mathbf{A}(\mathbf{w};\alpha)\cdot I_{\mathcal{L}}(z) + B(\mathbf{w};\beta) = \qquad (3)$$

where $I_{\mathcal{P}}$ and $I_{\mathcal{L}}$ denote the vectors of the indicator variables of the sets of the partitions $\mathcal{P}$ and $\mathcal{L}$; **A** and **C** are specified vector functions of **w** , the criterion parameter and the local parameter; and B is a specified function of **w** and the local parameter. For each node, (3) is specified by $\mathcal{P} = \mathcal{L} = Q$, where Q is the binary partition associated to a question Q. For a subtree T of the large tree, $\mathcal{P}$ is $\mathcal{L}_T$ and $\mathcal{L}$ is $\mathcal{L}_{T_{Max}}$. For a cluster defined by the amalgamation procedure, $\mathcal{P}$ is the corresponding partition and $\mathcal{L}$ is again $\mathcal{L}_{T_{Max}}$.

Two important special cases of (3) correspond to broad classes of problems that GLIMTREE can solve:

i) Outcome classification with confounder correction:
w is a confounder, v; C does not depend on **w**, and C(γ) = γ; **A** = 0; B(**v**) = βv. In this case a tree is built while correcting for the effect of the confounder, supposed global.

ii) Subgroup analysis:
w is a determinant, x, *i.e.* a variable whose relationship to outcome is especially important (*e. g.* treatment or control variable); **C** = xγ; **A** does not depend on x, and **A**(α) = α; B = 0. In this case the relationship between y and x, assumed linear, is of primary interest The tree is built to predict the slope of the line. The role of the local (vector) parameter α is to account for direct effects of the predictors on the outcome, supposed local.

Choice 2:
We recall that a natural measure of model fit in GLIM is the *deviance* of the model given by (3), *i.e.* the difference, devided by φ, of the maximized log-likelihood based on (3) and of the log-likelihood of the saturated model. The latter is the model for which each data point has a different value of θ. As is well known, the deviance has the advantage of being directly related to likelihood-based information measures, while being independent of φ.

In view of the desirable properties of the deviance, we chose $IC(\gamma|\mathcal{P}, U)$ to be the difference of the deviances of the model associated to the trivial (one-set) partition, and of the model associated to $\mathcal{P}$. The interpretation is then straightforward: the information content about γ of a partition, is simply the increase in the accuracy (litterally, the decrease in the inaccuracy!) of the prediction for γ, which can be attributed to the decision of making separate predictions for each set of the partition.

Choice 3:
The SDQ family at a node is chosen as the class of admissible simple questions. This is defined as follows. A *simple question* is a question involving one predictor at the time, and of the type: $[z_i \in A]$, where A is a subset of the possible values of z_i for discrete predictors, or: $[z_i \geq a]$ for ordered predictors. A question is admissible at a node if the sets of the associated split $\{Q_1, Q_2\}$, satisfy the condition: $\sum \{s(u^{(i)})$; sum over $Q_k\} \geq$ m, k=1,2, where s is a specified function and m a specified integer. Note that for $s(u) \equiv 1$, this is a a condition of minimum leaf size. Another possible choice for s, especially useful when y is a count, is $s(u) = y$, a minimum count condition.

Choice 4 and 5:
Thresholds are based on the AIC approach. The information loss of a subtree T with respect to T_{Max} is considered negligeable if it does not exceed twice the difference of the number of parameters associated to T_{Max} and to T, scaled by an estimate of ϕ.

4. An Example

We have analyzed epidemiological data from Appendix VII of Breslow & Day (1987; pag. 369-374). Number of nasal cancer deaths in a cohort of Welsh nickel refinery workers was the outcome variable; the other criterion variable, treated here as a confounder, was the exposure index (to nickel); predictors were Age of First Employment (AFE), Year of First Employment (YFE) and time since first employment (TFE). The figure shows the large tree constructed according to GLIMTREE, by specifying (2) as a Poisson distribution.

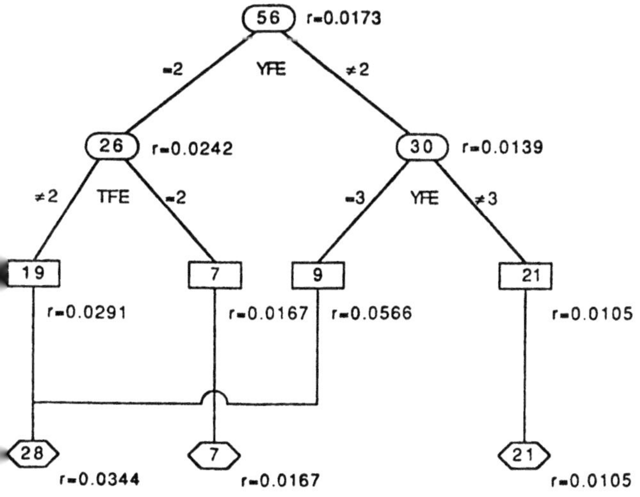

AFE	YFE	TFE
1=<20.0	1=1902-1910	1=0.019.9
2=20.0-27.4	2=1910-1914	2=20.0-29.9
3=27.5-34.9	3=1915-1919	3=30.0-39.9
4=35.0-54.4	4=1920-1924	4=40.0-49.9
		5=50.0+

$Log(r)=-7.761\cdot I_1-6.655\cdot I_2-7.384\cdot I_3+0.770\cdot EXP$

r --- death rate(per person-years)

i=1,2,3 --- leaf i starting from leftmost

5. Discussion

The approach presented here brings tree-growing to the exponential family given by (2). We have shown that the approach is feasible and extremely flexible, in that it permits one to treat, within the same framework, outcome prediction and subgroup analysis as well as adjustments for global *linear* confounding. In fact the model given by (3) can be considered as a mixture of a tree structure and a linear structure.

We conclude with a comment on the relationship of this work to Hastie & Tibshirani (1986). The two approaches can be seen as complementary, in that the non-linearities treated by generalized additive models are appropriate for modelling global non-linear effects, whereas GLIM trees are best at describing local effects or *interactions*, possibly in the presence of global linear effects.

REFERENCES

Breiman, L., *et al.* Classification and Regression Trees. Wadsworth, Belmont, Ca.

Breslow, N. E., Day, N. E. (1987) Statistical Methods in Cancer Research, Vol. 2, IARC, Lyon

Ciampi, A., Chang, C.H., Hogg, S. and McKinney, S. (1987) Recursive Partition: a versatile method for exploratory data analysis in Biostatistics. MacNeil and Umphrey (eds), **5**, Biostatistics, 23-50

Ciampi, A. and Thiffault, J. (1989) Pruning Regression Trees for Censored Survival Data: The RECPAM approach, Commun. Statist.- Theory Meth. **18(9)**, 3373-3388

Härdle, W. (1989) The interplay between statistics and Numeric Knowledge, E. Diday ed., Nova Science Publ., New York

Hastie, T. and Tibshirani, R. (1986), Generalized Additive Models (with discussion) Statistical Science, **1**, 297-318

McCullagh, P. and Nelder, J.A. (1989) Generalized Linear Models, Chapman & Hall, London

Algorithms and Statistical Software

BOJA: A Program for Bootstrap and Jackknife

A. Boomsma, Groningen

SUMMARY

Main features of a computer program for bootstrap and jackknife analysis are summarized. An application of a correlation model is presented to compare resampling methods, and different methods for estimating bias, standard errors and confidence intervals. Future program extensions are mentioned briefly.

Key words: bootstrap, jackknife, bias, standard errors, interval estimation, correlation, number of bootstrap replications.

INTRODUCTION

Over the last decade an increasing interest in resampling methods can be observed. In particular, the theoretical appeal of the bootstrap is expanding (Hinkley, 1988). Also, bootstrap and jackknife methods have been applied for a great variety of statistical models. This enhanced interest, both theoretically and practically, leads to a concomitant need for reliable, well documented programs for resampling procedures. This paper introduces the BOJA program, intended to offer the researcher a skilful scientific tool in performing familiar types of bootstrap and jackknife analyses.

SCOPE OF THE BOJA PROGRAM

The scope of the applications, and main program features are exposed first.

a. Resampling procedures. Given a sample **X** (n x p) of n independent observations on p variables, either bootstrap or jackknife resampling can be done.

For the bootstrap the resampling procedures include the standard bootstrap (Efron, 1982), the balanced bootstrap (Davison, Hinkley & Schechtman, 1986), the extended (balanced) bootstrap, and nested bootstrap resampling schemes. The standard bootstrap method generates random samples of size n from **X**, resulting in B bootstrap samples $\mathbf{X}_1^*, \ldots, \mathbf{X}_B^*$. By the balanced bootstrap each of the n observations from **X** are resampled equally often. The extended bootstrap method does not fix the number of bootstrap replications B in advance: resampling stops when standard deviations (or coefficients of variation) of bootstrap replications $\hat{\theta}^*$ do not differ in subsequent sets of bootstrap samples, according to a specified stop criterion. Nested (second-level) bootstrap resampling is of specific interest for the calculation of (variance stabilized) bootstrap t confidence intervals (see Tibshirani, 1988).

For the jackknife BOJA offers three types of resampling procedures: the standard, the random, and the grouped jackknife (Efron, 1982). The latter two

Compstat 1990

methods could be used when the sample size n is large.

For a new release of BOJA the implementation of importance sampling (Johns, 1988), antithetic resampling (Hall, 1989), and other up-to-date bootstrap computational devices aiming at efficiency (Efron, 1990), is considered.

b. Bootstrap and jackknife estimation. Main objectives of bootstrap and jackknife methods are estimation of bias (more precise estimates of unknown parameters), estimation of standard errors and confidence intervals, and for the bootstrap in particular the approximation of sampling distributions. BOJA calculates the bootstrap and jackknife estimates required for these purposes.

In theory, there are a large number of methods in constructing bootstrap confidence intervals (see DiCiccio & Romano, 1988; Hall, 1988). In BOJA the standard, percentile, bias-corrected percentile, bootstrap t, and variance stabilized bootstrap t interval estimation methods are implemented. For the jackknife only the standard interval is calculated. Other estimation methods (see, for example, Wu, 1986) are perpended for the next release of BOJA.

c. External and internal statistical applications. BOJA can handle the input and analysis of parameter estimates $\hat{\theta}^*$, based on bootstrap or jackknife samples, that have been computed outside the BOJA program (i.e. external statistical applications). For general applications a researcher might thus take the following steps of analysis: (a) do the resampling part with BOJA, and save the resulting resamples; (b) execute the required statistical application outside BOJA, and save the parameter estimates $\hat{\theta}$ and $\hat{\theta}^*$, based on the original sample $\mathbf{X}$ and the resamples $\mathbf{X}^*$, respectively; (c) return to BOJA, and calculate bootstrap or jackknife statistics, conditional upon the (re)sample parameter estimates that were obtained outside the BOJA program.

In addition, BOJA supports complete resample analyses for a number of statistical models. These so-called internal statistical applications are dealing with means, medians, correlation coefficients (product-moment, multiple, partial), (co)variances, and estimation in linear regression and ANOVA models (randomized groups and block designs). Examples of useful options for some of these applications are: Fisher's variance stabilizing transformation for correlations, bootstrap on observations or on residuals in regression, and accounting for heteroscedasticity in ANOVA. The user's guide of BOJA gives a detailed description of the statistical applications that can be handled.

d. Input, output, help, and technical features. The BOJA program is menu driven and easy to use. The user interface operates with a multilevel menu system and pop-up windows. The specifications needed to run a particular application are made by means of data entry forms with choice lists. They can be saved, easily modified by screen editing, and be loaded instantly.

The sample data $\mathbf{X}$ can be read from file or imported by keyboard. The friendly user interface has attractive editing facilities: sample data can be changed, variables as well as observations be deleted, labels of variables be modified. The changed sample data matrix can be saved to file for future use. In new versions of BOJA missing value problems will be regarded.

Bootstrap and jackknife samples can be written to file. For each resample, four types of matrices can be saved, which contain either the complete resample data $\mathbf{X}^*$, or summarized information from $\mathbf{X}^*$ (covariance matrices, sums of squares and cross-products matrices, or correlation matrices).

The view options of the program allow the user to inspect the sample data and its summary statistics, as well as estimates based on bootstrap or jackknife samples and its summary statistics (bootstrap and jackknife statistics, including confidence intervals). Separate parts of information can be examined by way of scrollable screen regions, and effectively saved.

An important attribute of BOJA is that it operates with an elaborate, context sensitive help device. The meaning of each entry of the program can be called upon instantaneously, by pressing a single function key. Explanations and consequences of all program options are offered on line by this utility. Thus in fact, a summary of the user's guide is directly accessible.

The program is available for IBM PC's and compatibles, as well as for DEC computers (VMS, ULTRIX); mainframe versions are currently in development. The system requirement for personal computers is a minimum memory of 640 Kb RAM. A mathematical coprocessor is not imperative. The programming language is C.

e. Monte Carlo evaluation and graphical options. If the user knows or has a hypothesis about the value of a population parameter, the results of a bootstrap or jackknife analysis may be evaluated according to that knowledge or hypothesis. Also, (standardized) empirical sampling distributions of bootstrap replications and jackknife pseudo-values can be inspected effectively. BOJA has two main graphical options: histograms and quantile (Q-Q) plots.

The graphics option works device independent. The next release of BOJA makes it feasible to redirect the graphical output to printers adequately: the image will be built at the printer's resolution.

If extensive Monte Carlo simulations are needed to support theoretical research, a separate utility of BOJA may be rather helpful: selected output of each Monte Carlo simulation can be written to file for further analysis.

AN APPLICATION BY REPLICATION

As an illustration of the potential capabilities of BOJA, artificial data were used to estimate bias, standard error, and confidence intervals regarding the product-moment correlation coefficient ρ. Partly as a test to BOJA, Monte Carlo devices comparable to those of Efron (1981) and Tibshirani (1988) were used: given slight design modifications and extensions, this research amounts to a replication as well as to an elaboration of their simulation studies. All BOJA calculations were done on a MicroVAX 2000 computer system.

With IMSL routine RNMVN, 200 samples of size n = 15 were generated from a bivariate normal distribution with $\rho = 0.5$, that are used in all comparisons. Fisher's variance stabilizing transformation $\zeta = \tanh^{-1}\rho$ is also studied.

The following resampling methods were compared: The standard jackknife, denoted as J15 (deletion of one observation at a time); Bootstrap methods, B15, B100, B400, B1000, where the number refers to the number of bootstrap replications; Balanced bootstrap methods, BA15, BA100, BA400, and BA1000. Normal theory results are based on the 200 random samples **X**, without doing resampling. True, theoretical values are calculated according to normal theory assumptions. Balanced bootstrap results are presented only if they differ substantially from those obtained by the unbalanced bootstrap.

1. Estimation of bias and standard error

The results in Tables 1 and 2 on resample estimation of bias and standard errors for $\hat{\rho}$ and $\hat{\zeta}$, resemble those acquired by Efron (1981). In this example the true, theoretical bias of $\hat{\rho}$ equals -0.014. Over the 200 samples, averages (standard deviations) of jackknife (J15) and bootstrap (B1000) estimates of bias for $\hat{\rho}$ are -0.015 (0.031), and -0.012 (0.020), respectively. In similar cases, it has been observed before that the expectations of bootstrap and jackknife bias estimates are small compared with their standard deviations.

Table 1. Estimates for ρ, $\zeta = \tanh^{-1}\rho$, and their bias, based on 200 samples, n = 15, from a bivariate normal distribution with $\rho = 0.5$.

	$\hat{\rho}$		bias $\hat{\rho}$		$\hat{\zeta}$		bias $\hat{\zeta}$	
Method	Mean	SD	Mean	SD	Mean	SD	Mean	SD
J15	.497	.214	-.015	.031	.545	.287	.021	.038
B15	.473	.214	-.008	.055	.589	.312	.023	.077
BA15	.471	.214	-.011	.025	.582	.309	.016	.034
B100	.472	.214	-.010	.027	.586	.310	.020	.038
BA100	.471	.212	-.011	.020	.585	.309	.019	.028
B400	.470	.213	-.012	.021	.585	.310	.019	.028
B1000	.470	.212	-.012	.020	.584	.310	.018	.028
Normal	.482	.212	-.018		.566	.298	.017	
True value	.486		-.014		.569		.020	

As normal theory estimates for the standard error of $\hat{\rho}$ and $\hat{\zeta}$ the estimates $\hat{\sigma}(\hat{\rho}) = (1 - \hat{\rho}^2)/(n - 3)^{1/2}$, and $\hat{\sigma}(\hat{\zeta}) = 1/(n - 3)^{1/2}$ were used. Along with means and standard deviations (SD), Table 2 gives estimates of Pearson's coefficient of variation CV = SD/Mean, and the root mean squared error (RMSE) from the true values, calculated as $\sigma(\hat{\rho}) = 0.210$ and $\sigma(\hat{\zeta}) = 0.287$.

If the number of bootstrap samples B is small, the balanced bootstrap performs with less variability than the unbalanced bootstrap. In this example, standard error estimation by resampling methods is less stable than normal theory estimation: the normal theory, jackknife (J15), and bootstrap (B1000) average estimates of $\sigma(\hat{\rho})$ are 0.209 (0.055), 0.216 (0.086), and 0.199 (0.064), respectively. For the estimation of bias and standard error B = 400 bootstrap samples seems adequate enough. An increase of B beyond 400 for standard error estimation seems pointless, indeed (cf. Efron, 1981, p. 531).

Table 2. Estimates of standard error for $\hat{\rho}$ and $\hat{\zeta}$; 200 samples, n=15, ρ=0.5.

	$\hat{\rho}$				$\hat{\zeta}$			
Method	Mean	SD	CV	RMSE	Mean	SD	CV	RMSE
J15	.216	.086	.400	.086	.298	.075	.252	.076
B15	.196	.080	.406	.081	.282	.084	.299	.084
BA15	.193	.075	.390	.077	.273	.073	.269	.075
B100	.198	.068	.343	.069	.285	.061	.213	.061
BA100	.199	.067	.338	.068	.285	.060	.210	.060
B400	.199	.065	.326	.066	.285	.056	.197	.056
B1000	.199	.064	.320	.065	.286	.054	.190	.054
Normal	.209	.055	.265	.055	.289	0	0	.002
True value	.210				.287			

2. Estimation of confidence intervals

For the estimation of two-sided 90% confidence intervals for ρ, a number of methods were compared. Standard jackknife and standard bootstrap intervals are $\hat{\rho} \pm 1.761\, \hat{\sigma}_J$, and $\hat{\rho} \pm 1.645\, \hat{\sigma}_B$, respectively; $\hat{\sigma}_J^2$ and $\hat{\sigma}_B^2$ are the jackknife and bootstrap estimate of variance. Bootstrap interval results are all based on B = 1000 (first-level) bootstrap replications. Regarding (variance stabilized) bootstrap t methods the number of replications for variance, and transformation estimation were 25 and 100, respectively (see Tibshirani, 1988). The (exact) normal theory interval inverts the $\tanh^{-1}\hat{\rho}$ approximation.

Table 3 gives results for left and right interval endpoints, interval length, and percentages of noncoverage. It shows that the bias-corrected (BC) percentile and variance stabilized (VS) bootstrap t interval methods do relatively well. They outperform the standard and the bootstrap t method, at least when the results are based on ordinary resample estimates $\hat{\rho}^*$. However, when Fisher's $\tanh^{-1}\hat{\rho}^*$ transformation is used (indicated by **F** behind interval labels in Table 3) the latter methods perform substantially better (all bootstrap t intervals are within [-1,1] then), as does the jackknife method. Where no **F** results are presented, no differences were found within methods between ordinary estimation procedures using $\hat{\rho}^*$, and those using $\tanh^{-1}\hat{\rho}^*$. The balanced bootstrap gave basically the same results as the unbalanced one.

In comparing results from Table 3 with those obtained by Tibshirani (1988, Table 1a), notice that the bootstrap estimate of variance $\hat{\sigma}_B^2$ was used here.

In this example, it could be shown that for interval estimation, in order to obtain reasonably accurate estimates of quantiles in tails of sampling distributions, 1000 bootstrap replications are no luxury (cf. Efron, 1987).

Table 3. Estimates of two-sided 90% confidence intervals for ρ; 200 samples, $n = 15$, $\rho = 0.5$. **F** indicates that Fisher's transformation was used.

Interval	Left		Right		Length			% Noncoverage		
	Mean	SD	Mean	SD	Mean	SD	CV	Left	Right	Overall
Jackknife										
Standard	.10	.34	.86	.15	.76	.30	.40	12.0%	3.0%	15.0%
Standard (**F**)	.04	.31	.77	.12	.73	.25	.34	5.5%	3.5%	9.0%
Bootstrap										
Standard	.14	.30	.80	.14	.65	.21	.32	13.5%	3.5%	17.0%
Standard (**F**)	.10	.30	.76	.13	.65	.21	.33	10.5%	3.5%	14.0%
Percentile	.11	.31	.75	.13	.64	.21	.33	11.5%	4.0%	15.5%
BC percent.	.08	.30	.74	.13	.66	.21	.31	7.0%	5.5%	12.5%
Boot. t	-.07	.39	.84	.19	.91	.42	.46	3.5%	4.5%	8.0%
Boot. t (**F**)	.04	.32	.75	.13	.72	.27	.38	5.0%	4.5%	9.5%
VS boot. t	.07	.29	.74	.13	.67	.21	.30	5.0%	5.5%	10.5%
Normal (**F**)	.08	.27	.75	.12	.67	.16	.23	5.5%	3.0%	8.5%

Acknowledgement: The author expresses his gratitude to J. Metselaar, programmer of BOJA, for his efficient computational help in preparing this paper.

REFERENCES

Davison, A.C., Hinkley, D.V., & Schechtman, E. (1986). Efficient bootstrap simulation. **Biometrika**, **73**, 555-566.

DiCiccio, T.J., & Romano, J.P. (1988). A review of bootstrap confidence intervals (with discussion). **Journal of the Royal Statistical Society, Series B**, **50**, 338-354.

Efron, B. (1981). Nonparametric estimates of standard error: The jackknife, the bootstrap and other methods. **Biometrika**, **68**, 589-599.

Efron, B. (1982). **The jackknife, the bootstrap and other resampling plans**. Philadelphia, PA: Society for Industrial and Applied Mathematics.

Efron, B. (1987). Better bootstrap confidence intervals (with discussion). **Journal of the American Statistical Association**, **82**, 171-200.

Efron, B. (1990). More efficient bootstrap computations. **Journal of the American Statistical Association**, **85**, 79-89.

Hall, P. (1988). Theoretical comparison of bootstrap confidence intervals (with discussion). **The Annals of Statistics**, **16**, 927-985.

Hall, P. (1989). Antithetic resampling for the bootstrap. **Biometrika**, **76**, 713-724.

Hinkley, D.V. (1988). Bootstrap methods (with discussion). **Journal of the Royal Statistical Society, Series B**, **50**, 321-337.

Johns, M.V. (1988). Importance sampling for bootstrap confidence intervals. **Journal of the American Statistical Association**, **83**, 709-714.

Tibshirani, R. (1988). Variance stabilization and the bootstrap. **Biometrika**, **75**, 433-444.

Wu, C.F.J. (1986). Jackknife, bootstrap and other resampling methods in regression analysis (with discussion). **The Annals of Statistics**, **14**, 1261-1350.

Prototyping Dynamic Graphics Functions in S

L. A. Clark and D. Pregibon, Murray Hill

Much of the work in *dynamic graphics* for data analysis has been done on specialized workstations using specialized software. The result is quite impressive, highly efficient, but non-portable code. Portability is especially important in this area as critical assessment of many of the ideas has yet to be undertaken. We provide crude versions of some existing dynamic graphics operations in the (new) S language [1]. S is UNIX® based and widely distributed. Our functions rely on a recently developed device driver [3] for a portable and nearly device-independent system for algorithm animation [2]. We foresee application of our approach by anyone interested in exploring the power of animation techniques for data analysis.

1. Introduction

Current work [4] on dynamic graphics for data analysis has focused on high performance workstations and specialized applications (*e.g.*, point cloud rotation, brushing scatterplot matrices, linked scatterplots). For performance reasons, workstation idiosyncrasies had to be exploited in the various implementations, precluding any routine port of the software to different environments. Also if the user had a different sort of application in mind, then unfamiliarity with the hardware and software usually provided a high enough barrier that few attempted to implement their ideas. Finally, most systems exist outside the statistical computing environment. Thus basic techniques such as linear regression, scatterplot smoothing, and data transformation are not available and must either be recoded in the target language or otherwise done without. For all of these reasons, dynamic graphics have not been used by the statistical masses. They exist as demonstration systems, and for the most part, have been used on examples showing them at their best. Thorough testing and evaluation still awaits these systems.

Our approach is at the opposite end of the spectrum: (nearly) device-independent quasi-dynamic graphics. The goal is to provide such functionality in the (new) S language [1]. We use the `animate()` device driver [2] which serves as a generator for the Algorithm Animation System [3] which we abbreviate as AAS. The `animate()` device driver is related to the `postscript()` and `pic()` device drivers distributed with S. All produce an ascii file containing graphics commands in a language to be post-processed outside of S. Users see no graphics being displayed; graphics interaction in S is not possible. How then can this model be used to implement dynamic graphics? The answer is that the post-processor provides for interactive control of the pre-computed scenes. We use the `movie` post-processor of AAS which accepts an ascii file in the `script/intermediate` languages and renders an animation sequence in a selected window on a bit-mapped terminal or workstation. The `movie` system allows various control mechanisms and ways to interact with the sequence; these are demonstrated in the next section.

Why would anyone be interested in using these tools? One answer is that they provide perhaps the only means of access to dynamic statistical graphics techniques. Another is that the user can concentrate on what is best for display, rather than what can be computed and displayed in real-time. This frees developers to spend their time on prototyping graphics capabilities rather than developing fast update formulae. Finally the ability to explore certain ideas, modify them, and extend them is the primary benefit of our approach.

Compstat 1990

2. An overview of the `animate()` device driver and AAS

We illustrate the `animate()` device driver and the `movie` rendering software by a single example. This allows the flavor of the interaction to be appreciated while suppressing much of the detail. Certain features of both pieces of software not explicitly discussed below will become apparent in the specific examples in the next section.

We consider data [6] on the length of eruption and duration between eruptions of the geyser, Old Faithful, between 8am and 5:30pm for eight consecutive days in the summer of 1978. These data were collected to develop a simple rule to predict the time of the next eruption to be posted at the visitors information center. The data are organized as lists, `Length` and `Duration`, which contain the consecutive readings for the eight study days. Consider the following set of S expressions:

```
animate("geyser")
view(1)
     background(T)
     plot(unlist(Length), unlist(Duration))
for(i in 1:8){
     click()
     view(1)
          points(Length[[i]], Duration[[i]], pch = "o")
     view(2)
          plot(Length[[i]], type = "l", ylim = range(unlist(Length)))
     view(3)
          plot(Duration[[i]], type = "l", ylim = range(unlist(Duration)))
}
```

The script above starts by initializing the device driver with the `animate()` function. The argument `"geyser"` names the sequence and creates the system file `geyser.s` where graphics output will be directed. AAS allows for multiple views of an animation sequence. The S function `view()` indicates which view succeeding graphics commands apply to. In this case there are 3 views, the first containing the scatterplot of `Duration` versus `Length`, the second and third being time-order plots of `Length` and `Duration` respectively. The scatterplot persists throughout the animation sequence with the explicit call to `background(T)`. All other S graphics commands are local to each frame of the animation sequence, these being delimited by the `click()` function.

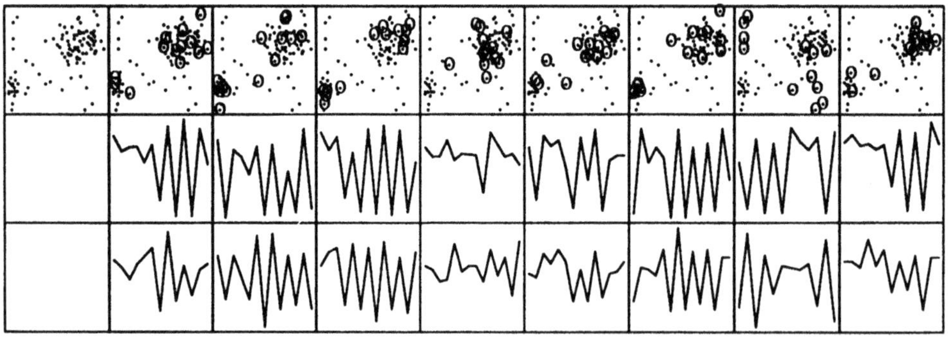

Figure 1. Stills from the animation sequence on predicting the next eruption of Old Faithful. Each row corresponds to a particular (numbered) view. Each column corresponds to a particular frame in the sequence. The first column is the scatterplot of `Duration` versus `Length` for all days. Successive columns display the data specific to each of the eight days.

At this point the user sees no graphics as all output is simply written to a file. The output itself consists of commands and labeled geometric objects in the `script` language of AAS. This file can then be processed (developed) by shell scripts either to produce hardcopy or for interactive viewing. The device driver supports interfaces to these shell scripts via the `stills()` and `movie()` functions respectively. Figure 1, which contains the 25 stills from the entire sequence, was produced by the S expression: `stills(framewid = .66, frameht = .66)`.

Although the interactive nature of animation is lost in hardcopy, certain features of these data are apparent in Figure 1. `Duration` tends to increase with `Length` for all days *except* Day 7, and within any given day, long `Durations` are followed by short ones and similarly for `Length`.

To demonstrate the interactive rendering software `movie`, we first embellish the animation sequence with labels using the S expression: `label(paste("Day", 1:8), background=T)`. The function `label()` creates an independent view named `label` indicating which day the readings correspond to. This view can manipulated like other named or numbered views. The `movie()` command creates a window on the screen (TTY5620/630 versions) in which the `movie` rendering software is downloaded together with the data for the current animation sequence. The time required for this process varies, but for this example the elapsed clock time was 45 seconds (at 9.6KB), 30 seconds of which is the overhead for downloading the `movie` software. Larger datasets or more complicated sequences can take upwards of a minute to download, this being viewed as (barely) acceptable. Once the `movie` window is created, other developed animation sequences can be downloaded without incurring the overhead of re-downloading the software.

A feature of the `movie` rendering software which requires no special allowance in the device driver is the ability to reshape, move, or overlay views in a multi-view animation sequence. These allow interactive changes to the aspect ratio of a view, enhancement of a particular view (see point cloud rotation), or construction of a complicated animation sequence from simpler ones, respectively. In Figure 2 we focus attention on Days 3 and 7, where the views for the `Length` and `Duration` series are overlaid. The pattern evidenced on Day 3, where the two series rise and fall together, is reproduced to a large extent for days not shown. Day 7 on the other hand exhibits the curious feature that as the `Length` series rises the `Duration` series falls and *vice versa*. Further inspection suggests that the two series are out of synchronization. This suggestion is corroborated by overlaying the `Duration` series on the `Length` series after offsetting it by one reading (panel 3). The offset results in an aligned pattern similar to that exhibited on the other days.

This anomaly in the data went undetected in the original analysis [6] but was subsequently pointed out [5]. The latter authors used overlays of vu-graphs to find and fix the problem. Since this technique is hardly routine, our device driver should make similar analyses less hit-and-miss.

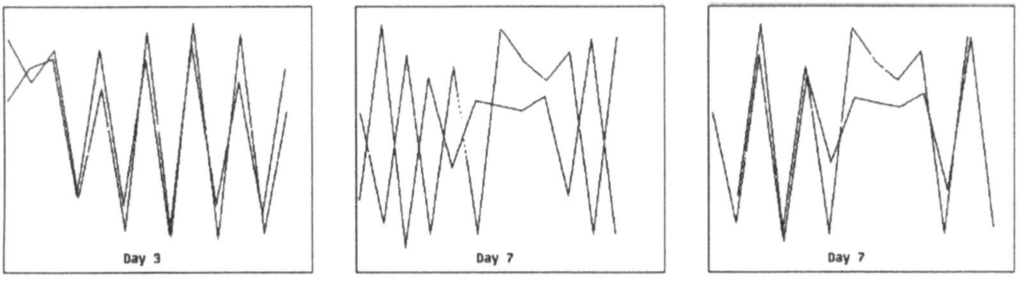

Figure 2. Example of whole and partial overlaying of plots. The first panel shows overlays of the `Length` and `Duration` series for Day 3. The second panels shows overlays of the `Length` and `Duration` series for Day 7. Overlays of the two series for other days are similar to that on Day 3 in that the series are fairly closely aligned. The third panel shows overlays of the `Length` and `Duration` series for Day 7 after the latter was visually offset by one reading.

This example was chosen to illustrate some of the flexibility of the `movie` rendering software. The key *control* features of the software, for speed, direction, and cruise control, were not and can not be demonstrated on paper. Imagination helps, but ultimately the advantages of animated graphs can only be appreciated by direct use of the software.

3. Some B Movies

In this section we illustrate the scope of the device driver with some functions in the S library `animate`. The illustrations are crude versions of dynamic graphical techniques which have been discussed in the literature, at conferences, or otherwise demonstrated. We say crude since our device driver imposes a far more restrictive type of interaction than is usually provided by existing systems for dynamic graphics. Specifically, since all scenes are precomputed in S, the only interaction possible is that provided by the animation control system on the precomputed scenes. In contrast, most tailor-made dynamic graphics systems compute on the fly using specialized software on a particular piece of hardware. Together with a flexible control system, this encourages improvisation. The cost, of course, is device-dependency, which we are attempting to avoid. The emphasis in this section is that at least for prototyping purposes, leveraging the S language and AAS results in a high degree of the functionality of the more specialized implementations.

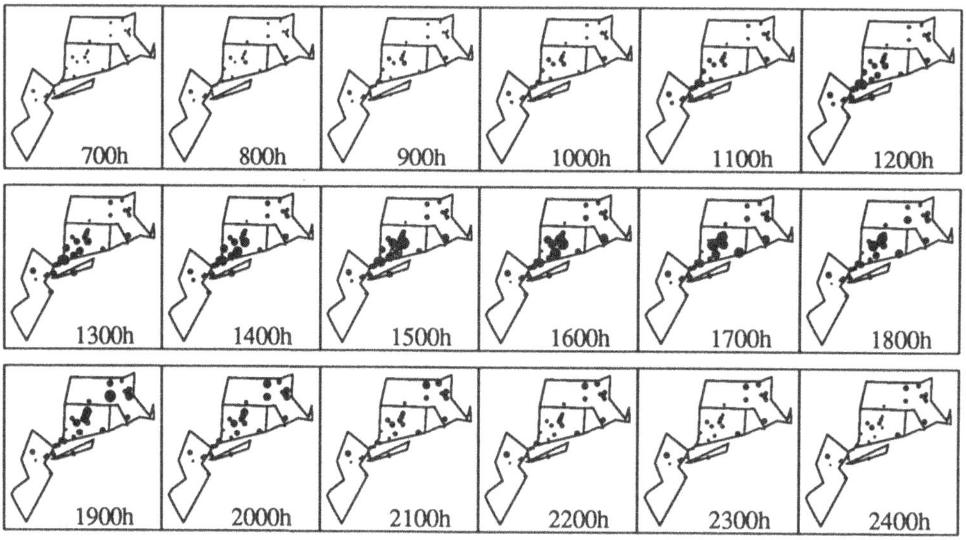

Figure 3. Time Series Maps. An animation sequence displaying the transport of airborne pollutants in the northeast corridor. Pollutant concentrations, coded by filled circles, are consecutive hourly maxima taken at 28 weather stations. It is apparent that high concentrations produced in the New York City Metropolitan area in the afternoon are effectively transported to the rural northeast several hours later. The S data structure `map` contains the coordinates of the exaggerated map of the northeast corridor, the data structure `stations` contains the coordinates of twenty-eight weather stations, and the matrix `ozone` the hourly maximum ozone concentrations recorded at these stations. The sequence was generated with the S expression `tsmap(stations$x, stations$y, ozone, map)`.

3.1. Time Series Maps

An early application of animation techniques concerns the display of temporal-spatial series. Specifically, consider a time series recorded at each of a fixed number of geographical locations. The usual time series plot consists of measurement versus time; there is no room for displaying the geographical information. The function `tsmap()` provides one solution by using the *x*-*y* plane to capture the geographic location, filled circles (at these locations) to code the measurement, and animation to capture time. The result is a two view animation sequence. The first view is the scatterplot of fixed geographic locations with (optional) background map. The second view consists of the labels associated with the time at which each measurement was taken. Figure 3 illustrates the idea. The most interesting aspect of this particular animation sequence is that it took only several minutes to produce. In comparison, the original movie of these data (circa 1970's) took several weeks to produce, while a more recent version using AAS directly [3] took several hours.

3.2. Scatterplot Linking

Consider the problem of tracking a subset of observations across several related but different plots. A particular case in point is a scatterplot matrix which displays all pairwise scatterplots of variables in a row-by-column layout. The dynamic graphic technique known as *brushing* is an example of how subsets of observations are linked across scatterplots. Standard implementations allow the user to mouse-select subsets of data points on one of the scatterplots with the corresponding points being highlighted on all other scatterplots. More general scenarios result in linking of selected scatterplots not necessarily resulting from a scatterplot matrix.

The function `link()` simulates subset highlighting in user-selected linked scatterplots. It takes an arbitrary number of arguments which specify the $x-y$ coordinates of each scatterplot, and a named argument `z` which specifies the subsets to be highlighted. If `z` is categorical, successive subsets correspond to the levels of `z`. If `z` is continuous, successive subsets correspond to deciles of `z`. The animation sequence consists of a (named) view for each scatterplot. Successive snapshots show the subsets being highlighted simultaneously in each view. Figure 4 provides an example.

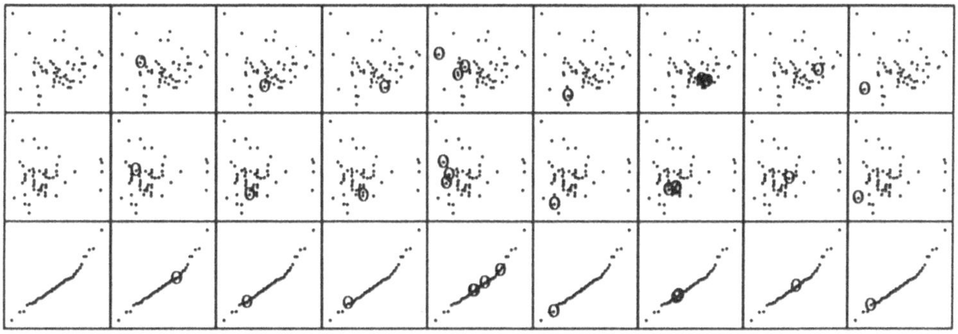

Figure 4. Scatterplot Linking. The data consist of average miles per gallon (mpg), `weight`, and `engine` size for 57 makes of automobiles classified according to `country` of origin. The first row contains the scatterplot of `resid` versus `weight`, the former obtained by regressing mpg on `weight`. The second row contains the scatterplot of `resid` versus `engine` size. The final row contains the normal probability plot of residuals. Successive columns identify automobiles from selected countries in each of the three rows simultaneously. For the most part, the residuals for these countries are below the median which means that the predicted mileage for these automobiles is greater than that observed. The sequence was generated with the S expression `link(scatter(weight,resid), scatter(engine,resid), qqnorm(resid), z=country)`.

3.3. Point Cloud Rotation

The single most popular dynamic graphical technique is arguably point cloud rotation. This permits a collection of points (x,y,z) in three dimensions to be viewed by utilizing the plane of the screen for two dimensions and motion for the third. Mouse interaction is the typical control mechanism whereby the speed, direction, and angle of rotation are varied.

We provide a crude version of point cloud rotation with the function `rotate()`, which rotates a point cloud around the vertical axis in equally spaced angles in $[0,2\pi]$. Each angle of rotation corresponds to a click. The animation system controls the direction and speed interactively.

Arbitrary rotations are possible, but not interactively. The argument `tilt` specifies an initial tilt of the data in the screen axis. Thus, rotation about the horizontal axis, `tilt=`$\pi/2$, or along any arbitrarily pre-specified tilting of the $x-y$ plane is possible.

A second view is included in `rotate()` which shows the coordinate axes spinning as the angle of rotation changes. This is an important visual cue to orient the scatterplot view. The scales of the two views are matched so that they can be overlaid if desired. The animation sequence in Figure 5 illustrates a rotating point cloud with an initial $\pi/8$ rotation in the $x-y$ plane.

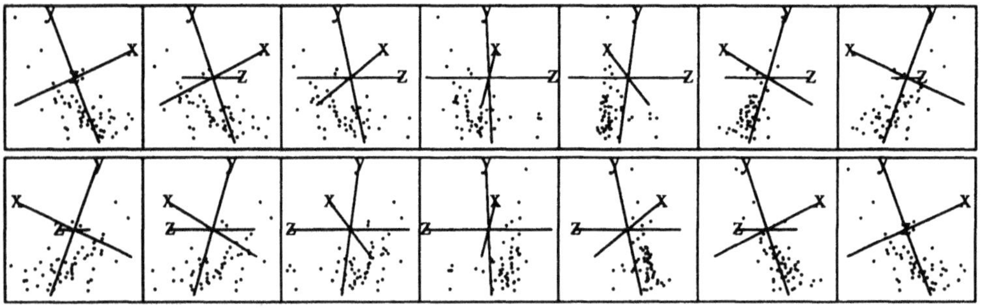

Figure 5. Point Cloud Rotation. The $x-y-z$ data are `weight-mpg-engine` respectively. Each frame corresponds to a specific angle of rotation from 0 (top left) to 2π (bottom right). The data were initially tilted by $\pi/8$ degrees in the $x-y$ plane prior to rotation about the vertical axis. The sequence was constructed with the S expression `rotate(weight, mpg, engine, tilt=pi/8, n=14)`.

4. User Interface

The functions presented in the previous section are designed to be as simple to use as possible. Arguments (x, y, z) are used consistently in the same order and manner. The functions behave satisfactorily whether or not z is categorical. Additional arguments are provided for more control.

For users who want something just a little different than we allow, the functions can be easily modified as they are S functions like any other. Editing functions is encouraged as the code is structured to be fairly self-documenting. Indeed, the primary motivation for our work is to allow users to develop and tailor dynamic graphics techniques to their own particular application. As a simple but illustrative example, we modified the `rotate()` function to accept an optional argument, `curve`, a curve in three dimensional space which generalizes the idea of a (two dimensional) scatterplot smooth. In our implementation the curve is assigned a view of its own which is rotated in parallel with the views for the data and axes. This adds two `if` expressions, each consisting of a few lines of code, to the original 20 or so. We have also developed related functions for animating point clouds, one for stereo viewing, and another based on slicing rather than rotation.

5. References

[1] RA Becker, JM Chambers, and AR Wilks (1988), *The New S Language*, Wadsworth, Pacific Grove, CA.

[2] JL Bentley and BW Kernighan (1986), *A System for Algorithm Animation*, Comp. Sci. Report 132, *AT&T*.

[3] LA Clark and D Pregibon (1990), *An Animation Device Driver for S*, Proc. Stat. Graphics ASA, Anaheim.

[4] WS Cleveland (1989), *Dynamic Graphics*, Wadsworth, Belmont CA.

[5] L Denby and D Pregibon (1987), *Example of the Use of Graphics in Regression*, American Statistician **41**.

[6] S Weisberg (1980), *Applied Linear Regression*, John Wiley, New York.

Programming Languages for Statistical Computation

M. Harman, S. Danicic and R. Gilchrist, London

Summary

Several workers in the field of statistical computation have pointed to the need for a statistical package which would include a programming language specifically designed for statistical applications[3].

Some interest has been shown in the possible applicability of the functional programming style. For example an implementation of a statistical language based on the functional programming language Lisp has been carried out by Tierney[8]. The general purpose language S has also been designed and written with statistical programming in mind by Chambers[1].

With the aim of providing a flexible approach to designing a language for statistical computation the authors have created a programming language prototyping system. This system allows rapid implementation of interpreters for new and mutually quite different languages, all of which are based on a core of statistical routines. Examples of the use of a working prototype, SOL, are given. SOL incorporates generalized linear modelling facilities in a Pascal-like framework extended to include functional language facilities.

Keywords : Statistical Programming, Functional Programming, Imperative Programming, Pascal, Generalized Linear Models.

Introduction

Experience with the FUNIGIRLS system[2] demonstrated the expressive potential of functional programming notation. However, in comparison with more conventional programming languages, functional languages have no global state and no sequencing of instructions. Moreover strong typing prevents the use of some attractive notation, such as the Wilkinson and Rogers' model definition syntax. This has been considered a drawback by some statisticians[2]. In contrast, the syntax of SOL closely resembles that of Pascal, with assignment, procedures and loop constructs. There is a built-in matrix data type and functions are provided for matrix manipulation and model-fitting. Wilkinson and Rogers' model definition is exactly as in GLIM3.77. SOL also provides an untyped list data structure.

Compstat 1990

In passing we note that our prototyping approach relies on result concerning isomorphisms between functional and procedural semantics, the details of which are contained in [5].

A complete description of SOL can be found in [4]. Here we give examples of SOL programs. Built-in functions are underlined, keywords are emboldened.

Functional Style Facilities

Reduce is an example of a higher-order function. It applies a function 'inbetween' the elements of a list with a given definition 'ABaseCase' for the empty list. For example if Add were the prefix version of Addition then Reduce(Add(),[1,2,3],0) would be 1+2+3+0. We can use Reduce to define a sort function succinctly. Unlike Pascal functions and procedures defined by a single statement require no **begin ... end** *bracketting.*

```
function Reduce(Afun,ABaseCase,AList);
if null(AList)
   then
   Reduce:=ABaseCase
   else
   Reduce:=Afun(head(AList),Reduce(Afun,ABaseCase,tail(AList))) ;
```

Insert adds an element x to a pre-sorted list l :

```
function Insert(x,l) ;
if null(l) then Insert := [x]
           else if x < head(l)
                   then Insert := cons(x,l)
                   else Insert := cons(head(l),Insert(x,tail(l)))

Sort := Reduce( Insert() , [] ) ;
write(Sort( [5,3,0,2323,6.9097,2] )) ;
writeln(Sort( ['Imelda','Cliff','Bob','Seb','Tony','Dave'] )) ;
```

Procedural Style Facilities

SOL also has more traditional programming constructs such as procedures, for-loops and variable parameters. To illustrate, we show how the SOL programmer could write the Updating algorithm in a style very close to Pascal, the main difference being the lack of type information in the procedure declaration. This algorithm updates the current mean vector XBAR *and the sums of squares and products matrix A, stored as a one-dimensional array, these quantities being based upon* k *previous observations of p-vectors* X. *A new p-vector* X *is added in, to return new values for* XBAR *and* A.

```
procedure Update( VAR A , VAR XBar, VAR k, VAR X, p ) ;
begin
   OneDIndex := 1;
   for i := 1 to p do begin
        X[i] := X[i] - XBAR[i] ;
        XBAR[i] := XBAR[i]+X[i] / (k+1) ;
        for j := 1 to p
          do begin A[OneDIndex] := A[OneDIndex]+k*X[i]*X[j]/(k+1);
                   OneDIndex := OneDIndex + 1 end (* for j *)
        end (* for i *) ;
   k := k+1 (* one extra reading so update reading count *)
end (* Update procedure *) ;
```

Fitting Generalised Linear Models

Generalized Linear Models[6] can be fitted to design matrices either using standard, built-in, model-fitting functions which include the standard distributions (Normal, Poisson, Binomial etc.) with appropriate link functions, or models can be fitted from first principles using the built-in ModelFit function.

Fitting Standard Distributions using Built-in functions

We can use standard error distributions and model definition syntax as in GLIM 3.77. For example the following code applies the Poisson, log-link fit to data in a two-way table classified by ulcer and blood type. In passing we show the use of the GL function which generates levels of a factor. Note that <<n,p,L>> constructs an n by p matrix the elements of which are drawn from the list L.

```
Mean := init(9,1,1.0) ;
Count := << 9,1,   [983,383,2892,697,416,2625,134,84,570] >> ;
Ulcer := GL(3,1,9) ;
Blood := GL(3,3,9) ;
(* Now define ulcer and blood types as factors *)
Ulcer := Fac(Ulcer) ; Blood := Fac(Blood) ;
Result1 := Poisson(LogLink, with yvar count, explained by mean);
(* Optional syntax with yvar, explained by similar to Minitab *)
Result2 := Poisson(LogLink, count, Mean + Blood + Ulcer) ;
```

Model-Fitting from First Principals

The following functions are used to define an Inverse Gaussian model with canonical link. The assignment to the variable InvGauss yields a function requiring parameters **yvar** *and* **design** *and returning model statistics for the Inverse Gaussian model. Init(38,1,1.0) creates an 38 by 1 matrix the elements of which are all set to the value 1.0. It is used here to provide a matrix for the prior weights, which are, in this case, all assumed to be 1.0.*

```
(* Variance function *)
function va(mui) ; va := mui^3 ; (* ^ is the power function *)
(* deviance increment *)
function di(yi,mui) ; di := (yi-mui)*(yi-mui)/(mui*mui*yi) ;
(* Link function *)
function fv(etai) ; fv := (0.5/etai)^0.5 ;
(* Starting value for linear predictor *)
function ls(yi) ; ls := 0.5/(yi*yi) ;
(* derivative of the link function *)
function dr(mui) ; dr := -1/mui^3 ;

InvGauss := ModelFit( [va,di], init(38,1,1.0), [fv,dr,ls]) ;
```

As an illustration, we may fit an Inverse Gaussian model to the failure times of air-conditioning in aircraft[7]; for simplicity we have only considered two aircraft. The first of which has 23 recorded failures and the second has 15.

```
y := <<38,1,  [413,14,58,37,100,65,9,169,447,184,36,201,118,4,
               31,18,18,67,57,62,7,22,34,74,57,48,29,502,12,70,
               21,29,386,59,27,153,26,326] >> ;
x := Fac(Gl(2,23,38)) ;
mean := init(38,1,1.0) ;
xmat := cat(mean,x) ; (* concatenate mean and x *)
Result1 := InvGauss(y,mean) ;
Result2 := InvGauss(y,xmat) ;
```

The Model Statistics

Fitting a model using ModelFit or one of the standard error distribution functions yields a list of model statistics containing : The Vector of deviances, the total deviance, the degrees of freedom, the vector of parameter estimates, the vector of values fitted according to the GLM, the scaling parameter and the co-variance matrix. The model statistics from the previous two examples could be printed by a user-defined procedure as shown below :

```
procedure ModelOutput(AModel) ;
begin
writeln('+---------------------------------------+') ;
writeln('|          Model   Statistics           |') ;
writeln('+---------------------------------------+') ;
writeln ;
writeln(' Total deviance ...................:',Index(2,AModel)) ;
writeln(' Degrees of freedom ...............:',Index(3,AModel)) ;
writeln(' Parameter estimates ..............:',Index(4,AModel)) ;
writeln(' Fitted Values ....................:',Index(5,AModel)) ;
writeln(' Scale Parameter ..................:',Index(6,AModel)) ;
writeln(' Co-variance matrix of estimates ..:',Index(7,AModel)) ;
writeln ;
writeln('       --- o ---') ;
end ;
```

Index(i,L) is a built-in function that returns the ith element of the list L.

Areas for Possible change in future prototypes

There are a number of features of the language which will be the subject of attention in the definition of any future prototypes :

Extra Statements

The main advantage of prototyping is that it will allow us to add and remove statements from the language definition, yielding a new prototype in a matter of days. There are many extra statements that statisticians may wish to try out in future prototypes.

User-defined Data types

Many modern programming languages allow the user to define their own data types. This may be desirable for statistical computations, and could be added into future language prototypes.

Aknowledgements

Mark Harman and Sebastian Danicic developed the prototyping system, Mark Harman designed and implemented SOL, Anthony Scallan wrote the statistical routines used by SOL and Robert Gilchrist added statistical input.

References

1. Chambers, J M : A Computing Environment for Statistics. *AT&T Statistical Research Reports* **40**. 1987.

2. Gilchrist, R & Scallan, A : FUNIGIRLS : A prototype functional programming language for the analysis of Generalized Linear Models. COMPSTAT Proceedings in Computational Statistics, Physica Verlag, Heidelberg, pp213-218, 1988.

3. Giusti, A N & Marchetti, M : Matrix languages for Statistics. *Bulletin of the ISI* **1**, Paris pp 381-2, 1989.

4. Harman, M & Danicic, S : SOL1 : A First Prototype Statistical Language. Internal Publication Department of Computing Mathematics and Statistics, The Polytechnic of North London. 1990.

5. Harman, M & Danicic, S : Functional Reasoning for Procedural Programs. Internal Publication Department of Computing Mathematics and Statistics, The Polytechnic of North London. 1990.

6. Nelder, J A & McCullagh, P : *Generalized Linear models*. Second Edition. Chapman and Hall, London, 1989.

7. Proschan, F : Theoretical explanation of observed decreasing failure rates. *Technometrics*, **5**, 375-385, 1963.

8. Tierney, L : XLISP-STAT : A Statistical Environment based on the XLisp language. *University of Minnesota tec. rept.* **528**. 1989.

Expert Systems in Statistics

Statistical Knowledge-Based Systems - Critical Remarks and Requirements for Approval

K. M. Wittkowski, Tübingen

1. What is a 'statistical knowledge-based system'?

The term 'statistical expert system' has been used extensively, although it is still ill-defined. We will call a program a *statistical expert system* (SES), if it can undertake actions specific for experts in the field of statistics (or a sub-field thereof) through application of knowledge-bases (WITTKOWSKI 1989).

Under this definition, it is necessary that the system uses a *knowledge-base* (KB) (of variables, study design, hypotheses, methods, programs, domain of application etc.) to draw conclusions. Thus a knowledge-based system is qualitatively different from more conventional programs (that might as well undertake some actions specific for experts) in that it handles data (numbers, characters ...), data structures (relations, matrices ...), and semantics (rules, frames, ...). In a knowledge-based system, these semantics are stored without redundancy as opposed to ordinary systems, where similar or even identical pieces of knowledge are either replicated several times within the program or have to be entered several times within a session.

If the 'expertise' consists primarily of application of relational operators and numerical algorithms, or the 'knowledge-base' is either redundant or not an integral part of the decision process, even a sophisticated program will not be considered 'knowledge-based'. In order to emphasize the role of the KB, we will use the term '*statistical knowledge-based system*' (SKBS) rather than the more ambigiuous 'statistical expert system'.

Compstat 1990

2. Which benefits could be expected from an SKBS ?

Some SKBS might be used by expert statisticians as a substitute of a highly specialized colleague. These systems would be comparable to 'conventional' expert systems (e.g. for medical diagnosis), in that the KB is confined to a single sub-field of statistical methodology and the user is assumed to be familiar with the study design, the statistical terminology and the mathematical background, lacking only some more detailed knowledge in this sub-field. Such systems will probably be developed soon and the first systems are already emerging. Unless algorithmic solution based on the mathematical foundation of statistics will finally prove more efficient than heuristic strategies, the value of these SKBS is beyond question.

Other SKBS might assist expert statisticians through accelerating tedious tasks, where a substantial proportion of decisions could be derived from a KB build during previous analysis tasks. Except for the initial step of knowledge acquisition, the expert could even delegate some tasks to the (statistically) naive user. These systems might be more difficult to develop, because different types of knowledge (and knowledge representation techniques) must be considered. On the other hand, the expert would be freed to concentrate on more important activities (including more extensive consultation).

These SKBS might also help solving a current problem in the use of statistical analysis systems. The frequency of type-III errors (application of inappropriate, though numerically correct programs) proves that these systems are often used by non-statisticians. Because faulty algorithms usually have less impact on the result than type-III errors, even use of 'approved' systems has only a minor impact on the quality of analyses. The more 'user-friendly' these systems become, the more frequent type-III errors seem to be. An SKBS, detecting these 'semantical' errors, might substantially improve the quality of analyses. Whether or not these benefits will outweigh the risk of having more non-statisticians performing analyses, will ultimately depend on the quality of the KB.

3. How could a SKBS be designed ?

In ELLIMAN, WITTKOWSKI (1987), an SKBS has been logically divided into INTERACTION, DOMAIN, STRATEGY, SEMANTICS, ACCESS, and EXECUTION layer (Fig. 1). The importance of these layers depends on the area of application and the intended user (WITTKOWSKI 1988a). The INTERACTION layer is especially important for (statistically) naive users, the DOMAIN layer is obsolete, if the system is independent of the area of application, etc.. We will first discuss the importance of the KB associated with these layers using the PANOS system (WITTKOWSKI 1985) as an example.

In PANOS, decisions are typically based on choices (alternatives), which might be either active or inactive, depending on the rules (STATISTICS KB) operating on facts (MODEL, DESIGN, DATA KB). Each rule is associated with a type of field in the PANOS screen form (WITTKOWSKI 1988b). If more than a single choice remains active (after application of DATA, DESIGN, and MODEL knowledge), the user has to make a decision. If the user's decision causes (temporary) inconsistencies, corresponding fields start blinking until all inconsistencies are resolved (the corresponding part of the KB is not shown in Fig. 2).

Linking rules to field types simplifies both evaluation and application of the KB, because only a small part of the KB has to be considered (by either the investigator or the system). Thus the STATISTICS KB is more similar to conventional programs than a 'typical' unstructured KB, where the systems decisions are time consuming and difficult to predict. On the other hand, PANOS differs from conventional programs in that the DATA, DESIGN, and MODEL KB are univer-

	U S E R	
dialog KB	INTERACTION	user KB
	DOMAIN	
model KB (D)	STRATEGY	statistics KB (P)
design KB (D)	SEMANTICS	methods KB (D)
data KB (D)	ACCESS	programs KB (P)
data	EXECUTION	*programs*

Fig. 1 Layers of a SKBS (D/P = declarative/procedural KB)

sally accessible (no information has to be entered more than once), that the STATISTICS KB has no redundancy (the rules apply to all types of experimental designs and domains of application), and that rules are based on *a-priori* knowledge and analysis goals (e.g. whether differences in expectation, tendency, or distribution are of interest) as opposed to rules based on the empirical/assumed distribution of data or residuals (see WITTKOWSKI 1986 and the discussion below).

```
FieldType     'RelatedToObsUnitType[A]' { ... [H] }
  AddChoice     'Unlinked'
    DeactIf       DESIGN.IsLinked and DATA.Exists
  AddChoices    'Crossed', 'Nested'
    DeactIf       DESIGN.IsUnLnkd and DATA.Exists
FieldType     'NumberOfObsUnits'
  AddChoice      DESIGN.NrOfUnts(MODEL.RelatedToObsUnitType)
FieldType     'Format'
  AddChoice     'Alpha'
    DeactIf        IsValidSIUnit(MODEL.Units) or
                   DATA.Exists and  MODEL.Format <> 'Alpha'
  AddChoices    'Numeric', 'Exponential'
    DeactIf        DATA.Exists and  MODEL.Format =  'Alpha'
FieldType     'Level'
  AddChoices    'Absolute', 'Interval'
    DeactIf        MODEL.Format = 'Alpha'
  AddChoices    'Ordinal', 'Nominal'
FieldType     'Type'
if not DATA.IsDerived then
  AddChoice     'Continuous'
    DeactIf        MODEL.Level = 'Nominal'
  AddChoice     'Discrete'
else
  AddChoice     MODEL.DerivedType(DATA.SourceVariables)
FieldType     'Influence'
  AddChoices    'DISTRI', 'TNDNCY', 'EXPECT'
    DeactIf        MODEL.Level < 'Interval'
  AddChoice     'MONTNY'
    DeactIf        MODEL.Level = 'Nominal'
FieldType     'Analysis'
  AddChoice     'TEST'
    DeactIf        MODEL.Level = 'Nominal' and
                   MODEL.SomeInfluenceIs ['TNDNCY','MONTNY']
                   or
                   MODEL.Level < 'Interval' and
                   MODEL.SomeInfluenceIs ['EXPECT','LINEAR']
  AddChoices    'SAVE', 'PLOT'
```

Fig. 2: Examples of rules from the PANOS STATISTICS KB.

4. Why have so many approaches to build a SKBS failed ?

RX (BLUM 1982) and REX/STUDENT (PREGIBON, GALE 1984; GALE, PREGIBON 1984) were among the first attempts to develop a SKBS. The goal of RX was to model the whole cycle of generating and testing hypotheses, integrating the results into the (medical) DOMAIN knowledge base, and, finally, generating further hypotheses. The goal of REX/STUDENT was to let the system learn statistical strategies (e.g. for linear regression) from the user. Neither of these approaches has resulted in a version that could be made available for use, probably because they were too ambitious even for the present (1990) state of the art. Nonetheless, they have helped in identifying actions that could be undertaken by a SKBS.

Another problem leading to failure was the type of knowledge implemented. Many systems were based on rules that may be misleading (WITTKOWSKI 1986, HAUX 1989, MOLENAAR 1989). For instance, what is a 'small proportion' (Fig. 3, RULE-1) and how affects eliminating records the overall result ? RULES-2/3 have been discussed in WITTKOWSKI (1986). If rules are based on 'significance' (e.g. RULE-4), it is often not clear, how power, level, and sample size are weighed.

<u>RULE-1 (BLUM 1982)</u>

IF only a small proportion of patients in the study is affected by a variable
AND the variable is present thoughout those records,
THEN eliminate those records from the study.

<u>RULE-2 (GALE, PREGIBON 1984)</u>

IF the distribution of y is unduly skew
AND the sign of y is positive
THEN assert that logarithms of y should be used

<u>RULE-3 (PORTER, LAI 1983)</u>

IF the scale is ordinal
THEN use a rank test

<u>RULE-4 (PFEIFFER 1988)</u>

IF the KOLMOGOROFF-SMIRNOF-Test rejects the hypothesis of a Gaussian distribution,
THEN consider a BOX-COX-Transformation

Fig. 3 Examples of rules commonly used in SKBS

5. Are there any SKBS available ?

GLIMPSE (NELDER, WOLSTENHOLM 1986) guides the user in using GLIM for fitting a (hypothetical) generalized linear model to some data (see WITTKOWSKI 1988b for an example). Given the complexity of understanding link-functions and interactions/high-order terms, the user will typically have to be an expert in the field of generalized linear models. Although the GLIMPSE KB contains rules like RULE-2 and is thus subject to the same criticism as the REX KB (MOLENAAR 1989), GLIMPSE could be considered as an example of an SKBS with a highly specialized KB.

KENS (HAND 1987) answers questions concerning non-parametric statistics and gives definitions, examples, and references. It is sometimes referred to as an 'expert system', although the author himself suggests that it might be better described 'as halfway between a question answering system and an information retrieval system' (l.c.). If KENS is not an 'expert system', then what is an 'expert system' or, more precisely, a SKBS ? We will try to apply the above definition to the following example.

6. What makes a statistical system 'knowledge-based' ?

CADEMO is aimed at 'consulting statisticians' and users with 'one week of training in experimental design' (RASCH, JANSCH 1989). Its KB consists of a tree of networks ('modules'). Each node is linked to an entry point to a hypertext-type dictionary, a list of references, and a screen-form with procedures (often computationally intensive) and 'rules' associated to its fields. Depending on the type of node, parameters are entered, a numerical result or message is presented, or the user selects the next node.

Is CADEMO an expert system ? Actions like computing sample sizes in non-trivial designs are specific for expert statisticians. Thus the first part of our definition is fulfilled. Is CADEMO knowledge-based ? Although the author was unable to obtain examples of the knowledge implemented, some conclusions can be drawn from the examples given in the manual (l.c.). Within a module (e.g. 'comparison of two means') CADEMO stores all parameters entered or computed (e.g. s[i], σ[i], δ, β, α, n[i]) and the path of previously passed nodes (i.e. the user's decisions) in a 'declarative KB' local to the module. The 'procedural KB' ('if the user chooses option X then goto node N') is local to the node. Neither 'KB' can be shared across modules, e.g. between 'comparison of two means' and 'clinical trials'. The 'explanation facility' displays the path and the list of parameters.

There are other systems with comparable functionality (RASCH 1989). If they are all 'expert systems', then <u>any</u> program is a 'knowledge base', any operating system an 'inference engine' and any input/output system an 'explanation facility'. Only those systems should be called 'knowledge-based' (or expert systems), where the system itself can handle semantics in a way that conforms to the (still less than perfect) definition given in Section 1.

7. How could a SKBS be approved ?

We have seen attempts to develop systems that leaped too far and we have seen ordinary programs advertised as 'knowledge-based'. The future of SKBS will probably lie somewhere in between. We have also seen that misleading rules might impair both the performance of the system and the adequateness of its decisions. In order to be able to separate the chaff from the wheat (is there a KB and is the knowledge appropriate ?), all rules (or equivalent representations) for each SKBS must be published in a similar manner as algorithms. All underlying concepts (assumptions, schools) have also to be given. In a second step, the performance of the system (e.g. how efficiently rules are linked via 'meta-rules') could be checked against human experts belonging to the same school by a 'TURING-test' (HAUX 1989, WITTKOWSKI 1989). Unless the KB is published and approved, we should be careful in assuming that any 'statistical system does the thinking for users' (THOMPSON E, *Digital Review*, Feb 6, 1989).

REFERENCES

BLUM RL (1982) *Discovery and representation of causal relationships from a large time-oriented clinical database: The RX project.* Berlin: Springer.

BOARDMAN TJ (ed., 1986) *Computer Science and Statistics.* Washington, DC: ASA

EDWARDS D, RAUN NE (eds., 1988) *COMPSTAT 1988.* Heidelberg, FRG: Physica

ELLIMAN AD, WITTKOWSKI KM (1987) The impact of expert systems on statistical database management. *Stat. Software Newsl.* **13**, 14 – 18

FAULBAUM F, UEHLINGER HM (eds., 1988). *Fortschritte der Statistik-Software 1*, Stuttgart: Fischer.

GALE WA, PREGIBON D (1984) Constructing an expert system for data analysis by working examples. In: HAVRANEK, SIDAK, NOVAK (eds., 1984) 227 – 236

HAND D (1987) A statistical knowledge enhancement system. *J. Royal Stat. Soc. A* **150**, 334 – 345

HAUX R (1989) Statistische Expertensysteme. *Biometrie und Informatik in Medizin und Biologie* **20**, 3 – 29

HAVRENEK T, SIDAK Z, NOVAK M (eds., 1984) *COMPSTAT 1984.* Wien, A: Physica

MOLENAAR IW (1989) Statistische Expertensysteme - Kommentar. *Biometrie und Informatik in Medizin und Biologie* **20**, 44 – 47

NELDER JA, WOLSTENHOLM DE (1986) A front-end for GLIM. In: BOARDMAN TJ (ed.) 113 – 117

PFEIFFER KP (1988) Entscheidungshilfen für die Modellauswahl am Beispiel der Diskriminanzanalyse. *Österr. Zeitschrift für Statistik und Informatik* **18**, 146 – 161

PORTER KM, LAI PY A statistical expert system for analysis determination. *Proc. ASA Stat. Comp. Sect.*, 309-311

PREGIBON D, GALE WA (1984) REX: an expert system for regression analysis. In: HAVRENEK T, SIDAK Z, NOVAK M (eds.) 242-8

RASCH D (1989) Statistische Expertensysteme – Kommentar *Biometrie und Informatik in Medizin und Biologie* **20**, 55 – 56

RASCH D, JANSCH S (1989) *Computer aided design of experiments and modelling – CADEMO version 2.1 – Handbuch.* Wiesbaden, FRG: HAND GmbH

WITTKOWSKI KM (1985) *Ein Expertensystem zur Datenhaltung und Methodenauswahl für statistische Anwendungen.* Stuttgart, FRG: PhD Dissertation; Internat. Bookseller Hartinger, Xantener Str. 14, D-1000 Berlin 15

— (1986) An expert system for testing statistical hypotheses. In: BOARDMAN TJ (ed.) 438 – 443

— (1988a) Building a statistical expert system with knowledge bases of different levels of abstraction. In: EDWARDS D, RAUN NE (eds.) 129 – 134

— (1988b) Dialogformen intelligenter Benutzerschnittstellen für statistische Auswertungen. In: FAULBAUM F, UEHLINGER HM (eds.) 212-225

— (1989) Statistische Expertensysteme – Kommentar *Biometrie und Informatik in Medizin und Biologie* **20**, 61 – 63

New Approach to Guha-Method from the Reliability Viewpoint

V. V. Kuzmenkov and O. I. Terskin, Moscow

Abstract. The goal is to show the importance of the sentence length account while the application of the GUHA-method takes place. We present the pattern of such account obtained by utilization of Bayes theorem.

Introduction

There are two independent ways of modern data analysis procedures design. On the one hand it is creation of statistical expert systems aligning well developed methods of multivariate analysis with the procedures of logic inference. On the other hand it is the development and justification of the new methods which comprise both logic and statistical inference.
The latter approach seems to be of great importance. The logic inference is of much more deductive nature and the predicate calculus has well validated axiomatic. At the same time statistical procedures generally possess distinctly inductive nature. One of such methodologies is realized in GUHA-method [1], the method of automatic hypotheses formation and testing.

Associative GUHA's peculiarities [2]

We understand the associative GUHA-method as a procedure for automatic listing of the hypotheses as follows :

$$k \sim_{\alpha} \lambda \qquad (1)$$

where k, λ - are the conjunctions of the derived features F_i which belong to the observational matrix $W_{M,k}$; $\sim$ - associative quantifier and α - the chosen level of significance.

Compstat 1990
© Physica-Verlag Heidelberg for IASC
(International Association for Statistical
Computing, 1990

The process of testing for the closed sentence amounts to the testing of the statistical hypothesis of mutual independence of k and λ in the four fold contingency table (FFCT). The observed frequencies in this table is obtained by the forming of the sentences like $k\sim\lambda$; $k\sim\bar{\lambda}$; $\bar{k}\sim\lambda$; $\bar{k}\sim\bar{\lambda}$, where "$\bar{a}$" means negation of property "a". The initial data for hypothesis formation is an observation matrix of " object-feature " type. Processing such data we obtained a set of sentences $\{Sent\}$ of different lengths which determine the correlation between causes and consequences. The number of such hypotheses may reach dozens or hundreds for commonly used significance level (α=0.05). This peculiarity is mentioned by the authors of GUHA-method [1]. The problem of GUHA's sentences set reduction is discussed in [3] and there we can find the attempt to solve this problem using the notation of "kernel" as more common part of GUHA's sentences. However the procedure based on "kernel" or "irreducibility"[1] principle is not adequate for many data analysis problems because it doesn't explain the reason of "kernel" existence and is based on the user intuitive ideas about the testing system.

Testing of GUHA's efficiency with reliability structural diagrams (RSD) restoration

We consider the scheme presented in Fig. 1 as RSD.

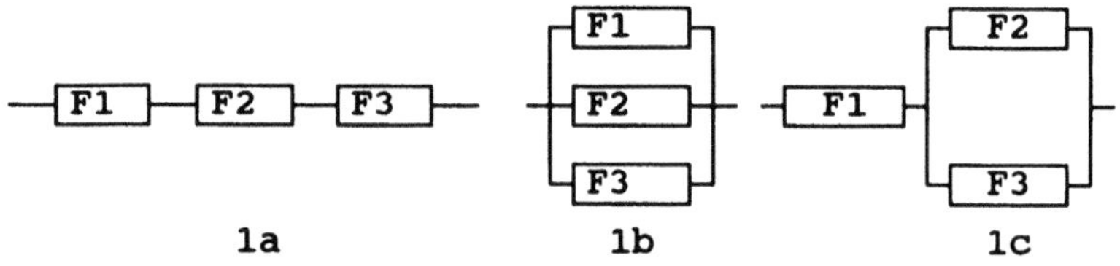

Figure 1. Possible RSDs with three elements.

The restoration problem of RSD is the probable determination of its shape according to the results of the system response analysis during its element testing. In that case the observational matrix is a matrix $W_{M,K}$ consisting of 0 and 1 elements,

where 0 - workability and 1 - failure of an element and M - number of experiments, K - total number of elements in the system. The response of the system is coded in the same way. In order to solve the problem of RSD restoration it is sufficient to determine its minimum path set (cuts), which are the conjunction of the elements which did not fail and with which the system is still working. From GUHA's point of view it means that the hypotheses ((0)F1,(0)F2,(0)F3 and all their possible combinations) are tested in order to find the most significant ones. For example: the scheme 1c has two paths (0)F1&(0)F2$\vee$(0)F1&(0)F3, two cuts (1)F1$\vee$(1)F2&(1)F3.
Calculations show that the restoration of the structures 1a and 1c by the GUHA is a trivial task which is solved by finding the sentences satisfying the criterion

$$\left\{H_{Fi}\right\} = \underset{i}{\operatorname{argmax}} \left\{ 1/\alpha_{Fi} \right\} \qquad \forall\, H_{Fi} \in \left\{\text{Sent}\right\}, i=1,K \qquad (2)$$

where F_i - the leading element in the chosen sentence; H_F - the tested hypotheses satisfying F_i. But such strategy of restoration 1c structure would result in the choice of "kernel" (0)F1 or (1)F1. This can be explained by the following reasons :
a) Structural importance [4] of F1 element in 1c which means simultaneous participation of F1 in two paths.
b) Different statistical chances for the shorter and longer sentences to be selected.
The former is post priori information which is not available during restoration. The latter is not taken into consideration by GUHA [1,2] although it is necessary to do that.

The procedure of sentence length tracking in GUHA's utilization

Let us consider the process of FFCT formation for any sentences similar to (1). If k and λ have the corresponding conjunction range l_a and l_s then all closed forms of (1) will result in $2^{l_a} \times 2^{l_s}$ contingency table (CT), where 2 - means the dichotomous observational matrix.

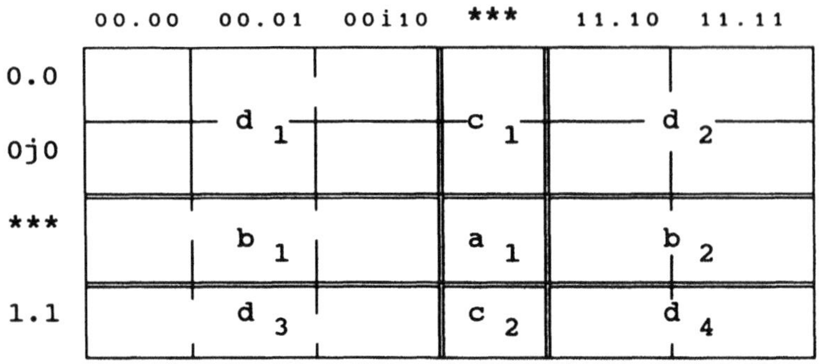

Figure.2 $2^{l_a} \times 2^{l_s}$ CT containing all closed forms for the sentences (1).

The cells of the CT contain frequencies corresponding to the observational matrix $W_{M,K}$ The process of FFCT formation becomes clear from Fig.2.The frequencies in FFCT will be formed according to the following equations :

$$a=a_1; \quad b=b_1+b_2; \quad c=c_1+c_2; \quad d=d_1+d_2+d_3+d_4 \tag{3}$$

Causes and consequences in the $2^{l_a} \times 2^{l_s}$ CT will be independent if we suppose all frequencies of all features to be distributed uniformly through the hole table, that is

$$m_{ij} = \frac{1}{2^{l_a+l_s}} \tag{4}$$

In this case the estimates of FFCT probabilities will be as follows :

$$\hat{p}_a = \frac{1}{2^{l_a+l_s}}; \quad \hat{p}_b = \frac{2^{l_s}-1}{2^{l_a+l_s}}; \quad \hat{p}_c = \frac{2^{l_a}-1}{2^{l_a+l_s}}; \quad \hat{p}_d = 1 - \frac{2^{l_a}+2^{l_s}-1}{2^{l_a+l_s}} \tag{5}$$

In order to obtain the post priori estimates of the parameter $\hat{\Theta}=(\hat{\theta}_1,\ldots,\hat{\theta}_4)=(\hat{p}_a,\hat{p}_b,\hat{p}_c,\hat{p}_d)$ we use the Bayes theorem where an a priori distribution of density function $p(\Theta)$ will be written as Dirichlet distribution :

$$p(\Theta)=\frac{\Gamma(\alpha_1+\alpha_2+\alpha_3+\alpha_4)}{\Gamma(\alpha_1)\Gamma(\alpha_2)\Gamma(\alpha_3)\Gamma(\alpha_4)} \hat{\theta}_1^{\alpha_1-1} \hat{\theta}_2^{\alpha_2-1} \hat{\theta}_3^{\alpha_3-1} (1-\hat{\theta}_1-\hat{\theta}_2-\hat{\theta}_3)^{\alpha_4-1} \tag{6}$$

The likelihood distribution function $f(y|\theta,M)$ should be written as :

$$p(\theta)=\frac{M!}{y_1!y_2!y_3!(M-y_1-y_2-y_3)!}\hat{\theta}_1^{y_1}\hat{\theta}_2^{y_2}\hat{\theta}_3^{y_3}(1-\hat{\theta}_1-\hat{\theta}_2-\hat{\theta}_3)^{(M-y_1-y_2-y_3)}, \quad (7)$$

where y_i - is an observed frequencies in the cell of FFCT.The presented form of $f(y|\theta,M)$ evolves from the initial multiplicative model derived from matrix $W_{M,k}$. We take the parameters $\alpha_i, i=1,4$ of Dirichlet distribution as:

$$\alpha_1=\hat{p}_a 2^{l_m}; \quad \alpha_2=\hat{p}_b 2^{l_m}; \quad \alpha_3=\hat{p}_c 2^{l_m}; \quad \alpha_4=2^{l_m}(1-\hat{p}_a-\hat{p}_b-\hat{p}_c), \quad (8)$$

where l_m - is the sum of the maximal lengths $l^m_a+l^m_s$ .So we can determine the post priori maximum likelihood estimates for the mean in the following way :

$$\hat{\theta}_1=(2^{l_m}\hat{p}_a+y_1)/(2^{l_m}+M); \quad \hat{\theta}_2=(2^{l_m}\hat{p}_b+y_2)/(2^{l_m}+M); \quad (9a,b)$$

$$\hat{\theta}_3=(2^{l_m}\hat{p}_c+y_3)/(2^{l_m}+M); \quad \hat{\theta}_4=1-\hat{\theta}_1-\hat{\theta}_2-\hat{\theta}_3; \quad (9c,d)$$

We use the obtained estimates for testing of hypothesis of k and λ independence based on exact Fisher Test or Chi-Square Test.

Numerical example

The observation matrix and its response are generated according to the RSD presented in Fig. 1 and consists of 2^3 experiments and four factors.

Possible sentences realized in matrix	RSD 1a		RSD 1b		RSD 1c	
	$10EFT_Q$		$10EFT_Q$		$10EFT_Q$	
GUHA	[+]	[1,2]	[+]	[1,2]	[+]	[1,2]
()F1	6.8 0.2	5.0 1.0	6.8 0.2	5.0 1.0	5.0 0.3	0!7 ?1:0
()F2	6.8 0.2	5.0 1.0	6.8 0.2	5.0 1.0	6.7 0.2	5:0 0.5
()F3	6.8 0.2	5.0 1.0	6.8 0.2	5.0 1.0	6.7 0*2	5.0 0*5
()F1&()F2	6.4 0.3	2.5 1.0	6.4 0.3	2.5 1.0	4!8 0*5	1.1 1*0
()F1&()F3	6.4 0.3	2.5 1.0	6.4 0.3	2.5 1.0	4!8 0.5	1.1 1.0
()F2&()F3	6.4 0*3	2.5 1*0	6.4 0$3	2.5 1$0	--	--
()F1&()F2&()F3	6!0 0.6	1!3 1.0	6!0 0:6	1!3 1:0	6.6 -- 0.5	3!8 -- *1.0
content of () is	paths (0)		cuts (1)		paths (0)	

Figure 3 Comparison of the GUHA-method [1,2] and presented

method (GUHA+) efficiency.Notations: * - minimal paths set; $ - minimal cuts set; ? - kernel; ! - minimal value of exact Fisher Test (EFT); Q - Yule's measure of association (Q=(ad-bc)/(ad+bc)) [5].

Conclusions

The theory presented and the numerical example show that account of a priori information in the length form is a useful technique which considerably increases the validity of results. Meanwhile with comparable significance levels which are used by GUHA+ the total amount of chosen sentences will be considerably less without loosing essential sentences. The problem of kernel is solved for RSD 1c.

This approach makes it possible to use Yule's measure of association which reaches the extreme (maximum) for likely paths as an information criterion.The latter conclusion became possible after we exclude zero frequencies from FFCT.

References

1. Hajek P. - Havranek T. Mechanizing Hypothesis Formation. Mathematical Foundations for a General Theory. Springer - Verlag Berlin (1978).
2. Hajek P. The New Version of the GUHA Procedure ASSOC (Generating Hypotheses on Associations)-Mathematical Foundations, in: COMPSTAT84, Physica Verlag, Wien, 1984, 360-365.
3. Renc Z. On interpretation of GUHA-results, Int. J. of Man-Machine Studies, 10, 1978, 37-46.
4. Barlow R.E. and Proschan F. Mathematical Theory of Reliability,Wiley, New York, (1965).
5. Upton J.G. The Analysis of Cross-tabulated Data, Wiley, New York, (1978).

Classifying Documents: A Discriminant Analysis and an Expert System Work Together

G. Hebrail and M. Suchard, Clamart

I - INTRODUCTION

The subject of the study here is the classification of documents into disjoint and pre-defined categories. The classification of these documents, that describe research projects, will be used in order to build a synthezised view of the activity of our research center.

Many studies on this subject have been carried out (see [BOBE63], [HAZA80], [TOFA87]) and textual data are usually processed by a discriminant analysis on the title's words or the key-words of the document. Non-textual data, like the review's name, the language, the category of the document are not classifying criteria.

We propose a new method which takes into account textual and non-textual data which are contained in the document :

- a representative sample of documents has been classified by a human expert,
- a preliminary process changes textual data into words and key-words,
- discriminant analyses are performed on words and key-words, and the results are injected into an expert system,
- the expert system is augmented by rules that it has automatically learned from the sample on non-textual data,
- a specialized inference engine determines whether the document can be classified or not.

We show that the results are better when using this method rather than only a discriminant analysis on words.

II - DISCRIMINANT ANALYSIS ON WORDS

We selected two different sets of words out of our projects :

- significant words of the title (without articles, pronouns, ...)
- key-words, extracted from the text of the project by an *automatic indexing technique* (see [BLMP90]).

Let us consider now the set of key-words extracted from the text. The sample is made of 1,500

Compstat 1990

projects in which nearly 2,500 different key-words appear. The average number of key-words extracted from the text of a project is 30.

Discriminant analysis on sparse qualitative data has been studied in the data analysis field. Two different approaches have been proposed :

- a linear discriminant analysis is performed directly or after a preliminary correspondence factorial analysis ([SAPO75]),
- a multinomial model of discriminant analysis, with independency between variables ([BIFH75], [BUKR80], [GODI78]).

In [SUCH89], we have experimented the two approaches and found that the multinomial model gives the best results (as reported in [BUKR80], [GODI78], [TITT81]).

We use a *multinomial discriminant analysis* (with independency between variables) where each variable figures the presence/absence of a word : we compute, for each project y to be classified and each possible class C_i, the probability for project y to belong to class C_i as follows:

$$p(y, C_i) \propto N_i \cdot \prod_{r=1}^{m} \left(\frac{y_r \cdot sup(N_i(r), 0.1) + (1-y_r) \cdot (N_i - sup(N_i(r), 0.1))}{N_i} \right)$$

where

$\propto$ for 'proportional to',
m is the total number of different words appearing in the sample,
$C_1, C_2, \ldots, C_k$ are the k classes,
$y = (y_1, y_2, \ldots, y_m)$ is the feature vector of project y ($y_r = 1$ if word r is in y, 0 otherwise),
$N_i(r)$ is the number of projects in the sample which are in class C_i and contain word r,
N_i is the number of projects in the sample which are in class C_i.

III - FROM DISCRIMINANT ANALYSIS TO RULES

We first split our sample into a *learning sample* (80% of the sample) and a *test sample*. The discriminant analysis is performed on the *learning sample*, and the *test sample* is classified. For each project and class, the result is a probability of membership.

We now replace the probability values by discrete values. The [0,1] interval is partitioned as follows :

- ONE for probability 1,
- L for probability in [0.9,1],
- M for probability in [0.5,0.9],
- S for probability in [0,0.5].

This partition was chosen after observing the distribution of probability values and remained relevant in all our experiments.

Then, we calculate the *confusion* (or *misclassification*) *tables* on the *test sample* for each preceeding level of probability. An element $T_{l,ij}$ of the confusion table T_l of level l is the number of projects which obtain a level l of probability for class C_i and are classified in class C_j in the *test sample.*

From these tables, we generate rules of the following form :

```
IF      DISCRIMINANT_ANALYSIS_CLASS = 'C1'
 AND    PROBA_LEVEL = 'L'
THEN    C1       FREQUENT
        C2       EXCLUDED
        C3       UNUSUAL
        C4       POSSIBLE
                 .....
        Ck       EXCLUDED
```

A rule is generated for each class and each level of probability (except level S).

In the rule dealing with class C_i and level l, we say class C_j is :

$$EXCLUDED \quad iff \quad T_{l,ij} = 0$$

$$FREQUENT \quad iff \quad T_{l,ij} > 0.5 \cdot \sum_{j=1}^{k} T_{l,ij}$$

$$UNUSUAL \quad iff \quad T_{l,ij} < \frac{\sum_{j=1}^{k} T_{l,ij}}{k}$$

$$POSSIBLE \quad otherwise$$

These rules form the first part of the expert system's rule base. One set of rules is generated from a discriminant analysis on title's words and another one from a discriminant analysis on text's key-words.

The two discriminant analyses are re-performed on the full sample, and the results on the projects to be classified are discretised to form the first part of the expert system's fact base.

For further details on this method, see [HESU89].

IV - ADDING RULES ON NON-TEXTUAL DATA

Our research projects are also described by non-textual data : the department which supports the project, and the customer of the project. These informations form the second part of the expert system's fact base.

By observing the sample, we generate rules with the same structure as in paragraph 3, for instance :

IF	DEPARTMENT = 'DEPT1'			IF	CUSTOMER = 'CUSTOMER1'		
THEN	C_1	FREQUENT		THEN	C_1	POSSIBLE	
	C_2	EXCLUDED			C_2	EXCLUDED	
	C_3	UNUSUAL			C_3	POSSIBLE	
		..				..	
	C_k	EXCLUDED			C_k	EXCLUDED	

These rules are generated by a process which is similar to the process described in paragraph 3. The learning process is quite simple (no 'generalisation' is made) compared to previous work achieved on the subject (see [BERZ88], [DIDA89], [LAUR88], [MICA86]), but generated rules allow a deduction by 'elimination'.

V - THE INFERENCE PHASE

Rules dealing with the results of the two discriminant analyses, the department and the customer, are gathered and applied to the projects to be classified.

The inference mechanism performs a deduction by elimination and assigns to each project, either a class when the different rules agree on one unique unexcluded class or on one unique more frequent unexcluded class, either no class with the label '*UNCLASSIFIED*' when no possibility is left or multiple choices remain.

After the inference phase, for each project, the list of the rules which have been used for its classification is available and provides an explanation of the classification process.

For further details on the inference engine, see [HEBR89].

VI - EXPERIMENTS

We have made experiments on several classifications. Here are the results of two of them : the first one is based on a sample of 2,000 projects and 1,000 projects to be classified into 7 categories, the second one is based on a sample of 1,100 projects and 1,300 projects to be classified into 28 categories.

Table 1 and 2 figures the results of our experiments : the column '*%CLASSIFIED*' represents the number of projects which have been classified by the method, the column '*%WELL CLASSIFIED*' represents the number of classified projects which have been assigned to the right class (according to the human expert).

Table 1 presents the results obtained with the discriminant analysis method on the title's words (the results with key-words are not better). For the classification into 7 classes, another discriminant analysis has been performed where the title's words, the department, and the customer have been gathered together : we obtained 86.5% of properly classified projects, the result being worse when key-words were added.

Table 2 contains the results obtained by the expert system using all the available data on the projects. The results are better than the discriminant analysis alone.

CLASSIFICATION	CORPUS	%CLASSIFIED	%WELL CLASSIFIED
7 classes	Sample Projects to be classified	100 100	93.6 78.7
28 classes	Sample Projects to be classified	100 100	97.5 68.4

Table 1 : Discriminant analysis on title's words

CLASSIFICATION	CORPUS	%CLASSIFIED	%WELL CLASSIFIED
7 classes	Sample Projects to be classified	96.5 90.6	96.8 92
28 classes	Sample Projects to be classified	96.8 79.9	98.7 73.8

Table 2 : Expert system results

VII - CONCLUSIONS AND PERSPECTIVES

The quality of the results provided by the expert system can be explained by three main reasons :

- it uses all the available information,
- it makes a confrontation of the different information sources, each source being used with an appropriate method,
- it classifies less projects but they are better classified.

We are currently studying the following directions to improve the method :

- use a crossvalidation technique for the discriminant analysis (rules on the results of discriminant analysis can be generated from the classification of the whole sample instead of the test sample),
- improve the discriminant analysis on key-words by using the semantic network that structures the key-words of our thesaurus,
- improve the learning process by using the hierachical structure of the research center and performing generalisation,
- try the method on other types of documents to evaluate its robustness.

VIII - REFERENCES

[BERZ 88] Berzuini C., "Generating Rules by Means of Regression Analysis", Compstat 88, Physica-Verlag, 1988.

[BIFH 75] Bishop Y.M., Fienberg S.E., Holland P.W., "Discrete Multivariate Analysis", The MIT Press, 1975.

[BLMP 90] Blosseville M.J., Monteil M.G., Penot N., "Indexation Automatique : Principe et Fonctionnement", Research Report to appear, 1990.

[BOBE 63] Borko H., Bernik M., "Automatic Document Classification", Journal of Association of Computing Machinery, vol.11, pp.138-151, 1963.

[BUKR 80] Butler W.J., Kronmal R.A., "Discrimination with Polychotomous Predictor Variables using Orthogonal Functions", Journal of the American Statistical Association, Vol.20, Nx390, June 1980.

[DIDA 89] "Data Analysis, Learning Symbolic and Numeric Knowledge", Proceedings of The Conference, Antibes, edited by E.DIDAY, Nova Science Publishers Inc., 1989.

[GODI 78] Goldstein M., Dillon W.R., "Discrete Discriminant Analysis", Wiley, 1978.

[HAZA 80] Hamill K.A., Zamora A., "The Use of Titles for Automatic Document Classification", Journal of the American Society for Information Science, pp.396-402, Nov. 1980.

[HEBR 89] Hébrail G., "Le Générateur de Systèmes Experts en Classements du Projet SPHERE : Traitement de Grands Volumes de Données et Apprentissage", Research Report EDF-DER HN-46/383, 1989.

[HESU 89] Hébrail G., Suchard M., "Classement Automatique de Documents par Apprentissage : une Analyse Discriminante et un Système Expert avec Apprentissage coopèrent", Research Report EDF-DER HN-46/386, 1989.

[LAUR 88] Laurière J.L., "Intelligence Artificielle", Tomes 1 and 2, EYROLLES, 1988.

[MICA 86] Michalski R.S., Carbonell J.G., Mitchell T.M., "Machine Learning : an Artificial Intelligence Approach", Vol.1 and 2, SPRINGER-VERLAG, 1983, 1986.

[SAPO 75] Saporta G., "Liaison entre plusieurs Ensembles de Variables et Codage de Données Qualitatives", Thèse Université Pierre et Marie Curie, Paris, France, 1975.

[SUCH 89] Suchard M., "Analyse Discriminante sur Données Qualitatives ou Textuelles", Research Report EDF-DER HN-46/334, 1989.

[TITT 81] Titterington D.M. and al., "Comparison of Discrimination Techniques Applied to a Complex Data Set of Head Injured Patients", JRSS, series A 144, 145-175, 1981.

[TOFA 87] Todeschini C., Farell M.P., "An Expert System for Quality Control in Bibliographic Databases", Journal of the American Society for Information Science, pp.2-11, Jan. 1989.

Key-words : Data Analysis, Discriminant Analysis, Expert Systems, Classification, Textual Data.

Estimation Procedures for Language Context: Poor Estimates are Worse than None

W. A. Gale and K. W. Church, Murray Hill

Abstract

It is difficult to estimate the probability of a word's context because of sparse data problems. If appropriate care is taken, we find that it is possible to make useful estimates of contextual probabilities that improve performance in a spelling correction application. In contrast, less careful estimates are found to be useless. Specifically, we will show that the Good-Turing method makes the use of contextual information practical for a spelling corrector, while attempts to use the maximum likelihood estimator (MLE) or expected likelihood estimator (ELE) fail. Spelling correction was selected as an application domain because it is analogous to many important recognition applications based on a noisy channel model (such as speech recognition), though somewhat simpler and therefore possibly more amenable to detailed statistical analysis.

1. Background

Statistical language models were quite popular in the 1950s, but interest faded rather suddenly when Chomsky argued quite successfully that statistics should not play a role in his competence model (Chomsky 1957, pp. 15-17). With the recent availability of large text corpora (of 100 million words or more), there has been a resurgence of interest in empirical methods, especially in recognition applications such as speech recognition (e.g., Nàdas, 1984), but also in many other areas of natural language research including machine translation (Brown et al, 1988). Since the sheer size of the available corpus data is largely responsible for the revival of these techniques, it is important to study the statistical estimation issues very carefully. Specifically, we will show that the Good-Turing (GT) method (Good, 1953) for estimating bigram probabilities makes the use of contextual information practical in our spelling corrector application, while attempts to use the maximum likelihood estimator (MLE) or expected likelihood estimator (ELE) fail.

Previous work led to the spelling correction program, *correct*, described by Kernighan, Church, and Gale (1990). In the course of that work, we observed that human judges were reluctant to decide between alternative candidate corrections given only as much information as was available to the program, the typo and the candidate corrections. We also observed that the judges felt much more confident when they could see a line or two of context around the typo. This suggests that there is considerable information in the context.

However, it is difficult to measure contextual probabilities. Suppose, for example, that we consider just the previous word, l. (A more adequate model would need to look at considerably more than just the previous word, but even this radically over-simplified model illustrates the problem.) Then we need to measure the conditional probabilities, $Pr(l|w)$, for all w and l in the vocabulary V. The problem is that V^2 is generally much larger than the size of the corpus, N. V is at least 10^5, so V^2 is at least 10^{10}. The largest currently available corpora are about 10^8 (100 million words). Thus, we have at least 100 times more parameters than data. In fact the problem is much worse because the data are not uniformly distributed.[1]

1. One might think that the sparse data problem could be solved by collecting larger corpora, but ironically, the problem only gets worse as we look at more data. The vocabulary is not fixed; both N and V grow as we look at more data. The rate of growth is still a matter of debate, but the evidence clearly shows that $V > O(\sqrt{N})$, and therefore, the sparse data problems only get worse as we look at more and more data.

Compstat 1990

Correct is reviewed in the next section, as it provides the framework within which this study is done. The section also discusses estimation techniques, rules for combining multiple sources of evidence, and evaluation procedures.

2. Correct

The *correct* program takes a list of misspelled words (typos) as input (as might be produced by the Unix® *spell* program), and outputs a set of candidate corrections for each typo, *along with* a probability. These probability scores distinguish *correct* from other spelling correction programs, that output a (long) list of candidiate corrections, many of which are often extremely implausible.

Here is some sample output:

Typo	Corrections
laywer	lawyer (100%) layer (0%) lawer (0%)
negotations	negotiations
notcampaigning	???
progession	progression (94%) procession (4%) profession (2%)

The entry *???* indicates that no correction was found.

The first stage of *correct* finds candidate corrections, c, that differ from the typo t by a single insertion, deletion, substitution or reversal. For example, given the input typo, *acress*, the first stage generates candidate corrections in the table below. Thus, the correction *actress* could be transformed by the noisy channel into the typo *acress* by replacing the t with nothing, @, at position 2. (The symbols @ and # represent nulls in the typo and correction, respectively. The transformations are named from the point of view of the correction, not the typo.) This unusually difficult example was selected to illustrate the four transformations; most typos have just a few possible corrections, and there is rarely more than one plausible correction.

Typo	Correction	Transformation			
acress	actress	@	t	2	deletion
acress	cress	a	#	0	insertion
acress	caress	ac	ca	0	reversal
acress	access	r	c	2	substitution
acress	across	e	o	3	substitution
acress	acres	s	#	4	insertion
acress	acres	s	#	5	insertion

Each candidate correction is scored by the Bayesian combination rule $Pr(c)\ Pr(t|c)$, and then normalized by the sum of the scores for all proposed candidates. Care must be taken in estimating the prior because of sparse data problems. It is possible (and even likely) that a proposed correction might not have appeared in the training set. Some methods of estimating the prior would produce undesirable results in this case. For example, the maximum likelihood estimate (MLE) would estimate $Pr(c) = 0$, and consequately, many candidate corrections would be rejected just because they did not happen to appear in the training set. We will encounter even more severe forms of the sparse data problem when we consider context.

We will consider four estimation methods for dealing with the sparse data problems. All of these methods attempt to estimate a set of probabilities, p, from observed frequencies, r. It is assumed that the observed frequencies are generated by a binomial process with N total observations. The estimation methods generate an adjusted frequency r^*, where r^* is a function of r. Once r^* has been determined, then p is estimated as $p \approx r^*/N^*$. $N^* = \sum r^*\ N_r$ where N_r is the frequency of frequency r, assuring that the estimated probabilities add to one. The maximum likelihood estimator (MLE) sets $r^* = r$. The MLE estimate is particularly poor when $r = 0$, since the true probabilities are almost certainly greater than 0.

Following Box and Tiao (1973), we can assume an uninformative prior and reach a posterior distribution for p. Using the expectation of this distribution amounts to using $r^* = r+.5$. We call this the expected likelihood estimate (ELE). Gale and Church (1990) discuss weaknesses of the ELE. The third method is the minimax (MM) method (Steinhaus, 1957), which sets $r^* = r+.5\sqrt{N}$. Its derivation is based on a risk analysis; it minimizes the maximum quadratic loss. The fourth method is the Good-Turing (GT) method (Good, 1953). This method sets $r^* = (r+1)\ N_{r+1}/N_r$. Unlike the MLE, all three other methods assign nonzero probabilities, even when $r = 0$. This is probably a desirable property.

We use the ELE for the probabilities of single words as they are frequent enough not to require elaborate treatment. The channel probabilities, $Pr(t|c)$, are computed from four confusion matrices: (1) $del[x,y]$, the number of times that the characters *xy* (in the correct word) were typed as *x* in the training set, (2), $add[x,y]$, the number of times that *x* was typed as *xy*, (3) $sub[x,y]$, the number of times that *y* was typed as *x*, and (4) $rev[x,y]$, the number of times that *xy* was typed as *yx*. Probabilities are estimated from these matrices by using $chars[x,y]$ and $chars[x]$, the number of times that *xy* and *x* appeared in the training set, respectively, as the total number of observations appropriate to some cell of a matrix. The probabilities are estimated using the Good-Turing method (Good, 1953), with the cells of the matrices as the types.

Returning to the *acress* example, the seven proposed transformations are scored by multipling the prior probability (which is proportial to 0.5 + column 4 in the table below) and the channel probability (column 5) to form a raw score (column 3), which are normalized to produce probabilities (column 2). The final results is: *acres* (45%), *actress* (37%), *across* (18%), *access* (0%), *caress* (0%), *cress* (0%). This example is very hard; in fact, the second choice is probably right, as can be seen from the context: *...was called a "stellar and versatile* acress *whose combination of sass and glamour has defined her....* The program would need a much better prior model in order to handle this case. The next section shows how the context can be used to take advantage of the fact that that *actress* is considerably more plausible than *acres* as an antecedent for *whose.*

c	%	Raw	freq(c)	Pr(t|c)
actress	37%	.157	1343	55./470,000
cress	0%	.000	0	46./32,000,000
caress	0%	.000	4	.95/580,000
access	0%	.000	2280	.98/4,700,000
across	18%	.077	8436	93./10,000,000
acres	21%	.092	2879	417./13,000,000
acres	23%	.098	2879	205./6,000,000

Many typos[2] such as *absorbant* have just one candidate correction, but others such as *adusted* are more difficult and have multiple corrections. The table below shows examples of typos with candidate corrections sorted by their scores. The second column shows the number of typos in a seven month sample of the AP newswire, broken out by the number of candidate corrections. For example, there were 1562 typos with exactly two corrections proposed by *correct.* Most typos have relatively few candidate corrections. There is a general trend for fewer choices, though the 0-choice case is special.

#	Freq	Typo	Corrections
0	3937	admininistration	
1	6993	absorbant	absorbent
2	1562	adusted	adjusted dusted
3	639	ambitios	ambitious ambitions ambition
4	367	compatability	compatibility compactability comparability computability
5	221	afte	after fate aft ate ante
6	157	dialy	daily diary dials dial dimly dilly
7	94	poice	police price voice poise pice ponce poire
8	82	piots	pilots pivots riots plots pits pots pints pious
9	77	spash	splash smash slash spasm stash swash sash pash spas

We decided to look at the 2-candidate case in more detail in order to test how often the top scoring candidate agreed with a panel of three judges. The judges were given 564 triples and a few concordance lines:

absurb absorb absurd

financial community . *E* *S* " It is absurb and probably obscene for any person so engaged to und

The first word of the triple was a *spell* reject, followed by two candidates in alphabetical order. The judges were given a 5-way forced choice. They could circle any one of the three words, if they thought that was what the

2. For the purposes of this experiment, a typo is a lowercase word rejected by the Unix® spell program.

author had intended. They could also say "other" if they thought that some other word was intended, or "?" if they were not sure what was intended. We decided to consider only those cases where at least two judges circled one of the two candidates, and they agreed with each other. This left only 329 triples, mainly because the the judges often circled the first word, indicating that they thought it had been incorrectly rejected by *spell*.

The following table shows that *correct* agrees with the majority of the judges in 87% of the 329 cases of interest. In order to help calibrate this result, three inferior methods are also evaluated. The *channel-only* method ignores the prior probability. The *prior-only* method ignores the channel probability. Finally, the *neither* method ignores both probabilities and selects the first candidate in all cases. As the following table shows, *correct* is significantly better than the three alternative methods. The table also evaluates the three judges.[3] The table shows that the judges significantly out-perform *correct*, indicating that there is room for improvement.

Method	Discrimination	%
correct	286/329	87 ± 1.9
channel-only	263/329	80 ± 2.2
prior-only	247/329	75 ± 2.4
chance	172/329	52 ± 2.8
Judge 1	271/273	99 ± 0.5
Judge 2	271/275	99 ± 0.7
Judge 3	271/281	96 ± 1.1

3. Context

As previously noted, the judges were extremely reluctant to cast a vote without more information than *correct* uses, and they were much more comfortable when they could see a concordance line or two. This suggests that contextual clues might help improve performance. However, it is important to estimate the context carefully; we have found that poor measures of context are worse than none.

In this work, we use a simple n-gram model of context, based on just the word to the left of the typo, l, and the word to the right of the typo, r. Although n-gram methods are much too simple (compared with much more sophisticated methods used in AI and natural language processing), even these simple methods illustrate the problem that poor estimates of contextual probabilities are worse than none. The same estimation issues are probably even more critical when the simple n-gram models of context are replaced by more sophisticated AI models.

The variables l and r are introduced into the Baysian scoring function by changing the formula from $Pr(c)Pr(t|c)$ to $Pr(c)Pr(t,l,r|c)$, which can be approximated as $Pr(c)Pr(t|c)Pr(l|c)Pr(r|c)$, under appropriate independence assumptions. The issue, then, is how to estimate the two new factors: $Pr(l|c)$ and $Pr(r|c)$. We have four proposals: MLE, ELE, MM and GT. Let us consider one way of using the ELE method first. It is straight-forward and similar to our best method, but hopelessly wrong.

$$Pr(l|c) = Pr(lc)/Pr(c) \approx \frac{(freq(lc)+0.5)/d_1}{(freq(c)+0.5)/d_2} \propto \frac{freq(lc)+0.5}{freq(c)+0.5}$$

where $d_1 = N+V^2/2$ and $d_2 = N+V/2$. We use the proportionality, since d_2/d_1 is constant over all candidate corrections. Similarly, $Pr(r|c) \propto \frac{freq(cr)+0.5}{freq(c)+0.5}$. When these estimates for $Pr(l|c)$ and $Pr(r|c)$ are substituted in the formula, $Pr(c)Pr(t|c)Pr(l|c)Pr(r|c)$, we have:

$$\frac{Pr(t|c)\ (freq(lc)+0.5)\ (freq(cr)+0.5)}{(freq(c)+0.5)} \qquad \text{E/E}$$

This new formula produces the desired results for the *acress* example, as illustrated in the following table. (The

3. Judges were only scored on triples for which they selected one of the proposed alternatives, and for which the other two judges agreed on one of the proposed alternatives. A triple was scored "correct" for one judge if that judge agreed with the other two and "incorrect" if that judge disagreed with the other two.

column labeled raw is 10^6 times the formula E/E, as only proportionalities matter.) Note that *actress* is now prefered over *acres* mostly because *actress whose* is more common than *acres whose* (8 to 0). Presumably the difference in frequencies reflects the fact that *actress* is a better antecedent of *whose*. Note also though, that *cress* is now considered a plausible rival because of errors introduced by the ELE method. The high score of *cress* is due to the fact that it was not observed in the corpus, and therefore the ELE estimates $Pr(l|c) = Pr(r|c) = 1$, which is clearly biased high.

c	%	Raw	freq(c)	Pr(t|c)	freq(lc)	freq(cr)
actress	69%	1.85	1343	55./470,000	2	8
cress	27%	.719	0	46./32,000,000	0	0
caress	3%	.091	4	.95/580,000	0	0
access	0%	.000	2280	.98/4,700,000	2	0
across	0%	.011	8436	93./10,000,000	0	20
acres	0%	.003	2879	417./13,000,000	0	0
acres	0%	.003	2879	205./6,000,000	0	0

We will consider five methods for estimating $Pr(l|c)$. The method just described is called the E/E method, because both $Pr(lc)$ and $Pr(c)$ are estimated with the ELE method. The M/E method uses the MLE estimate for $Pr(lc)$ and the ELE estimate for $Pr(c)$. The E method takes $Pr(lc)$ proportional to the ELE estimate $(freq(lc)+0.5)$, but the denominator is adjusted so that $\sum_c Pr(l|c)=1$. The MM method adjusts the minimax suggestion in (Steinhaus, 1957) in the same way. The G/E method uses the enhanced Good-Turing (GT) method (Church and Gale, 1990) for $Pr(lc)$ and the ELE estimate for $Pr(c)$.

method	num	den	Pr(l|c) ∝
E/E	ELE	ELE	$(freq(lc)+0.5)/(freq(c)+0.5)$
M/E	MLE	ELE	$freq(lc)/(freq(c)+0.5)$
E			$(freq(lc)+0.5)/(freq(c)+0.5V)$
MM			$(freq(lc)+0.5\sqrt{freq(c)})$ $/(freq(c)+0.5V\sqrt{freq(c)}))$
G/E	GT	ELE	$r^*/(freq(c)+0.5)$

The first two methods are useless, as shown by the performance of the context alone:

Poor Estimates of Context Offer Little or No Help

	chance	M/E	E/E
wrong	164.5	15	169
uninformative	0	136	4
right	164.5	178	156

The other three are better. The performance of G/E is significantly better than the other four.

Better Estimates of Context Exist

	E	MM	G/E
wrong	62	59	45
uninformative	0	0	4
right	267	270	280

For the Good-Turing estimates, we use an enhanced version of the Good-Turing estimator. The basic estimator is applied to subgroups of the bigrams. The subgroups have similar values of $Np_x p_y$, where p_x and p_y are the probabilities for the individual words. The grouping variable is the expected frequency of the bigram if the words occurred independently. Its use is discussed in detail by Church and Gale (1990). It results in about 1400 significantly different estimates for bigrams not seen in the training text, and in about 150 different estimates for words seen once.

When combined with the prior and channel, G/E is the only one of the five estimation methods that improves significantly[4] on the performance of *correct*. The following table shows *correct* in column 1, followed by the two disastrous measures M/E and E/E, then the two useless measures E and MM, and finally the one useful measure G/E.

Context is Useless Unless Carefully Measured

	no context	disastrous +M/E context	disastrous +E/E context	useless +E context	useless +MM context	useful +G/E context
wrong	43	11	61	39	40	34
uninformative	0	136	0	0	0	0
right	286	182	268	290	289	295
% right	86.9 ± 1.9	55.3 ± 2.7	81.5 ± 2.1	88.1 ± 1.8	87.8 ± 1.8	89.7 ± 1.7

4. Conclusions

We have studied the problem of incorporating context into a spelling correction program, and found that the estimation issues need to be addressed very carefully. Poor estimates of context are useless. It is better to ignore context than to model it badly. Fortunately, there are good methods such as G/E that provide a significant improvement in performance. However, even the G/E method does not achieve human performance, indicating that there is considerable room for improvement. One way to improve performance might be to add more interesting sources of knowledge than simple n-gram models, e.g., semantic networks, thesaurus relations, morphological decomposition, parse trees. Alternatively, one might try more sophisticated statistical approaches. For example, we have only considered the simplest Baysian combination rules. One might try to fit a log linear model, as one of many possibilities. In short, it should be taken as a challenge to researchers in computational linguistics and statistics to find ways to improve performance to be more competitive with human judges.

Acknowledgements

Mark Kernighan's work on *correct* laid the groundwork for this study of context modelling. We thank Jill Burstein for judging.

References

Box, G. E. P., and G. C. Tiao, 1973 *Bayesian Inference in Statistical Analysis*, Addison-Wesley, Reading, Massachusetts.

Brown, P., J. Cocke, S. Della Pietra, V. Della Pietra, F. Jelinek, R. Mercer, and P. Pietra, 1988, "A Statistical Approach to French/English Translation," *Proceedings RIAO88*, Conference on User-oriented Content-based Text and Image Handling, Cambridge, Mass , March 21-24.

Chomsky, N., "Three Models for the Description of Language," *IRE Transactions on Information Theory*, vol. IT-2, Proceedings of the Symposium on Information Theory, 1956.

Church, K. W. and W. A. Gale, 1990, "Enhanced Good-Turing and Cat-Cal: Two New Methods for Estimating Probabilities of English Bigrams," submitted to *Computer, Speech, and Language*, Academic Press.

Gale, W. A. and K. W. Church, 1990, "What's Wrong with Adding One?" submitted to *IEEE Transactions on Acoustics, Speech, and Signal Processing*.

Good, I. J., 1953, "The population frequencies of species and the estimation of population parameters," *Biometrika*, v. 40, pp. 237-264.

Kernighan, M. D., K. W. Church, W. A. Gale, 1989, "A Spelling Corrector Based on Error Frequencies" *Proceedings of the Thirteenth International Conference on Computational Linguistics*

Nàdas, A., (1984), "Estimation of probabilities in the language model of the IBM speech recognition system," *IEEE Transactions on Acoustics, Speech, and Signal Processing*, v. ASSP-32 pp. 859-861.

Steinhaus, H., (1957), "The problem of estimation," *Annals of Mathematical Statistics*, v. 28, pp. 633-648.

4. The GT method changes the program's preference in 25 of the 329 cases; 17 of the changes are right and 8 of them are wrong. The probability of 17 or more right out of 25, assuming equal probability of two alternatives, is .04. Thus, we conclude that the improvement is significant.

Knowledge Modelling for Statistical Consultation Systems; Two Empirical Studies

G. M. van den Berg and R. A. Visser, Leiden

In discussions on the design of computerized support in statistics, it is often contended that individual differences between experts' ideas on statistical consultation and on the application of analysis methods are so large, that it prevents general acceptance of statistical consultation systems. Of course, one could design a support system based on knowledge acquisition with a single expert. However, we preferred to consider first how serious the problem of individual differences is. Ultimately, this is an empirical question.

Two empirical studies are reported in this paper. In the first exploratory investigation, we asked twenty expert statisticians with which analysis methods they were familiar and how they chose between these methods, especially during consultation with a client. From this study we gained knowledge about the concepts and reasoning that are applied during statistical consultation. In the next study, we examined the similarities and differences the same statisticians perceive among analysis methods. This produced a representation of methods, called a *method catalog*, for each expert. Subsequently, the method catalogs of all experts have been compared. Both studies clarify the elements on which all experts seem to agree, and the elements on which there is less or no agreement of opinion. Comparison of the results of the two studies leads to the conclusion that the representation of the problem during statistical consultation differs from the representation of the solution in terms of the analysis method to be applied. This may have serious implications for the design of a system to support the choice of an analysis method.

1. A study of reasoning during statistical consultation

Statistical consultation

Statistical consultation can be described as a process in which two agents cooperate. One is a client with data and a question about a specific empirical subject, the other a statistician with procedural knowledge about abstract methods that could produce an answer to the question. Eventually, the client role and the statistical consultant role may coincide in one person in which case the process may simply be called data analysis. A common feature in descriptions of data analysis and statistical consultation is the cyclical succession of stages, usually a preparatory, a computational and an interpretative stage (e.g. Hand, 1986; Gale, 1987). When data analysis is viewed as a cyclic process, a distinction can be made between a research problem and an analysis problem. The research problem controls the overall process, since it is the question that a specific study as a whole is supposed to answer. The concept of analysis question is restricted to the question to be answered by one separate analysis. Knowledge about the characteristics and context of the research problem is indispensable for the specification of the separate analysis questions. This is the kind of knowledge the client-agent is supposed to contribute.

Compstat 1990

Once an analysis question has been formulated, the data-analytical agent has to provide information on potential solutions. From this point of view we study the concepts used in statistical problem solving.

Results on reasoning during statistical consultation

A study was done to learn more about the data-analytical concepts used by statistical consultants in reasoning about a problem and about the relations between these concepts. A report of this study can be found in Van den Berg (1990a). Twenty experts in data analysis and consultation have been asked which considerations are relevant for the determination of a proper analysis method. The experts were selected on the basis of the number of articles they had published in recent years and on recommendations by colleagues. They were active in a variety of research fields, notably social sciences and economics, and each of them had ample experience in statistical consultation.

The interviews were processed by classifying the remarks in a number of categories and by arranging these into related groups. The following types of considerations have been identified:

1. the analysis question: presence of independent and dependent variables, describing versus testing, emphasis on research units versus emphasis on variables, preliminary analyses, special data like time series, models;
2. characteristics of the data collection: research design, sampling;
3. characteristics of the data: measurement level, number of variables, partitioning of the variables, distribution;
4. conditions for & external influences on data analysis: budget, available computer programs, weight of the research question, available information;
5. conditions for & influences on the consultation process: knowledge of the client, traditions and practice.

Most of the ideas of the experts could be classified in the categories described: in the end only 3 of the 159 remarks made by the experts did not belong to a specific category or a general type of consideration. Since hardly any remarks were thus left out, these groups probably constitute an exhaustive list of factors that are relevant to the choice of an analysis method. Essentially these factors are related to the kind of information that must be present to be able to make a well-founded decision. They can be thought of as the elements of the processes of data analysis and statistical consultation, but they do not constitute a model of the way in which these processes actually take place.

The first three types of considerations determine which analysis method is most suitable from a statistical point of view; the last two types determine which method is feasible from a practical point of view. In addition, it is obvious that the separate categories that have been arranged into five types of considerations, are not all of the same nature and the same impact. For some considerations, like the question to be answered by the analysis and the characteristics of the data, it is clear that information is absolutely necessary to be able to choose an analysis method. Other concepts, like the characteristics of the data collection, seem to be more related to the interpretation of the results of an analysis than to the choice of an analysis method.

Differences between experts in reasoning during statistical consultation

Disagreements between experts mainly take two forms. In the first place, some disagreement over the meaning of concepts has been found. In particular the interpretation of a concept was not necessarily the same for every expert who mentioned it. In general, these disagreements between experts were not very big, with one exception: there are quite some differences of opinion on the nature of measurement levels. Especially on the treatment of ordinal variables contrasting opinions are voiced. These differences in opinion could not be traced back to the disciplines in which the experts were active, nor to other systematic variations between them.

The second type of disagreement between experts is related to the concepts they have mentioned. Remarks concerning "analysis question" and "characteristics of the data" were made by almost every expert. "Characteristics of the data collection", "conditions for & external influences on data analysis", and "conditions for & influences on statistical consultation" were less frequently reported. There were no experts who made remarks on every concept. Analysis of the concepts that were and that were not mentioned by every expert, shows that especially statisticians involved in economic research distinguish themselves from the other experts, who were mainly active in the social sciences. In particular, economists more often made remarks on modelling and on the availability of information, and they made no remarks at all on characteristics of the data collection, on the distribution of variables, and on the conditions for & influences on statistical consultation.

Conclusions on reasoning during statistical consultation

The exhaustive list of concepts that results from this study can be used as a guideline for the design of computerized support. These are the considerations that are relevant to the choice of an analysis method, so these are the items about which clients have to provide information to enable a support system to reach a decision. The fact that some of these concepts are essential from a statistical point of view, while others are important from a practical point of view, and yet other concepts only influence the interpretation of results, should be taken into account in the structure of a support system. Statisticians involved in economic research seem to occupy a somewhat special position. If this result can be further corroborated, it may be necessary to restrict efforts to design a support system for statistical consultation to a single discipline.

2. A study of method catalogs

Method catalogs

The term method catalog is used for the conceptual network in which each statistician mentally represents the relations among analysis methods. We assume that experts perform a structured search in their method catalog, when they choose a method. Differences between experts may be caused by differences in their method catalogs and therefore we want to determine to what extent these catalogs match. We specifically want to know whether the catalogs of all or most of the experts can be related to a common basic structure that may be somehow viewed from different perspectives causing disagreements in opinions. What is needed therefore, is a representation of method catalogs that shows both the parts that all experts agree upon and the parts that are unique to individual statisticians or to a group of statisticians.

Results on method catalogs

Eighteen of the twenty statisticians interviewed for the first study, have participated a second time. Detailed results of this second study have been reported in Van den Berg (1990b). The analysis methods they had mentioned in the first interview plus seven methods mentioned by the majority of the experts, were presented to them on cards and they were asked to sort these cards in as many piles as they wished and as many times as they could. Each time the cards were sorted, the experts were asked to give a description of the categories that had been formed, and of the concept underlying the classification as a whole.

Seven methods or groups of techniques have been judged by every expert: (1) regression analysis, (2) analysis of variance, (3) LISREL, (4) loglinear modelling, (5) principal components analysis, (6) one-dimensional scaling, and (7) nonparametric statistics. In addition, sixteen other analysis methods have been judged by at least two of the experts. The number of methods per expert ranged from 9 to 21 with a mean of 14. On average, 5 classifications were made with 3 categories in each classification. Some classifications made by different experts, are clearly similar:

* type of research question for which an analysis method can be used (13 experts): usually two categories, viz. exploration or description versus testing, confirmation, or causality;
* level of measurement (11 experts): methods that can be applied when the data are categorical (also called: qualitative or discrete) versus methods for metric (quantitative, continuous) data;
* theoretical background of analysis methods (9 experts): one category always indicates that some methods are based on the (generalized) linear model, but everyone of the 9 experts has made different categories for the other methods to be sorted.

Other classifications are concerned with things like the number of variables in the analysis, the distribution of the variables, the level of difficulty of analysis methods, the design of the data collection. Overall, 94 classifications have been made, 69 of which were generated by two experts or more.

Next, each classification was interpreted as a nominal variable. Through applying generalized principal components analysis (De Leeuw, 1984) a representation in three dimensions was obtained for every expert separately. The scores of the analysis methods on each of the principal components of an expert's solution were considered to be coordinates in a three-dimensional space. The 18 matrices with methods and coordinates have been analyzed using the program MATCHALS (Commandeur, 1990), an improved and generalized version of the PINDIS program of Lingoes & Borg (1978), designed to match configurations according to a number of models. The model used in this study starts with generalized Procrustes analysis and proceeds by allowing differential dimension weighting. In effect this means that the program searches the shared characteristics in the 18 configurations, and fits this common configuration to every separate configuration by stretching or shrinking each of the dimensions. The over all fit of the model to the data of this study accounts for 65% of the variance in the data. In the common configuration two dimensions can be described. The first one is related to the presence of dependent variables and puts methods without dependent variables, like cluster analysis, multidimensional scaling, principal components analysis, against methods with dependent variables, for example regression analysis, analysis of variance, logistic regression. The second dimension shows methods for one or two variables, like non-parametric tests, t-test, χ^2, correlation coefficients, versus methods for a larger number of variables, for example discriminant analysis and principal components analysis.

Differences between experts in method catalogs

Differences between experts should first of all be sought in the weights they attach to the dimensions. A tentative conclusions is that it seems that the statisticians involved in economical research somewhat cluster together. They weight the first dimension, that is related to the absence or presence of dependent variables, heavier than the second dimension. The psychologists in the sample also display a tendency to cluster. These experts consider both dimensions to be equally important. However, further study is certainly necessary to see whether these results will be confirmed.

Conclusions on method catalogs
Three classifications have been made by the majority of the experts. The first two of these are quite clear: similarities and differences between analysis methods are related to the type of research question for which an analysis method can be used, and to the measurement level of the variables to be analyzed. The third classification poses more of a problem, since only the category of methods based on the linear model is unambiguous. There is no agreement among the statisticians about the other categories of this classification (after all: what is the opposite to the linear model?).

An unexpected result of this study is the large number of classifications that were unique to a single expert. In general, there seems to be a lot of variability in the concepts experts use to think about analysis methods. This is reflected in the fit of the MATCHALS-model, from which can be inferred that only approximately two thirds of the variance in the data is accounted for by the common configuration and the dimension weighting.

3. Representing the problem versus representing the solution

Looking at the substance of the ideas generated in the first study and the classifications made in the second study, there are two concepts that show up clearly in both studies. The type of research question (description versus confirmation) and the measurement level of the variables are found both in reasoning during statistical consultation and in the method catalogs of the statisticians. In addition, the common configuration that results from MATCHALS, orders the analysis methods according to the absence or presence of dependent variables and according to the number of variables. We may conclude, that information about these characteristics of an analysis problem is always required to be able to choose an appropriate analysis method. Note, however, that this information is not sufficient for a definite choice.

Research question, measurement level, type and number of variables, seem to be the only common part in the results of the two studies. Other concepts are applied during statistical consultation than in structuring method catalogs. This difference may be explained from a difference in focus in the two studies that corresponds to a real difference in the way an analysis problem may be approached. One could state that during statistical consultation the focus is on description of the client's analysis *problem*, while in method catalogs attention is centered on the *solution* to such problems. In other words, two perspectives are possible on the choice of an analysis method: either one focuses on the problem description or on the solution. It is not hard to imagine that the perspective chosen in the design of a statistical support system is essential to the use of such a system. Focusing on the problem may facilitate the communication with a user who is not necessarily very informed on statistical concepts and terminology, and thus not on solutions. Focusing on the solution may have the advantage of enabling finer distinctions between different analysis methods. Further study of the consequences of these different perspectives is certainly inevitable. In addition, it seems worthwhile to investigate whether the same contrast between problem representation and solution representation shows up in other fields of application.

When we compare the results of the first study with those of the second, another thing stands out clearly: there is much more agreement among statisticians about the concepts applied in reasoning during statistical consultation than about the concepts applied to describe similarities and differences between analysis methods. This does not necessarily mean that all statisticians agree about all the ideas relevant to statistical consultation, but there are almost no unique individual concepts. In contrast, approximately one fourth of the classifications was unique to a single statistician in the second study. This is surprising when we consider the fact that all kinds of formal differences between analysis methods are documented in data-analytical and statistical handbooks, while statistical consultation is much less frequently described and studied. Actual practice seems to impose more common structure than theoretical representation.

As far as systematic disagreements between the experts are concerned, both studies point in the same direction which indicates that statisticians in economic research may use other concepts than statisticians involved in research in the social sciences. If this is confirmed in future studies, it may be necessary to focus on a single discipline when the design of a statistical support system is contemplated.

Acknowledgements: The authors like to thank Willem Heiser for important advice about the methods used in these studies and Jacques Commandeur for his help with the application of MATCHALS. We also thank the twenty anonymous statisticians who lent us several hours of their time.

References

Commandeur J.J.F. (1990). *Matching n Configurations*. Leiden: DSWO Press (thesis Leiden University).

De Leeuw J. (1984). The GIFI system of nonlinear multivariate analysis. In E. Diday, M. Jambu, L. Lebart, J. Pagès & R. Tomassone (eds.) *Data analysis and informatics* (III, pp. 415-424). Amsterdam: Elsevier Science Publishers B.V.

Gale W.A. (1987). Statistical applications of Artificial Intelligence and knowledge engineering. *Knowledge Engineering Review,* **2**, 227-247.

Hand D.J. (1986). Patterns in statistical strategy. In W.A. Gale (ed.) *Artificial intelligence and statistics* (pp. 355-387). Reading, Mass.: Addison-Wesley Publishing Company.

Lingoes J.C. & Borg I. (1978). A direct approach to individual differences scaling using increasingly complex transformations. *Psychometrika*, **43**, 491-519.

Van den Berg G.M. (1990a). Ideas of statistical experts on the choice of an analysis method. *Journal of Applied Statistics*, in press.

Van den Berg G.M. (1990b). *Comparing method catalogs of statisticians*. Leiden: Leiden University (Dept. of Behavioral Computer Science, Research Report 90-01).

Keywords: statistical consultation, statistical reasoning, comparison of experts, comparison of analysis methods.

An Expert System Strategy for Selecting Interesting Results

F. Gebhardt, Sankt Augustin

Abstract. The system EXPLORA extracts from a set of data a collection of interesting statements utilizing as far as possible the semantics of the data. The system may find too many such statements. A procedure is described that suppresses among them those that are sufficiently inferior to, and at the same time sufficiently similar to other statements that are retained. Some properties of the procedure, in particular in a statistical environment, are described.

Keywords: Expert systems; machine learning; generalization; classification.

1 Extracting interesting statements from data

In our project EXPLORA [Klösgen 1989, 1989a], the goal is roughly to analyze large data sets and to extract interesting results displaying them in a textual form. One of the sub-goals is to make precise what "interesting" should mean.

EXPLORA analyzes recurrent data sets since for a single set the effort of setting up the system would generally be too high. Obviously the system has to get a lot of knowledge on the meaning of the data in order to select the interesting results prudently.

An essential part of this environmental knowledge is a set of "statements". These are prototypes of potential results in a textual form together with computable predicates. Conceptually, the first step in analyzing the data is to search for those sets of argument values that render the predicates true; thus one finds the true statements among the possible statements.

The next step is to eliminate redundant results. "Redundant" depends on the application and is therefore part of the environmental knowledge. A typical case is that a type of results that holds for a set of objects automatically also holds for subsets; displaying the statement for subsets would be redundant.

Next certain non-interesting results are eliminated. A typical case here is that a result for a set of objects does not imply the same result for subsets but makes them plausible or otherwise uninter-

Compstat 1990

esting; in this case it would be interesting if for a particular subset this statement does not hold (or if in fact an opposite statement does hold).

Of course, the actual sequence of computations is different. EXPLORA exploits knowledge on the parameters occurring in the statements, in particular different kinds of ordering that may be valid such as hierarchical, linear, set inclusion. Thus, if a statement is true and this implies the truth for certain other parameter values, these need not enter computation since the results are redundant anyways. There are several other cases implemented that may reduce the search tree: If there exist opposite statements (as "significantly above / below average"), only sets that cover at most about half the object space need be checked since otherwise the opposite statement for the complement is more interesting; there may be a lower limit for the size of object sets that are interesting or for which a statistically reasonable statement can be made; if a parameter has a linear order, only intervals where the first and the last value produce deviations from the expected result (e.g. the average) into the same direction need be considered.

The search algorithm as implemented in the EXPLORA system is based on the theory of categorical grammars. The application of this theory to our situation is documented in [Latocha 1989].

2 Selection procedure

The system might still produce too many statements. As an example consider the results of an election. Goal objects are the 50 best election districts out of 248 (German Federal election) for a party. We are looking for subsets described by a few demographic variables and containing a high portion of goal objects. In our case, about a dozen different demographic variables are available.

The description of such a subset will be called a "partial generalization" since it is composed of common properties of a part of the goal set, admitting noise, however. A typical result is "Most (14 out of 15) of the constituancies with very high population density and low or moderate unemployment rate belong to the best 50 for 'Die Grünen' ".

After eliminating redundant results, there may be several dozen, even a few hundred different partial generalizations that each cover a portion of the goal objects and, generally, a few non-goal objects. For further reducing this mass, a supplementary selection procedure has been developed [Gebhardt 1989].

In a simplified form, one has a collection of partial generalizations characterized by the sets of objects covered and the task is now to discard from this collection all those that are sufficiently inferior to, but also sufficiently similar to, another partial generalization. That is, an inferior result has to be kept only if it is sufficiently distinct from all better ones. "Sufficiently inferior" and "sufficiently similar" have to be made concrete.

The basic ingredients for the selection procedure are a measure for the quality of a partial generalization G (called evidence V(G)) and a possibly unsymmetric measure for the similarity of two partial generalizations G_1 and G_2 (called affinity $S(G_1, G_2)$) where $S \leq 1$. G_1 is discarded if $V(G_1) \leq S(G_1, G_2)\, V(G_2)$.

There are several quite diverse environments for this procedure. The evidence may be derived from the portion of goal objects among all objects of the set. Then the union of several disjoint partial generalizations with nearly equal evidence has again about the same evidence.

The evidence may be a statistical measure, say the deviation of a statistical variable from its expectation (under a suitable null hypothesis) in terms of multiples of its standard deviation. Then the union of disjoint sets has a higher evidence (typically proportional to the square root of the size).

Although both cases behave differently in many respects, it turns out that the same type of affinity leads to good results:

$$S(G_1, G_2) = [\, |M_1 \cap M_2| / |M_2| \,]^{\kappa},$$

where M_i is the set of objects associated with G_i and $|\,.\,|$ is the weight of a set, in the simplest case the number of objects, in the statistical environment typically the variance of the underlying statistical variable.

The affinity is unsymmetric (thus it is not a similarity); this has a simple reason. Consider the case $M_1 \subset M_2$. If M_1 is worse than M_2 (has a lower evidence), then M_2 should always suppress M_1, but if M_1 is better, it should depend on the relative sizes and on the relative evidences if M_2 is suppressed by M_1 or not. Thus, one needs $S(G_1, G_2) = 1$ and $S(G_2, G_1) < 1$.

The procedure is not transitive: G_3 may be suppressed by G_2, G_2 by G_1, without G_1 suppressing G_3.

Incidentally, if the evidence is based on the number of objects, one should choose the number of goal objects as weight (not the number of all objects). Two partial generalizations are roughly equivalent, if they contain nearly the same goal objects; a varying number of non-goal objects will effect the evidences accordingly and thus induce a preference.

Finally, κ is an adjustable parameter that may (besides other possible variations of this basic procedure) be used to satisfy certain changing user needs, in particular the amount of output. In the examples that have been checked so far, values of κ between 0.25 and 0.5 show a good behaviour. The smaller κ, the smaller is in general the set of remaining statements.

The procedure is loosely related to strategies of machine learning [Ellman 1989; Diday and Brito 1989; Lebowitz 1986] where, however, search procedures hardly admit noisy data; an exception are systems based on a classification algorithm such as MAGGY [Kodratoff, Manago and Blythe 1987].

3 The selection procedure in a statistical environment

A typical situation in statistics is that one has a stochastic variable that measures the deviation of a set of objects from 'normal' and this variable is additive. More exactly:

With each set G there is associated a stochastic variable $\xi(G)$ with $E(\xi) = 0$ under a suitable null hypothesis of 'nothing conspicuous, everything as expected'; if $G \cap H = \emptyset$, then $\xi(G \cup H) = \xi(G) + \xi(H)$ and $\mathrm{var}(\xi(G \cup H)) = \mathrm{var}(\xi(G)) + \mathrm{var}(\xi(H))$. Then one can prove the following

Theorem. Assume that the conditions just stated hold; let $|G| = \mathrm{var}(\xi(G))$ and $V(G) = \xi(G) / |G|^{1/2}$. If $\xi(G) > 0$, $\xi(G \cup H) > 0$, but $\xi(H) < 0$, then G suppresses $G \cup H$ for $\kappa \leq 0.5$.

This gives in a statistical setting a natural boundary for κ: κ should not exceed 0.5. If $\kappa < 0{,}5$, then G suppresses $G \cup H$ also for $0 < \xi(H) \ll \xi(G)$, in particular if G deviates considerably from 'normal' but H only slightly, well within chance fluctuations. The exact limits up to where $G \cup H$ is suppressed depend of course on the ξ's, on $|H| / |G|$ and on κ. With growing $\xi(H)$, G as well as $G \cup H$ will be presented to the user, then $G \cup H$ only, next $G \cup H$ and H; finally H suppresses $G \cup H$.

In practical applications, the assumptions may not hold exactly. If e.g. ξ measures the deviation from a grand mean that is itself estimated from the population, then the variances are no longer additive (the total sample has e.g. variance 0). But this need not bother us. The selection procedure is essentially a heuristic procedure and the theorem just gives an impression of its behaviour.

For a fixed $\xi(G) > 0$ and $\kappa < 0.5$, we get this behaviour: If $\xi(H) > 0$ is small, H is suppressed; with One of the areas where we have used this procedure successfully is an opinion poll on banking habits of people. We are looking for instance for demographic groups where a particular bank is over- or underrepresented. Some parameters such as the portion of the population being customer of a particular bank or being engaged in a particular kind of banking business are taken from the sample and the distributions involved are binomial (the variances are not exactly additive); thus the assumptions of the theorem are not exactly true. Nevertheless, the procedure produced meaningful and useful results.

4 Variants of the selection procedure

The selection procedure can be modified to take into account further properties of the search problem or user preferences. We give here just a few examples.

It may be desirable to concentrate on partial generalizations that have a simple description. Given a conspicuous generalization, its evidence can often be increased a little bit by adding extra terms. But this is not what we want. Therefore one can decrease the evidence somewhat according to the complexity of the description, e.g. by a factor < 1 (say, 0.9 or 0.95) for each AND-operator in the description.

If a describing variable has a hierarchical structure, higher levels should be more meaningful than lower ones; this can be reflected by devaluing evidences for descriptions involving lower levels. If a variable has an ordinal structure, one-sided intervals can likewise be preferred to two-sided ones.

In a similar way one can treat variables in a description that are hard to interpret in a given context. Devaluing them may result in more meaningful solutions as long as they are only slightly better using the original evidences.

Of course there is the danger of introducing too many adjustable parameters; therefore such variants should be used with care. On the other hand, there is no reason not to use them where they are meaningful. Statistical procedures such as classification or factor analysis abound with adjustable parameters and free choices, and even in a simple statistical test one has usually at least an error probability whose value is set by custom and by our use of the decimal system.

References

Diday, Edwin; Brito, M. Paula: Symbolic cluster analysis. In: *Conceptual and numerical analysis of data : proc. of the 13th Conf. of the Gesellschaft für Klassifikation* (Augsburg, April 1989) / O. Opitz (ed.). Heidelberg : Springer, 1989, pp. 45 – 84.

Ellman, Thomas: Explanation-based learning: a survey of programs and perspectives. In: *Computing Surveys 21* (1989), pp. 163 – 176.

Gebhardt, Friedrich: *Choosing among competing generalizations*. Sankt Augustin : GMD (Arbeitspapiere der GMD 421), 1989, 16 pp. (preprint).

Klösgen, Willi: The generalization step in a statistics interpreter. In: *Data Analysis, Learning Symbolic and Numerical Knowledge* / E. Diday (ed.). New York : Nova Science, 1989, pp. 473 – 480.

Klösgen, Willi: EXPLORA: content interpretation of statistical data. In: *Fortschritte der Statistik-Software* / F. Faulbaum (ed.); H. M. Kehlinger (ed.). To be published (probably Stuttgart and New York : G. Fischer, 1989a).

Kodratoff, Yves; Manago, Michel; Blythe, Jim: Generalization and noise. In: *International Journal of Man-Machine Studies 27* (1987), pp. 181 – 204.

Latocha, Peter: *Exploration von Aussagenräumen: ein semantischer Ansatz*. Sankt Augustin : GMD (GMD-Studien 164), 1989, 144 pp.

Lebowitz, Michael: Concept learning in a rich input domain: generalization-based memory. In: *Machine learning: an artificial intelligence approach, vol. II* / R.S. Michalski (ed.); J.G. Carbonell (ed.); T.M. Mitchell (ed.). Los Altos : Kaufmann, 1986, pp. 193 – 214.

Computer Assisted Interpretation of Conditional Independence Graphs

M. J. Cottee, Buckinghamshire

Abstract

Conditional independence graphs are one of the few graphical techniques available for the representation of a model which has been fitted to a given set of data. Through the use of CIGE, the computer software described in this paper, it is now possible to represent without ambiguity a much larger number of log-linear models fitted to discrete data, as well as to represent a certain class of models, the covariance selection models, fitted to continuous data. It is suggested that CIGE provides a useful tool for the graphical communication and interactive interpretation of such models, which may be of some use when working with researchers with limited statistical knowledge.

Keywords: conditional independence graphs, graphical modelling, conditional interaction graphs, log-linear models, covariance selection models, generating class, interactive computer graphics.

1 Introduction: Conditional Independence Graphs

Conditional independence graphs, or independence graphs for short, were expounded by Darroch, Lauritzen and Speed (1980), as an effective visual representation technique for log-linear models. Since then, independence graphs have been used extensively in graphical modelling (for example, in graphical log-linear modelling for discrete data; in graphical Gaussian modelling for continuous data; and in conditional Gaussian modelling for mixed data). A good introduction to the use of independence graphs in graphical modelling is contained in Whittaker (1990).

This paper is concerned with how the independence graph may be used for the representation of fitted models, for which it is one of the few graphical techniques in existence (compare this situation with the profusion of graphical techniques available for Exploratory Data Analysis (EDA) and, to a lesser extent, for the analysis of residuals). As a tool for graphical modelling, the independence graph has a number of mathematical properties which makes it very profound. However, as a representation technique for fitted models, it has a number of limitations.

Compstat 1990

In the following section, a package will be described which has been developed for the representation and communication of fitted models using ideas based upon the independence graph. The package is called the "Conditional Independence Graph Enhancer" (CIGE). Description of CIGE will include further discussion of independence graphs and their limitations. The package has been written in C, for monochrome SUN work-stations, using standard library routines. Colour has deliberately not been used, so that any display obtained using the software can be printed out and reproduced readily in technical reports and journals without colour printing.

2 Conditional Independence Graph Enhancer

2.1 Input and Construction of Graphs

In the case of discrete data, the package can take as its input the generating class of the fitted log-linear model, or the parameters (both zero and non-zero) of the terms in the model (from which the generating class can be determined). The generating class of a (hierarchical) model is the set of non-zero interactions whose presence in the model is not implied by any other interactions contained in the model, but which together imply the presence of all other non-zero interactions in the model.

In the case of continuous data, the package takes as its input the 'strengths' of the pair-wise associations in the fitted covariance selection model, as determined using the partial correlation coefficients or the edge exclusion deviances, which are monotonically related (see Knuiman (1978), Whittaker (1990)).

The corresponding independence graph is constructed with one vertex per variable, which may be located freely on the plane. In CIGE, the vertices are drawn, by default, equidistant around the circumference of a circle. In the case of discrete data, a single undirected edge is drawn between a pair of vertices if, and only if, a two-way association between the corresponding pair of variables is contained within, or implied by, the elements of the generating class of the model. In the case of continuous data, if the inputted strength of a pair-wise association between variables is non-zero, then a single undirected edge is drawn between the corresponding pair of vertices.

2.2 Manipulation of Graphs

Location of the vertices equidistant around the circumference of a circle will not necessarily provide the most 'pleasing' representation of the independence graph. For example, if the vertices were to be located in some other position, some or all of the cross-overs involving the edges may be eliminated, making the graph easier to read. There is no known graph theoretic algorithm to determine and construct any given graph with the minimum number of cross-overs. Therefore, CIGE permits the viewer to 'drag' the vertices around the window, using the mouse device, until a visually pleasing representation is obtained. This ability to relocate the vertices also enables the user to cluster together subsets of variables: for example, if they are strongly associated, etc..

2.3 Display of Conditional Interaction Graphs (discrete data only)

One limitation of independence graphs is that two or more hierarchical models, having different generating classes, may have the same independence graph, owing to the pair-wise nature of the edges in the graph. In the context of graphical modelling, a given independence graph is regarded as representing the corresponding 'graphical model' –that is, the model for which the elements of the generating class correspond to the cliques (the maximally complete sub-graphs) of the graph. Thus it is not possible to represent non-graphical models uniquely using the independence graph approach.

Cottee and Hand (1989) propose the adoption of an edge coding scheme for independence graphs, giving rise to the conditional interaction graph, or interaction graph for short. By adopting different edge codes to correspond to different orders of interaction (for example, a dashed edge corresponds to a two-way association; a continuous edge to a three-way association; and a double edge to a four-way association, etc.) it becomes possible to represent a larger number of models, including some non-graphical models, uniquely. For example, Figure 1 shows an interaction graph which represents the given non-graphical model uniquely. However, it is still not possible to represent all generating classes without ambiguity, and there remains a need for further information in order to ascertain which model is represented –this is dealt with in section 2.4.

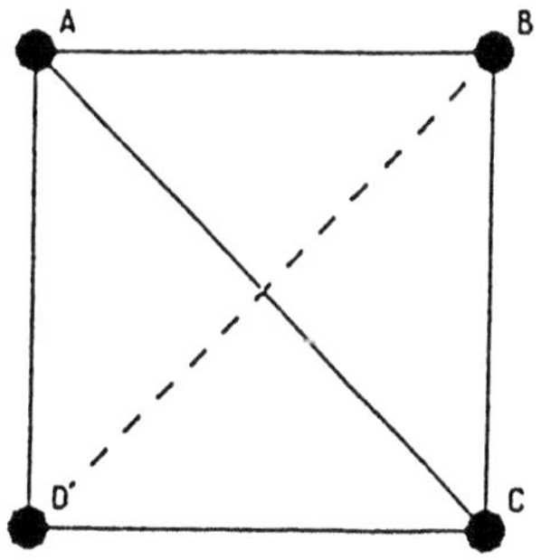

Figure 1: Interaction graph corresponding to the model with generating class $\{[ABC], [ACD], [BD]\}$.

Because CIGE uses the generating class as input, it is quite simple to display the interaction graph in preference to the independence graph, at any stage.

2.4 Display of Generating Class (discrete data only)

The most straight-forward solution to the problem of ambiguity in independence or interaction graphs is to display individually the sub-graphs corresponding to the elements of the generating class of the model represented. Using CIGE, this can be done in a second window, in either a sequential or simultaneous manner. The edges in the sub-graph will correspond to the type of graph displayed in the main window (ie. single edges for the independence graph, and edge codes corresponding to the order of interaction for the interaction graph), and the location of the vertices will correspond to the location of the vertices in the main window. If the vertices in the main window are subsequently relocated, this will be mirrored in the second window.

In Figure 2, the individual elements of the generating class are shown sequentially for the independence graph drawn to represent a given (non-graphical) model.

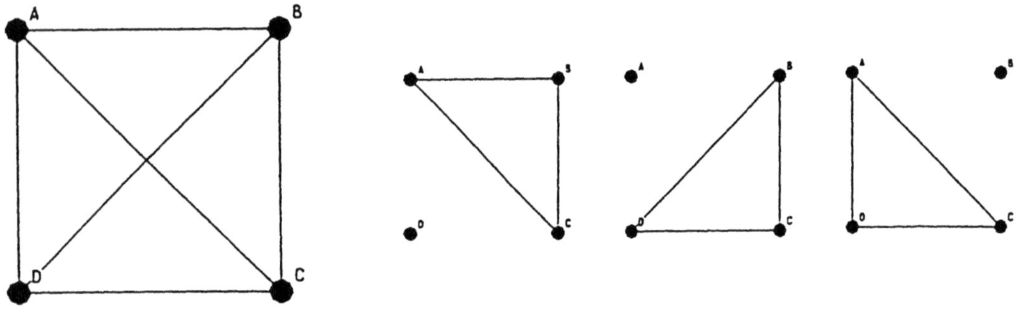

Figure 2: Independence graph corresponding to the model with generating class $\{[ABC], [BCD], [ACD]\}$, together with the individual elements of the generating class.

2.5 Incorporating Strength of Association (continuous data only)

The amount of information communicated by the independence graph can be increased by incorporating some information about the strength of the pair-wise associations on the graph.

In the past, some attempt has been made to do this by applying principal components analysis (PCA) to the matrix of partial correlations and plotting the vertices/variables in the space defined by the first two principal components (Whittaker (1988)). Using the absolute values of the partial correlation matrix, it is found that, in general, variables which are strongly associated are located close together, whereas weakly associated variables are located farther apart.

Although PCA results could readily be incorporated into the software, and would thus determine the layout of the vertices in the graph, in practice this is not desirable. Firstly, the user is no longer free to reposition the vertices, either to reduce the number of cross-overs or to make the graph easier to read where vertices may be tightly clustered together, without destroying the relationship between strength of association and distance. Secondly, since the vertices are represented in the first two dimensions only, it may be the case that the relationship between strength and distance is non-monotonic.

We therefore suggest two alternative techniques for the encoding of information about the strength of association in independence graphs, which are not dependent on distance. One possibility is to encode the strength of association by the width of the corresponding edge, so that weak associations are represented by thin edges, and strong associations by wide edges. the other is to encode the strength of association by the grey-tone shading of the corresponding edge, so that weak associations are represented by faint edges, and strong associations by dark edges. These two approaches could be combined to provide yet more emphasis. All of these methods of encoding the strength of association by edge style (width, grey-tone shading, or both combined) are available within CIGE.

2.6 Dynamic Display of Strength of Association (continuous data only)

In practice, because of the finite size of a pixel, it is possible to represent only about six distinct levels of width or grey-tone shading. This is usually sufficient to give a feel for the strengths of the associations in the data. However, by the use of a slider, it is possible to examine the strengths of the associations in more detail.

The slider is used to select a lower threshold for the strengths of interest, between zero and the maximum value (inclusive), and all edges which fall below the chosen threshold will be dropped from the display. Use of the slider means that it is not necessary to determine a priori which edges with near-zero associations will be omitted from the graph. These can now be determined by the user post hoc, perhaps assisted by theoretical knowledge. Edges to be deleted can then be removed from the model, by setting them to zero.

2.7 Derivation of Numerical Values

2.7.1 Continuous data

By clicking on an edge of the graph, the user can obtain the value of the strength of the corresponding pair-wise association. This is in addition to the information conveyed graphically about the strength of the association by the width and/or grey-tone shading of the edge, or by the use of the slider (see Sections 2.5 and 2.6).

2.7.2 Discrete data

By clicking on a single vertex (corresponding to a main effect), on two vertices (corresponding to a two-way interaction effect), or on three or more vertices (corresponding to higher order interaction effects), the user can obtain the parameter estimates, and their standard errors, corresponding to the selected effect. The estimates are displayed in the form of a table, with one entry per level of the selected variable, in the case of a main effect, or one entry per combination of the levels of the variables, in the case of an interaction effect.

This necessitates input of the parameter estimates to the software.

2.8 Incorporation With Model Fitting Routines

At present, CIGE is primarily of use for the graphical display and communication of fitted log-linear and covariance selection models. It does this by providing a graphical interface between the user and various items of information pertaining to the model (such as the elements of the generating class, and the strengths of associations), which need to be inputted by the user.

A logical extension to the usefulness of CIGE would be to incorporate it with model fitting routines so that the user could input the raw data directly, rather than fitting the model outside of CIGE using a separate package, and then inputting the fitted model, and also so that the model can be changed and refitted if appropriate, perhaps by the addition and/or deletion of edges.

It is suggested that GLIM or MIM (developed by Edwards (1987)) are the most appropriate packages for fitting the model outside of CIGE. To extend the usefulness of CIGE, a package such as MIM could perhaps be developed which employs CIGE as an interface its model fitting routines.

Acknowledgement

This research was supported by the Economic and Social Research Council.

References

Cottee M.J. & Hand D.J. (1989) Edge coding in conditional independence graphs. *Technical Report STAT-89-1.* Department of Statistics, The Open University

Darroch J.N., Lauritzen S.L. & Speed T.P. (1980) Markov fields and log-linear interaction models for contingency tables *Ann. Statist.*, 8, pp522–539

Edwards D. (1987) A guide to MIM. *Technical Report* #1. Statistical Research Unit, University of Copenhagen

Knuiman M. (1978) Covariance Selection. *Biometrics*, 28, pp157–175

Whittaker J. (1988) E.S.R.C. Workshop: "Graphical modelling: Transparencies". *Technical Report.* Department of Statistics, University of Lancaster

Whittaker J. (1990) *Graphical Models in Applied Multivariate Statistics* John Wiley & Sons, Chichester

WAMASTEX - Heuristic Guidance for Statistical Analysis

W. Dorda, K. A. Froeschl and W. Grossmann, Vienna

SUMMARY: The current state and the direction of further development of the WAMASTEX system are described. The main portion of the paper discusses the empirical assessment of several decision heuristics WAMASTEX's internal workings are based upon.

KEYWORDS: statistical guidance, empirical validation of statistical heuristics

1. WAMASTEX 1 – Heuristic Statistical Guidance

We have started to develop statistical analysis software within the broad framework of "expert systems" in 1986 focusing at the statistical needs of clinical physicians working and researching in hospital departments. Intended primarily as a means of relieving the burden of statistical consultancy placed on a few overcharged "experts", the project by now has resulted in a couple of interactive programs – termed WAMASTEX – typically confined to retrospective analysis of data sets gained by clinical experiments. Coarsely speaking, WAMASTEX organizes dialogue sessions guiding its users conditionally through a route map of analyses based on a set of *heuristics* applied to both, observation data, and user responses to queries spawned by the program. In any case, WAMASTEX generates "assumed to be best" suggestions without denying the user a possibility to proceed otherwise. WAMASTEX is entirely integrated into the SAS®package by means of that system's macro language facilities.

WAMASTEX 1 is a fully functional program covering *descriptive* statistical analysis (currently favoring univariate methods; [2]). Though being a natural choice to begin with, this part of data analysis consists of highly informal strategies the formalisation of which would run into severe troubles when based on firm decision theoretic foundations of statistical inference (viz., multiple/conditional tests). Hence, "weak" methods insensitive to most of possible peculiarities of data samples, depending – at best – only on sample size, are practicable. The internal workings of the system are therefore based on a *heuristic taxonomy* of data samples defined in terms of decisive properties of a set of presupposed probability distributions (e.g., normal/heavy–tailed distributions). Thereupon, statistical methods supplying answers to particular research goals are selected with respect to this heuristic taxonomy (this is the procedure we believe to be the one usually followed).

WAMASTEX 1 conducts each dialogue session through the stages (i) data validation, (ii) determining proper level of scale, (iii) distribution analysis (outliers, skewness, modality), (iv) inferring an appropriate set of descriptive measures, (v) output of results (in general, split in three parts: qualitative, graphical, and numerical presentation).

2. WAMASTEX 2 – Modelling a Computational Environment

Experiences with the various versions of WAMASTEX have shown that what state–of–the–art expert systems can offer to a wide range of statistical users does not really address the crucial problems with, and requirements of, (exploratory) statistical data analysis. Instead of forcing the flow of analytical progress in investigating statistical problems, or

Compstat 1990

scrutinizing statistical data sets, into the straight jacket of "canned" formal strategies, it would be highly desirable to have systems supporting the *creative process* of statistical thinking by computational versatility. Consequently, the functionality of analysis systems will have to cover the whole range of *on-line* activities of contemporary statistical computing (including, e.g., planning of experiments, maintaining analysis protocols, etc., but also text editing, symbolic computing, Monte Carlo methods, ...; see e.g. [5]). Correspondingly, the model of man/machine interaction replaces linear, system-driven dialogue sequences by a rather flexible arrangement of access to the system's functions allowing individual "overall" approaches to statistical analysis (comprising, in particular, interlaces and successions of functions not foreseen by the system's designers) as well as spontaneous digressions from a broad course of an analysis pursued. Yet, in order to be comprehensible such interaction models have to obey the structure and constraints of human perception, and way of thinking (i.e., *heuristic* problem solving).
Currently, WAMASTEX is undergoing several revisions in the light of these arguments. As an initial effort, essential parts of WAMASTEX 1 are integrated in a window-based software environment. This remake (on PC hardware) is developed along the lines of combining high computational power with familiar paper-and-pencil methods.

3. Empirical Foundation of Statistical Heuristics

Whatever may be the general tendency of statistical expert system development – the competence of the system will be related intimately to the quality and, hence, acceptance of the heuristics incorporated. Both, WAMASTEX 1 and 2, will rely basically on the same set of heuristics. Therefore we have decided to calibrate a couple of essential heuristics used very thoroughly: among those concerned (with an eye towards normal distribution hypothesis) are in the first place distributional properties such as skewness, modality, outliers, and automatic grouping (histograms). In particular, the treatment of *outliers* and the assessment of *skewness* (→ data transformations) as well as *modality* (→ presumption of – hidden – grouping factors) of sample distributions are scrutinized. In order to check the validity, acceptance, and utility of these heuristics (univariate case) they have been subjected to three stages of assessment: (i) response of heuristics to random samples generated according to theoretical distributions; (ii) a survey shedding some light on the rate of concordance of human experts' judgments with system's judgments; (iii) another survey collecting the reactions of WAMASTEX 1 users; these investigations will be discussed in the sequel (a more detailed presentation will be given in [3]).

3.1 Performance of Heuristics: Simulation Results

Monte Carlo-simulations have been carried out in order to check the reliability of some of the heuristic decision rules incorporated into WAMASTEX. In particular, the behaviour of skewness and modality heuristics was tested in the following setup:

- Each heuristic was subjected to a series of random samples generated according to several types of distributions such as
  - standard normal distribution $N(0,1)$
  - gamma distribution (with $\alpha = 1$)
  - mixtures of two normal distributions of shape $qN(0,1) + (1-q)N(\mu_2, 1)$ where $q = 0.5, 0.6, 0.7, 0.8$ and $\mu_2 = 2, 3, 4$.
- From each type of distribution 100 random samples with sample sizes $n = 20, 50, 100$, respectively, were generated.
- The skewness heuristic has been applied only to samples already decided to be unimodal by means of the modality heuristic.

The following tables summarize obtained results. The figures in the tables' triple columns refer to sample sizes 20, 50, 100 (in this order). Since in every case 100 samples were generated figures are absolute as well as percent rates. Occasionally, corresponding figures do not sum up to 100; those "missing" samples are classified by WAMASTEX as *uniformly distributed.*
Table 1 shows that larger samples are recognized fairly well as symmetric (and, hence, unimodal) in case of standard normal distribution, whereas the skewness criterion (which is based on the interquartile range) reacts too conservative: 52% of the Gamma(1)-samples (with $n = 100$) are considered symmetric (however, it should be noted that WAMASTEX merely prefers the symmetry assumption, giving the system user a chance to override). On the contrary, the modality criterion works eagerly: smaller $N(0,1)$ samples are likely to be (mis-)classified as multimodal.

distribution	*symmetric*			*left–skewed*			*right–skewed*			*multimodal*		
$N(0,1)$	41	71	89	21	5	0	15	2	3	22	22	8
Gamma(1)	38	45	52	2	0	0	51	47	42	9	8	6

Table 1: Standard Normal and Gamma Distribution Results

Table 2 covers both, bimodal and (right-)skewed distributions in addition to fairly "regular" ones. In case of distinct bimodality ($q = 0.5, \mu_2 = 4$) the heuristic apparently performs well. Since the theoretical distributions of mixed samples with $q = 0.5$ are symmetric whereas they become skewed with increasing q (especially for $q = 0.7$), the pattern of figures in the right half of the table indicates that the decision criteria respond accordingly; the bimodal cases ($\mu_2 = 3, 4$) are discerned better, of course, with increasing μ_2. Distributions with $q = 0.8$ are – again – almost symmetric (and unimodal regardless of μ_2); this, too, is recognized at least for larger sample sizes.

q	μ_2	*symmetric*			*left–skewed*			*right–skewed*			*multimodal*		
	2	44	54	74	13	6	3	9	5	1	29	35	22
0.5	3	31	22	28	1	1	2	2	3	1	65	74	69
	4	9	4	1	2	0	0	2	0	0	85	96	99
	2	42	60	74	10	5	0	18	6	4	29	28	22
0.6	3	34	19	23	1	0	0	18	25	11	46	56	66
	4	8	4	3	0	0	0	19	19	2	73	77	95
	2	44	55	78	9	3	0	28	14	9	18	28	13
0.7	3	35	25	35	1	0	0	41	39	32	22	36	33
	4	6	7	0	0	0	0	60	53	68	34	40	32
	2	51	72	87	8	3	0	27	10	4	14	15	9
0.8	3	50	64	70	8	1	1	32	27	23	10	8	6
	4	47	56	71	5	4	0	43	35	28	5	5	1

Table 2: Mixtures of Normal Distributions Results

Altogether, this simulation study entails the following conclusions:

- The chosen skewness criterion does not reflect the visual impression of "skewness" conveyed by ordinary histograms; in particular, the heuristic reacts too conservative in favor of the symmetry assumption. Basically, this is due to the insensibility (robustness) of the interquartile range to the shape of distribution tails.

- As already mentioned, the modality heuristic tends to reject unimodality rather than accept it; the criterion is particularly sensitive to intervening empty classes in the histogram. Hence, the heuristic is rather unreliable in case of smaller sample sizes. In general, diagnosed multimodality may not be taken literally, but should encourage a closer investigation of data.

3.2 How Experts Respond

In order to assess the correspondence of our designed heuristics with experts' decisions in *real-life* situations 37 experts (from Austria and FRG) have been asked for their opinion about an appropriate description of real data samples. The selected group of experts addressed was composed of more theoretically oriented as well as applied statisticians, some of them with a bias to either methodological work or a prevailing consulting activity.

Features of the questionnaire:

- Data: 16 samples, comprising between 10 to 400 cases, with various scales (metric, percent, log) and with a scope of distributions representative for the kind of data WAMASTEX will be confronted with (i.e. ranging from symmetric/unimodal to – as we feel – bluntly skewed or multimodal; occasionally including outliers, too). Information provided per sample comprised a (WAMASTEX-) histogram, sample size, measuring unit, resolution of the measuring device, and a set of commonly used statistics, but no observation data were supplied.
- The questions (to be answered for each data sample in turn) aimed at:
  (i) classifying the suitability/utility of each proposed statistic (from a given set)
  (ii) evaluating the histogram (WAMASTEX uses standard SAS®–histograms albeit with number of groups, class widths, and class midpoints computed differently)
  (iii) an overall assessment of the sample regarding skewness, modality, and normality of the distribution as well as presence of outliers.

The inquiry form provided also extra space for each item to add comments, suggestions, censures, etc.

Survey results in qualitative terms:

Altogether 11 statisticians responded. Though these participating colleagues form a well–balanced subset of the group addressed originally, this number should be kept in mind in interpreting the following statements.

- Apparently, each statistician applies his *personal* analysis strategy fairly regardless of the characteristics of a particular data sample. These attitudes (e.g. choosing some preferred set of statistics) are adjusted in rather extreme circumstances only, e.g., rather small samples ($n < 15$ or so), obvious outliers, etc. In general, our investigation suggests the conclusion that statisticians may be classified (roughly) as either "routine" (preferring model–based analysis with less emphasis on data description) or "thorough" (devoting attention to even minor peculiarities of observation data).
- There is an evident tendency to use only a few common descriptive statistics; kurtosis, modus, and interquartile skewness seem to be especially unpopular.
- Statisticians seem to select single statistics for their own merits of interpretation instead of considering the adequacy of whole groups of related statistics (e.g. location measures). Taking scale measures as an example, it turned out that in case of assumed bimodality the interquartile range is often preferred to variance, presumably because of an ascribed surplus of interpretation capability (w.r.t substantial terms).
- Some *heuristic* rules were found to be commonly employed:

- multimodality forces to reject mean value, variance, skewness and kurtosis measures, and modus
- a small sample size ($n < 15$) rules out meaningful interpretation of any statistic
- kurtosis information is of no use in case of skewed distributions
- in order to be classified as multimodal a distribution is required to possess local peaks of almost equal height which are not too close together.

- Statisticians' judgments are highly influenced by the graphical presentation of data; somewhat unexpectedly this still holds if the shown histogram is disqualified explicitly (because of, e.g., improper grouping, or scaling). Numerical measures are referred to for evaluating skewness and normality but, in general, are ignored by and large in assessing modality or possible outliers. In case of larger samples there is a tendency to take increased notice of the statistics' values.
- Information about measuring units (e.g. percents) was rarely used at all, e.g. to propose data transformations to cope with skewness.
- Histograms are a major source of criticism: class midpoints and width should be "round" numbers, with a coarser resolution (by factors 10^k, for some appropriate k, or some simple fraction thereof, e.g. $\frac{100}{4}$, $\frac{1000}{8}$ etc.,if need be) than the one data are recorded with. Unfortunately, alternative proposals appear to be quite divergent (different class widths etc.). In several cases the chosen number of classes was remarked to be too high.
- Histograms aside, statisticians accept the usual scope of output of statistical program systems without complaint; notably almost no augmentations (exception: variation coefficient) have been suggested.
- Due to the insufficiences of "standard" histograms, alternative graphics have been demanded quite frequently:
  - Q–Q plots as a means to verify normality assumption
  - plots of empirical distribution function, especially in case of smaller samples
  - stem and leaf diagrams (for sample sizes $n < 100$)
  - smoothed histograms (for larger sample sizes)
  - box plots (in almost every situation)
  - plots of transformed data (root and log transformations).

Summary: Concordance of experts and WAMASTEX

This section summarizes the numerical results of the survey conducted. Essentially, the tendency of the experts' judgments corresponds with the results obtained by the simulation study. Table 3 illustrates the rate of agreement between the system's and experts' decisions (in the latter case *majority votes* are recorded only) across the 16 samples being considered. Presence (absence) of a feature is marked with "+" ("–"); "?" means neither "+" nor "–" (i.e., *ambiguous*), or simply *unknown*.

System	*symmetric*				*unimodal*			*no outliers*		
Experts	+	–	?	Σ	+	–	Σ	+	–	Σ
+	**4**	0	1	*5*	**12**	0	*12*	**7**	4	*11*
–	5	**4**	0	*9*	0	**2**	*2*	0	**3**	*3*
?	0	0	**2**	*2*	0	2	*2*	0	2	2
Σ	*9*	*4*	*3*	*16*	*12*	*4*	*16*	*7*	*9*	*16*

Table 3: System vs. Expert Decisions

Note: Since WAMASTEX assumes normal distribution implicitly, "extreme" observation values are more likely to be classified as outliers.

3.3 Users' Reactions

WAMASTEX 1 is publicly available for use at Vienna's University Hospital and is actually employed by physicians and biometricians. In order to get some *feedback* from system users, to the end of each dialogue session an on-line inquiry form has been appended. This form is partitioned into three sections which, in turn, deal with

- quality of dialogue (questions clearly phrased ?, flow of analysis intelligible ?)
- result presentation (notably, the distinction between recommended/not recommended descriptive measures)
- program usage (key stroke sequences, screen layouts, help system, etc.).

In addition, an overall judgment of WAMASTEX is asked for.
Up to now, around about 60 users have carried out more than 100 WAMASTEX sessions, whence 32 inquiry forms have been completed (mostly by physicians).
Generally speaking, reception of WAMASTEX is quite encouraging. Notably dialogue management and system usage have got high scores throughout. The statistical results produced have been judged as correct, sound, and presented in an easily comprehensible layout. Certainly the weakest and, hence, less appreciated feature of WAMASTEX is its low quality histogram. Altogether, 21 (out of 32) users attested WAMASTEX being "good" (13) or even "very good" (8); only one (1) user signalled total dissatisfaction. (For further details cf. [1].)

3.4 Concluding Summary

The principal findings of our investigation can be summarized as follows:

- WAMASTEX's knowledge base has been designed in accordance with the customary (though, in general, not articulated expressis verbis) attitude to consider statistical measures rather isolated from each other. If the – at times subtle – properties of even "simple" statistics are examined closely at all, the treatise is likely to be approached in an item-by-item manner (e.g., cf. [4]). However, the results presented above indicate that *interactions between various statistics are decisive and cannot be ignored* (e.g., outliers accompanied with clearly negative kurtosis). Hence, "good policy" judgments have to be based on more complex description patterns.
- Because of the – unduly – high impact of visual impressions on even the experts' perception of sample information the graphical presentation of data turns out to be rather a delicate (since controversial) theme. As a consequence, substantial efforts are required to improve both, the design of histograms and the choice of suitable (i.e., not misleading) kinds of statistical graphics.

As a final remark, we would like to thank all colleagues taking part in our survey; unfortunately, a good deal of addressed experts denied their collaboration.

References

[1] *Dorda W.*, "Erfahrungen aus Entwicklung und Probebetrieb eines statistischen Konsultationssystems", Biometrie & Informatik, **21**, 1990 (in press).
[2] *Dorda W. et al.*, "WAMASTEX: Ein statistisches Expertensystem-Modell am Beispiel deskriptiver Statistik", Statistik und Informatik, **18**, 1988, pp. 172–180.
[3] *Dorda W. et al.*, "WAMASTEX: Empirische Analyse einiger statistischer Entscheidungsheuristiken" (in preparation).
[4] *Madansky A.*, "Prescriptions for Working Statisticians", Springer, New York–Berlin, 1988.
[5] *Thisted R.A.*, "Computing Environments for Data Analysis", Stat. Science, **1**(2), 1986, pp. 259–275.

Multivariate Data Analysis and Model Building

On Model Search Methods

T. Havránek, Praha

Abstract: In the present paper questions linked with parallel using of symbolic and numeric computations in model search tasks are discussed as well as possibilities of high level parallelisation of simultaneous evaluation of sets of models in implementing model search algorithms.

Key words: model search, symbolic computations, parallel computations.

INTRODUCTION

We shall consider a class of model search methods that are designed to find a set of acceptable models (supported by data) from a finite, but possibly large, set of models. Such situation is typical in multivariate analysis; consider e.g. multivariate contingency tables and log-linear models, variable subset selection in multivariate regression or discrimint analysis etc. (see Havránek, 1988). The underlying idea is that a model search procedure should not to give a user only one "best" model, but a set of alternative models reasonably supported by data since in interpreting such results, user's knowledge in the field can be involved.

In realization of such methods, models are considered as **formulas of a symbolic (logical) calculus** and deduction between models can be applied. This point of view was developed in a rather formal way in (Hájek & Havránek, 1978, Havránek, 1981) and applied more deeply for solving various practical tasks in (Havránek, 1984, Edwards & Havránek, 1985, 1987). Originally, all methods suggested were implemented using standard sequential computers and standard languages riented towards

Compstat 1990

numerical tasks (FORTRAN, Pascal). As recognized in the last time this point of view could be advantageous if the implementation involving specialized (symbolic or numeric) as well as parallel (distributed) hardware and software is considered. The view to distributed computing presented here is closer to that of Sherwish (1988) than that of Stewart (1988); the later is oriented more to linear algebra (fine granularity tasks). In (Sherwish, 1988) the reader can find a simple explanation of necessary notions from this field.

Let us mention that logical dependencies between models (hypotheses) are in statistics typically used in construction of effective multiple test procedures (Bergman & Hommel, 1988, Shaffer, 1986).

MODEL SEARCH TASK

Consider a finite set of statistical models $\mathcal{M}$. Each model $m \in \mathcal{M}$ is endowed by a parametric space Θ_m. As an example, one can imagine a set of hierarchical log-linear models for a multidimensional contingency table as considered in (Edwards & Havránek, 1985).

Usually a deduction rule I is linked with $\mathcal{M}$ enabling to deduce from the fact that m_1 is true ($m_1 \in Tr$) the fact that $m_2 \in Tr$ for some $m_2 \in \mathcal{M}$. In the simpliest case, I is a relation on $\mathcal{M} \times \mathcal{M}$ and $Cn^+(m_1) = \{m_2 : m_2 \in \mathcal{M}, m_1 I m_2\}$ is a set if *positive* consequences of m_1. One can consider a set of *negative* consequences as well: $Cn^-(m_1) = \{m_2 : m_2 \in \mathcal{M}, m_2 I m_1\}$. If m_1 is false ($m_1 \in Fs$) then each $m_2 \in Cn^-(m_1)$ is false too.

As an example of a deduction rule consider the simplicity ordering $\leq$ of log-linear models in (Edwards & Havránek, 1985). Note that $m \in Tr$ means in fact that the model is true for a parameter values from Θ_m.

Now, if $S \subseteq \mathcal{M}$ is a set of models, than we can say that $m_1 \in S$ is I-*minimal* in S if there is no $m_2 \in S$, $m_1 \neq m_2$ such that $m_2 I m_1$. Similarly, m_2 is I-*maximal* it there is no m_1, $m_2 \neq m_1$ and $m_2 I m_1$. Two models m_1, m_2 are *independent* if neither $m_1 I m_2$ nor $m_2 I m_1$. A set $S \subseteq \mathcal{M}$ is independent it each two m_1, $m_2 \in \mathcal{M}$, $m_1 \neq m_2$ are independent.

Consider now the case, that in a stage of evaluation of $\mathcal{M}$, one has as the only knowledge at disposal an independent set of models $S=(S \cap Tr) \cup (S \cap Fs)$; i.e. he/she knows which models from S are true or false. Then $\mathcal{M}^? = \mathcal{M} - (Cn^+(S \cap Tr) \cup Cn^-(S \cap Fs))$ is the set of "yet not desided" models. The greatest information gain can give now evaluating of $m \in \mathcal{M}^?$ with either the maximal $\#(Cn^+(m) \cap \mathcal{M}^?)$ on maximal $\#(Cn^-(m) \cap \mathcal{M}^?)$.

SYMBOLIC AND NUMERIC COMPUTING

What means to evaluate a model $m \in M^?$. In statistics we induce $m \in Tr$ or $m \in Fs$ from a data set M, using a decision rule d, such that $d(m,M) = a$ symbolizes the decision to accept $m \in Tr$ and $d(m,M) = r$ the decision to reject $m \in Tr$ (clearly, it one uses a goodness-of fit test, the acceptance could be doubtfull, cf. Havránek & Soudský, 1989). A decision rule is *coherent* if the situation in which $m_1 I m_2$, $d(m_1,M) = a$, $d(m_2,M) = r$ cannot occur. If a decision rule is coherent, then the symbolic and numeric levels of computations are linked in such a way that if $d(m_1,M) = a$, one can, instead of numerically compute $d(m_2,M)$ for another model $m_2 \in M$, to ask whether $m_2 \in Cn^+(m_1)$ (similarly for the case of $d(m,M) = r$).

Clearly, for a single model m_1 and a single model m_2, the **cost** of symbolic computation linked with deciding whether $m_2 \in Cn^+(m_1)$ (or $m_2 \in C_n^-(m)$) is, usually, much less than the **cost** of numeric computation of $d(m_2,M)$.

On the other hand, if one has at disposal a large set $A(M) = \{m; m \in M, d(m,M) = a\}$ or $R(M) = \{m; m \in M, d(m,M)=r\}$ then the cost of deciding whether $m_1 \in Cn^+(A(M)) = \cup_{m \in A(M)} Cn^+(m)$ (or $m_1 \in Cn^- R(M)$) could be comparable with or greater than the cost of $d(m_1,M)$. Hence a comprehensive representation of A(M) or R(M) should be considered, using I-minimal elements of A(M), $\min_I A(M)$, or I-maximal elements of R(M), $\max_I R(M)$.

From this point of view (the cost of symbolic) computations realizing the search for "greatest information gain" model should be reconsidered.

SEQUENTIAL AND PARALLEL THINKING

In (Hájek & Havránek, 1978), purely sequential approach was used. The underlying idea was apriori to order the models in *M* linearly by some ordering $\leq$. Then, in inspecting *M*, one start with the first model m_1 w.r.t $\leq$. If $m_1 \in Tr(M)$, then he/she or a computer finds the uniquelly defined $m_2 = \min_{\leq} \{m_2 \notin Cn^+(m_1)\}$. Now, if in a stage of computation, a set $S = (S \cap Tr) \cup (S \cap Fs)$ is at disposal (or an appropriate representation via I-minimal and I-maximal elements of $S \cap Tr$ or $S \cap Fs$ is available), then $m = \min_{\leq}\{m \notin M; m \notin Cn^+(S \cap Tr), m \notin Cn^-(S \cap Fs)\}$ should be evaluated. Using a reasonable relation between ≤ and I such a way can be in some cases computationally feasible, but by no means optimal in the number of evaluated models, i.e. in the number of numerical steps "compute d(m,**M**)".

If our primary aim is to put down the number of numerical steps, another approach should be considered: given independent $S = (S \cap Tr) \cup (S \cap Fs)$, take one of the **maximum information gain** models, i.e. one of the models from $m \in M^?$ with maximal $\#(Cn^+(m) \cap M^?)$ or maximal $\#(Cn^-(m) \cap M^?)$. Since *M* is finite, such models can be found by symbolic computations, but the **cost** can be high.

Think, e.g. $\#M = 2^{30} = 1073741824$ in variable selection tasks and the way of inspecting $Cn^+(m)$ and $Cn^-(m)$ for all $m \in M^?$.

Now, substantial is to note that a necessary condition for a model to be maximum information gain model is to be either I-minimal on I-maximal ir $M^?$.:

If $m \notin max_I(M^?)$ then there is a model m_1, mIm_1, $m \neq m_1$. Then $Cn^-(m) \subsetneq Cn^-(m_1)$. Similarly for $m \notin min(M^?)$. This leads to the way chosen in (Edwards & Havránek, 1985, 1987): Given $(S \cap Tr)$, $(S \cap Fs)$ for an independent S, consider in the next stage either $max_I(M^?)$ or $min_I(M^?)$ (say choose one of them with the less number of models contained). After evaluating d(m,M) for all models in the chosen sec S_1 update S, $S \cap Tr$, $S \cap Fs$ (omitting some redundant models to obtain independency) and find the new $M^?$ etc.

A substantial feature here is, that a whole set of models S_1 should be evaluated simultaneously. This evaluation can be done in **parallel**.

Note that under mild assumptions effective symbolic computation methods for finding of $max_I(M^?)$ and $min_I(M^?)$ are available (Edwards & Havránek, 1987).

MORE DEEP VIEW TO PARALLEL PROCESSING

The set up is now the following: a symbolic process calls in parallel the numeric processes evaluating d(m,M) for each $m \in S_1$:

(*) symbolic **process**: find S_1

↓ **in parallel; explicit fork**

numeric **processes**: evaluate d(m,M) for each $m \in S_1$

↓

symbolic **process**: update S, $S \cap Tr$, $S \cap Ts$ and go to (*) if not $M^? = \phi$.

In a software implementation, a *pcall* instruction (see Kober, 1988) can be used: pcall $(d(m,S_1);\ m \in S_1)$. Clearly an effective software implementation should match the computer architecture. A hierarchical architecture with a "symbolic" master processor, slave numeric processors and a shared memory for data could be considered:

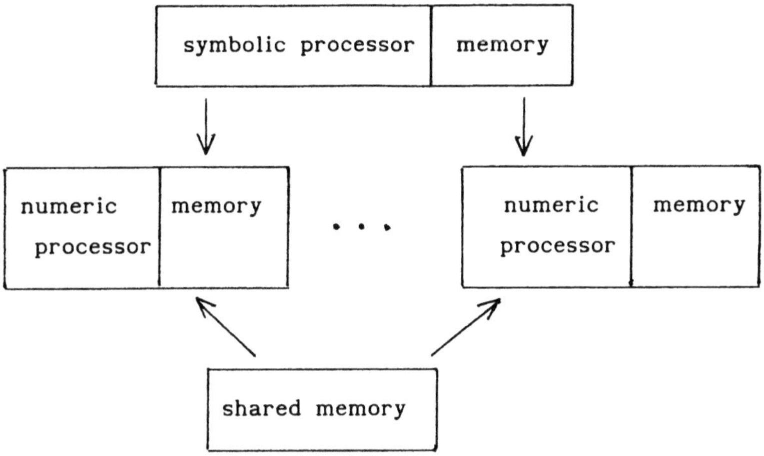

In the shared memory, data M are stored. If a numeric processor evaluates $d(m,M)$, only the necessary part of data is copied into his attached memory (think of necessary marginal tables in evaluating log-linear models).

What is substantial here is that data in the shared memory are not affected and numeric processors can work **asynchronously**.

In this set up, the symbolic process waits until obtaining the value $d(m,M)$ for all $m \in S_1$ and after that, a new S_1 is found.

In many cases, if $M^?$ is, after obtaing $d(m_1,M)$ for a model $m_1 \in M^?$, shanged to $M_1^?$, the sets $\min_I(M^?)$ and $\max_I(M^?)$ can be "easily" updated to $\min_I(M_1^?)$ $\max_I(M_1^?)$. In such a case, the symbolic process can update $\min_I(M^?)$, $\max_I(M^?)$ (for obtaining new S_1) immediately after obtaining a value $d(m,M)$ for a model $m \in S_1$ (not waiting for the whole vector of $d(m,M)$, $m \in S_1$):

(*) symbolic **process**: ask to evaluate S_1

↓ **explicit fork**

numeric **processes**: evaluate $d(m,M)$ for each $m \in S_1$

↓ **data flow**

symbolic **process**: after obtaining d(m,M) for a $m \in S_1$, update S,$S \cap Tr$, $S \cap Fs$, new S_1. It for each $m \in S_1$, d(m,M) was obtained go to (*) if not $M^? = \phi$.

Clearly, the computational model here is not very complicated Master-Slave mode (Scherwish, 1988), but two features are typical: first, the symbolic nature of the master process (realized on the master processor perhaps using a dialect of LISP) and second, the granularity of slave processes. The slave processes could be numerically rather complex: in such a process, it is, typically, necessary first to copy (part of) data M in the attached memory, then estimate parameters from Θ_m (think of iterative proportional fitting, nonlinear regression or robust multivariate regression as suggested by Antoch, 1986) and, finally, to compute a test statistic. Hence, as a processor realizing a slave numeric process, one can imagine a 32 bite processor with 4MB memory (and with well tested programs in FORTRAN77).

For the performance of the whole system, the number of slave processors is substantial. Number of models in S_1 should be comparable with the number of processors; since $\#S_1$ is usually tens, then a system with 16 or 32 etc. slave processors should be appropriate.

The symbolic mode of computation can be in this connection used with some advantage: it $\#S_1$ is greater than the number of processors, then in realizing the **pcall** instruction models (slave processes evaluating them) should be scheduled to free slave processors (a processor is free after sending the message d(M,m) to his master - after finishing the slave process). In such scheduling estimated, by a symbolic analysis, complexity of computing of d(m,M) could be used (think on decomposable and non decomposable log-linear models). Simple d(m,M) - tasks could be sheduled first - after their completing, the symbolic process can immediately start with updating $M^?$ etc.

At the end, two additional features should be mentioned. First, in symbolic mode not only "primitive" algorithmic techniques for finding, say $\max_I(M^?)$, could be used. This level is surely appropriate for realizing advanced AI-techniques for model search or for control of (exploratory) data analysis in general.

Second, communication between slave processes (processors) could be usefull. In many times some computations, say in estimating parameters from Θ_{m_1} and Θ_{m_2}, are not "independent". For example, some results obtained for a model m_1 could be updateds easily for m_2. Then a communication between the corresponding processes could be usefull, but, clearly, there some synchronization is needed - and since usually such communication could not be done in a uniform mode (it is necessary to recognize which models are "dependent"), the application of such high level parallelism is complicated and risky task (due communication/synchroni- zation there could be in fact a negative speed-up).

CONCLUSIONS

We tried to draw attention to various aspects of linking symbolic and numeric computations in model search tasks. To summarize, from the computational point of view two questions seems to be vital. First, it is necessary to carefully compare relative costs of symbolic and numeric computations (considering for both levels "optimal" software and hardware) and then to decide, when possible, which numeric tasks can be substituted or improved by using symbolic computations. Second, it is necessary to consider not only parallelism of symbolic and numeric computations, but medium or high level parallelism inside the numeric level (controlled by the symbolic process).

In parallelisation, the fit between the nature and granularity of the numeric processes on one side and the computer architecture available has to be achieved; in the model search tasks, the granularity seems to be in general more rought than the granularity considered in Sherwish (1988) examples. The parallelization inside the symbolic level together with employing AI techniques not only for the control of model search process but more deeply for model synthesis should be considered in the future as well.

ACKNOWLEDGEMENTS

Since in Czechoslovakia up to now there are no real parallel systems, I am gratefull to my friend and colleague Petr Hájek for convince me of the usefullness of abstract thinking.

References

Antoch, J. (1986): Algorithmic development in variable selection procedures, COMPSTAT 86, Physica-Verlag, Heidelberg, 83-90.

Bergman, B. & Hommel, G. (1988): Improvements of general multiple test procedures for redundant systems of hypotheses, Multiple Hypotheses Testing, Medizinische Informatik und Statistik, Vol. 70, Springer-Verlag, Berlin.

Chambers, J.M. (1989): Discussion of software environments for statistical computing, Statistical Software Newsletter 15, 81-84.

Edwards, D. & Havránek, T. (1985): A fast procedure for model search in contingency tables, Biometrika 72, 339-351.

Edwards, D. & Havránek, T. (1987): A fast model selection procedure for large families of models, J. Amer.Statist.Assoc. 82, 205-213.

Hájek, P. & Havránek, T. (1978): Mechanizing hypothesis formation, Springer-Verlag, Heidelberg.

Havránek, T. (1981): Formal systems for mechanized statistical inference, Int.J.Man-Machine Studies 15, 333 -350.

Havránek, T. (1984): A procedure for model search in multidimensional contingency tables, Biometrics 40, 95- 100.

Havránek, T. (1988): On a general algorithm for model choice in multivariate analysis, Statistics 19, 465-475.

Havránek, T. & Soudský, O. (1989): Model choice in the context of simultaneous inference, Y.Dodge (ed.), Statistical Data Analysis, North-Holland, Amsterdam, 165-176.

Kober, R. (1988): Parallelrechner-Architekturen, Springer- Verlag, Heidelberg.

Schervish, M.J. (1988): Applications of parallel computation to statistical inference, J.Amer.Statist.Assoc. 83, 976-983.

Shaffer, J.P. (1986): Modified sequentially rejective multiple test procedures, J.Amer.Statist.Assoc. 81, 826-831.

Stewart, G.W. (1988): Parallel linear algebra in statistical computations, COMPSTAT'88, Physica-Verlag, Heidelberg.

Principal Components Analysis with Respect to Instrumental Variables Via Univariate Splines

J. F. Durand, Montpellier

Introducing univariate splines provides an approach for solving the PCAIV problem in the non linear case: That is to build a representation of the observations as near as an another study, presupposing that the link between the two studies'variables takes an additive form. The linear PCAIV (Bonifas et al., 1984) is then a particular case of the spline-PCAIV (Durand, 1989) in the following sense: The solution of the linear PCAIV belongs to the set of feasible solutions of the spline-PCAIV and will constitute the first step of the associated iterative algorithm. Moreover this method presents an extension to regression using additive splines and can be considered as a particular canonical analysis in one sense which is specified.

A numerical application related to the treatment of an *anamorphosis* is given at the end of the paper.

1. THE SPLINE-PCAIV

Let $S(D)$ be the set of the D-symmetric matrices, i.e., the set of matrices A such that $DA = A'D$ where A' is the transposed matrix of A. It is known, (Robert & Escoufier, 1976), that the symmetrical bilinear form $< , >$ defined by $< A, B >= tr(AB)$, where $tr(.)$ is the trace operator, defines a scalar product on $S(D)$. It will be noted $||.||$ the norm deduced, more precisely $||A|| = (tr(A^2))^{\frac{1}{2}}$.

Let (Y,Q,D) be a study of q variables measured on n observations, *completely defined*: Y is $n \times q$ and the matrix Q, $q \times q$ symmetric and positive definite, which provides a metrics to the space of the observations, has been chosen by the statistician. The same has determined the matrix D of the weights associated to observations which allows the definition of a metrics on the space of the variables (D is $n \times n$, diagonal positive with tr(D)=1).

Let (X,R,D) be a second study of p variables measured on the same observations with the same weights, X being $n \times p$. It is wished to move the representation of the observations provided

Compstat 1990

by (X, R, D), choosing a transformation of X using a basis of r B-splines which gives $T = T(X)$ and choosing a metrics R $p \times p$, in order to fit at best the representation of the observations provided by (Y, Q, D), called the reference study. The matrix T, $n \times p$, can be written

$$T = MA\,. \tag{1.1}$$

The matrix M, whose j^{th} block of r columns is obtained by the transformation of the jth column of X, is $n \times rp$. The matrix A containing the unknown parameters a_{ij} is $rp \times p$. These parameters are gathered in a vector a of $\mathbb{R}^{rp}$.

Generally speaking, the representation of the observations is provided by the eigenvectors of the operators $YQY'D$ and $TRT'D$ which belong to $S(D)$. Note f the objective function deduced from the discrepancy between these two operators, i.e. from $\|YQY'D - TRT'D\|^2$

$$f(R, a) = \frac{1}{4}\,\|TRT'D\|^2 - \frac{1}{2}\,tr(YQY'DTRT'D)\,. \tag{1.2}$$

The problem consists in determining a metrics $\overline{R}$ and a vector $\overline{a}$ (i.e. an array $\overline{T} = M\overline{A}$) such as

$$f(\overline{R}, \overline{a}) = \min_{R,a} f(R, a)\,. \tag{1.3}$$

If it is supposed that a is fixed, we can obtain an optimal metrics by solving

$$\nabla_R f(R, a) = \frac{1}{2}(S_{TT}RS_{TT} - S_{TY}QS_{YT}) = 0\,.$$

with $S_{TT} = T'DT$, $S_{TY} = T'DY$, $S_{YT} = S'_{TY}$ .That gives

$$R = S_{TT}^{-1}S_{TY}QS_{YT}S_{TT}^{-1} \tag{1.4}$$

where S_{TT}^{-1} is the generalized Moore-Penrose inverse of S_{TT}.

In particular, the choice $a = \xi$ of the nodal values computed from the knots, is not necessarily optimal but gives $T = X$ and by the expression of the optimal metrics allows the consideration of the linear PCAIV's solution as belonging to the set of the feasible solutions of the spline-PCAIV.

Once $\overline{a}$ and thus $\overline{T} = M\overline{A}$ is determined, $\overline{R}$ is given by (1.4). A representation of the observations is obtained by performing the PCA of $(\overline{T}, \overline{R}, D)$. This PCA is equivalent, because the equality of the operators, to that of the triple $(P_{\overline{T}}Y, Q, D)$, where $P_{\overline{T}}$ is the D-orthogonal projector on $\mathbb{E}_{\overline{T}}$ the linear subspace of $\mathbb{R}^n$ spanned by the columns of $\overline{T}$.

$$P_{\overline{T}} = \overline{T}S_{\overline{T}\overline{T}}^{-1}\overline{T}'D\,. \tag{1.5}$$

The reconstruction of the representation of the observations is exact if $P_{\overline{T}}Y = Y$, i.e. if Y belongs to $\mathbb{E}_{\overline{T}}$. This will happen when, for every $k = 1, \ldots, q$, there exists p numbers λ_{kj} so that the kth variable of Y

$$Y_k = \sum_{j=1,\ldots,p} \lambda_{kj} \overline{T}_j(X_j) .$$

Thus when $q = 1$, the spline-PCAIV is a part of the regression techniques using additive splines (Hastie & Tibshirani, 1986).

2. THE SPLINE-PCAIV AND THE CANONICAL ANALYSIS

Without any restriction it can be supposed that Y is centred with respect to D. It is shown that the spline-PCAIV of $(Y, (Y'DY)^{-1}, D)$ and X is the best linear canonical analysis between Y and any T defined by (1.1) in the following sense

$$tr((P_Y P_{\overline{T}})^2) \geq tr((P_Y P_T)^2) .$$

3. ALGORITHMICAL CONSIDERATIONS

The explicit expression of the gradient of the function f with respect to a is given so that the knowledge of the gradients of f with respect to R and a allows the implementing of a numerical method of relaxation:

An initial value for a is chosen. The natural idea consists in alternating a step of computing an optimal R for fixed a, thanks to (1.4), with a step of a descent algorithm with respect to a, R being fixed.

If the initial choice is $a = \xi$, very easy to compute starting from the knots, the first step is nothing else but taking with (1.4) the optimal metrics of the linear PCAIV. In the absence of constraints on a, the chosen direction of descent could be that of a quasi-Newton method. In the case where the vector a of the parameters submits to a constraint, the direction of descent will be given by the projection of the gradient on the linear subspace spanned by the fulfilled constraints (projected gradient method). Unless the objective function defined by (1.2) is convex with respect to a, and it will be seen that is not the case, this algorithm leads to a local optimum. However the nature of f function of a is studied and an explicit choice at the time of the unidimensional minimization is provided.

4. ONE APPLICATION TO AN ANAMORPHOSIS RECONSTRUCTION

The representation of the observations provided by the triple (Y, Q, D) constitutes, in the case where the number of variables q is equal to 2 a plane form called the *reference form*. In the following example, the reference form, marked Y for simplicity, will be constitued of $n = 100$ observations (100 points of the plane), the metrics used here will be $Q = I_2$ and $D = 100^{-1} I_{100}$. Y has the form of a dog well known by the cartoons lovers, see figure 1.

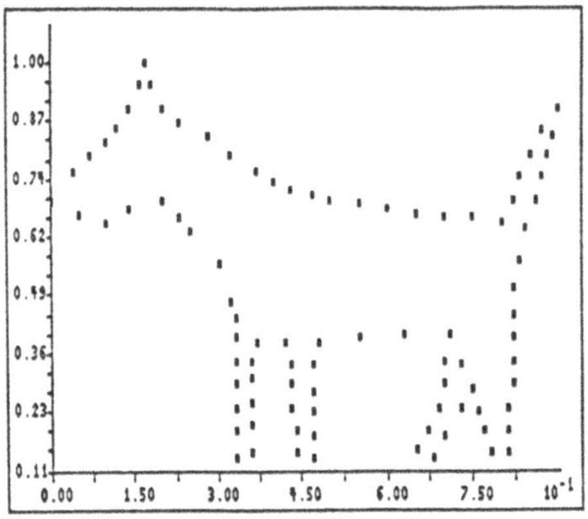

Figure 1 : The reference form.

The representation of the same 100 observations provided by the triple (X, I_2, D) with $p = 2$, is the plane form called the *anamorphosis* of Y or the *form to be distorted* taking Y for a model, see figure 2.

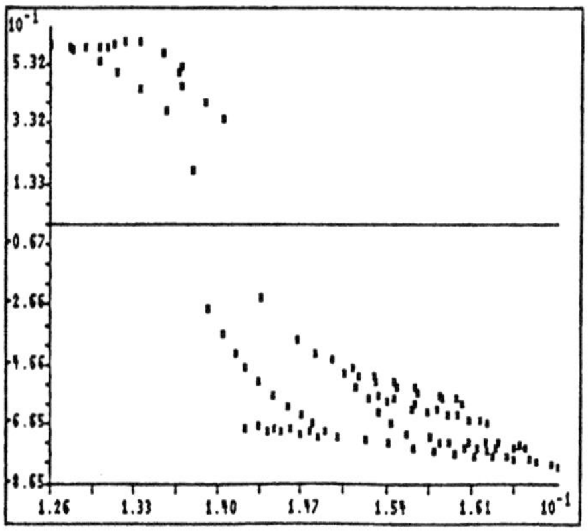

Figure 2 : The anamorphosis.

The anamorphosis of Y, marked X for simplicity, is *built* using transformations given by

$$\begin{cases} X_1 = \frac{1}{10} Acos(\frac{1}{3}Y_2 - \frac{1}{6}Y_1) \\ X_2 = (\frac{1}{3}Y_2 - \frac{2}{3}Y_1)^{1/3}. \end{cases}$$

These transformations had been chosen so that the link between the variables of Y and X takes the additive form

$$\begin{cases} Y_1 = 2cos(10X_1) - 2(X_2)^3 \\ Y_2 = 4cos(10X_1) - (X_2)^3 . \end{cases}$$

Two reconstitutions are implemented:

The first, corresponding to the linear PCAIV, consists in building the graphical display of the array $P_X Y$, see figure 3, stopping the spline-PCAIV algorithm at the first step

$$P_X Y = X(X'DX)^{-1}X'DY .$$

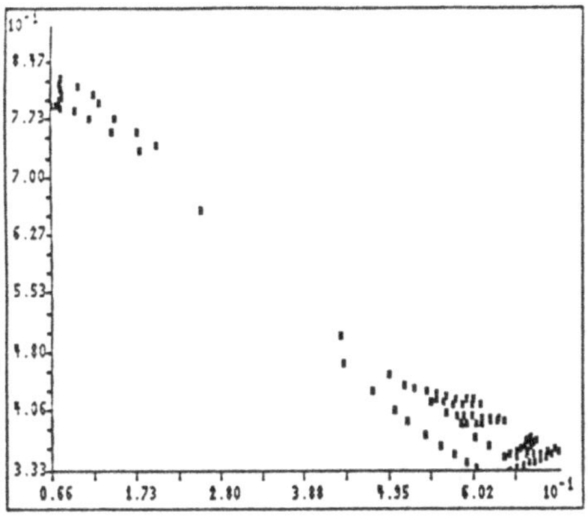

Figure 3 : The reconstruction using the linear PCAIV.

The second, which implements the algorithm described in the preceding pages, leads to the graphical display of the array $P_{\overline{T}}Y$, see figure 4. The splines used here are B-splines of order 3 with 4 interior knots.

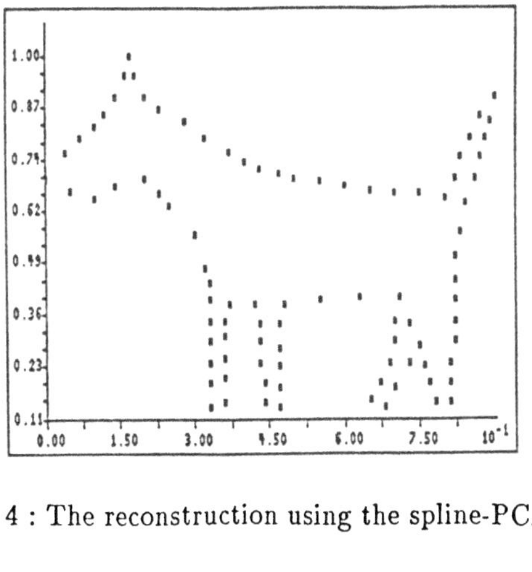

Figure 4 : The reconstruction using the spline-PCAIV.

The fact of having chosen statistical studies based on two variables only, avoids having to carry out PCA's at the end of the treatment in order to simplify the representations of the observations. The graphic comparison of the two analyses give prominence to the savings obtained in the reconstruction by the second approach, i.e. that using the splines.

ACKNOWLEDGEMENTS: I would like to express my thanks to professor Y. ESCOUFIER for his valuable advice and the quality of the welcome in his laboratory. I thank R. SABATIER and A. BERLINET for their careful reading of the manuscript and their helpful remarks.

REFERENCES

BONIFAS L., ESCOUFIER Y., GONZALEZ P.L., SABATIER R. 1984, *"Choix de variables en analyse en composantes principales"*, Revue de Statistique Appliquée, vol.XXXII, $n^o 2$, p.5 – 15.

DURAND J.F., 1989, *"Principal Components Analysis with respect to Instrumental Variables via univariate spline transformations"*, R.T. 8902, Unité de Biométrie, ENSAM-INRA-USTL, Montpellier.

HASTIE T., TIBSHIRANI R. 1986, *"Generalized additive models"*, Statistical Science, Vol. 1, $n^o 3$, p. 297 – 318.

ROBERT P., ESCOUFIER Y. 1976, *"A unifying tool for linear multivariate methods: The RV-coefficient"*. Applied Statistics, 25, 257 – 265.

SCHUMAKER L.L. 1981, *"Spline functions : Basic theory"*, J.Wiley New York.

Selecting the Best Subset of Variables in Principal Component Analysis

P. L. Gonzalez, Evry, R. Cléroux and B. Rioux, Montréal

The problem of variable selection, in Principal Component Analysis (PCA) has been studied by several authors [1] but as yet, no selection procedures are found in the classical statistical computer softwares. Such selection procedures are found, on the other hand, for linear regression or discriminant analysis because the selection criteria are based on well known quantities such as the multiple correlation coefficient or the average prediction error.

In PCA, the selected subset of variables must reproduce in the best possible way, the general characteristics of the observations. In particular, we wish to approximate in the best possible way, the distances between individuals as computed from the initial data matrix. The model of Principal Component Analysis with respect to Instrumental Variables (PCAIV) together with the optimization of the RV coefficient [4] provides a reasonable setup for variable selection.

The paper is organized in the following way. In Section I we recall the RV coefficient and the PCAIV model. In Section II, an algorithm for selecting the best variables subset in PCA is proposed. It is adapted from the Furnival and Wilson algorithm for multiple linear regression. Two examples are considered in Section III and a conclusion follows in Section IV.

I. PRINCIPAL COMPONENT ANALYSIS WITH RESPECT TO INSTRUMENTAL VARIABLES

1. The RV Coefficient

Consider the experiment denoted by (Y,Q,D) where Y is a *nxp* data matrix, p quantitative variables being measured on each of n individuals, Q is a *pxp* positive definite matrix used in computing distances between individuals and D is a *nxn* diagonal matrix of weights on the individuals which trace equals one.

We assume that the columns of Y have 0 mean vector. In order to compare the experiment (Y,Q,D) with other experiments, the operator WD=Y'QYD, which characterizes

Compstat 1990

(Y,Q,D) is used. Its eigenvectors, properly normalized, are the principal components. More precisely two experiments E_1 and E_2 concerning the same individuals are compared using the RV coefficient between their respective operators W_1 and W_2:

$$RV = RV\ (E_1, E_2)\ = \frac{tr\ (W_1 D\ W_2 D\ )}{\left[tr\ (W_1 D)^2\ tr\ (W_2 D)^2 \right]^{1/2}}\ .$$

By definition, $0 \leq RV \leq 1$. It is equal to 1 if and only if W_1 and W_2 are proportional (homothetic distances between individuals). It is equal to 0 if all the principal components of E_1 are orthogonal to all those of E_2.

2. PCAIV of X with respect to the experiment (Y,Q,D)

Let E = (Y,Q,D) be an experiment and X be a *nxq* data matrix, q variables being measured on the same individuals having the same weights. The problem is to find the metric *M*: *qxq* (positive definite matrix) such that the experiments (Y,Q,D) and (X,M,D) are as close as possible in the sense of the RV coefficient. In other words, we wish to find M such that the distances between individuals computed from the variables forming X, together with M, are as close as possible to those computed from the variables forming Y, together with Q. One then talks about the PCAIV of X with respect to (Y,Q,D). The solution of this problem is known to be

$$M = (X'DX)^{-1}\ (X'DY)\ (Y'DX)\ (X'DX)^{-1} \qquad (1)$$

It is seen that the PCAIV of X with respect to (Y,Q,D) is equivalent, as far as individuals representation is concerned, to the PCA of (X,M,D). It is seen also that, if $P_x = X(X'DX)^{-1} X'$ D is the D - orthogonal projector onto the subspace generated by the columns of X, then the PCAIV of X with respect to (Y,Q,D) is also equivalent to the PCA of $(P_x Y, Q, D)$.

In the sequel of this paper, RVREG will denote the maximum value of RV, that is

$$RVREG\ \left[X/(Y,Q,D) \right] = RV \left[YQY'D,\ XMX'D \right] \quad \text{with M given by (1).}$$

II. VARIABLE SELECTION IN PRINCIPAL COMPONENT ANALYSIS

1. The method

Consider again an experiment (Y,Q,D), where Y: nxp, and a data matrix X: nxq . The variables defining X may be identical to those defining Y, or partially or totally different. The problem here is to obtain, for any given r < q, a subset of r columns from X, denoted $X_{[r]}$, and to obtain the corresponding metric $M_{[r]}$ such that the two experiments (Y,Q,D) and $\left[(X_{[r]}, M_{[r]}, D\right]$ are as close as possible. In other words, we wish to obtain the subset of r variables which maximizes RVREG $\left[X_{[r]} / (Y,Q,D)\right]$ where r < q is given. When both r and p are moderately large, the exhaustive computation of all possible values of RVREG becomes intractable and a selection algorithm becomes necessary. In this paper, such an algorithm is proposed.

2. The algorithm

The objective is thus to find the best subset $X_{[r]}$ of X. Let r be the number of variables in the desired subset, m=q-r be the number of variables eliminated from the initial data matrix X and Θ_i be the value of RVREG computed from all the columns of X except X^i, the ith one. To initialize the procedure, we compute the Θ_i ' s and label them in such a way that

$$\Theta_1 \geq \Theta_2 \geq , \ldots , \geq \Theta_q \qquad (2)$$

The best subset of size q-1 is thus the one obtained by omitting the first variable. The principle of the algorithm consists of comparing the value α of RVREG, obtained from a subset of r variables, after eliminating m=p-r variables in which the highest index is j, to Θ_{j+1}. If $\alpha \geq \Theta_{j+1}$, then no subset resulting from the elimination of variables having indices higher than j will lead to a bigger value of RVREG. The details of the algorithm are the following:

Step 1: Compute the value of RVREG after the elimination of variable $X^1, X^2, \ldots , X^m$ where the indices are ordered as in (2). If this value is greater than Θ_{m+1}, then the optimal subset of size r contains variables $X^{m+1}, X^{m+2} , \ldots , X^q$. Otherwise we proceed to Step 2.

Step 2: Variable X^{m+1} is candidate for elimination. The m values of RVREG resulting from the elimination of m variables including X^{m+1}, among the m+1 first variables, are obtained. If the biggest of the m+1 values of RVREG, obtained in Steps 1 and 2, is not smaller than Θ_{m+2} then the corresponding subset is optimal of size r . Otherwise we proceed to Step 3 where X^{m+1} and X^{m+2} are candidates for elimination. And so on.

The preceeding algorithm can produce, for any r, the k best subsets of size r . It can also be adapted for the use of other association measures such as, for example, the redundancy index introduced by Stewart and Love [5].

III. EXAMPLES

1. Best subset selection in PCA

The data matrix consists of n=49 countries and p=7 variables recorded in each country. The variables are the following: GNP= gross national product per capita in 1957 US dollars, INFD= member of infant deaths per 1000 live births, PHYS= member of inhabitants per physician, DENS= population per square kilometer, AGDS= population per 1000 hectares of agricultural land, LIT= percentage literate of population aged 15 and over, HIED= member of students enrolled in higher education per 100,000 population. The complete data set can be found in [2]. The best subsets of respective sizes $r = 1,2, \ldots, 6$ are given by

RVREG	Variables					
0,6802	LIT					
0,8387	DENS	LIT				
0,9187	DENS	LIT	GNP			
0,9558	AGDS	LIT	INFD	MIED		
0,9813	DENS	PHYS	GNP	INFD	HIED	
0,9967	AGDS	LIT	PHYS	GNP	INFD	HIED

The best subsets of sizes r=1,2,3 are also obtained by a forward procedure but not the others. It was noted in the data set that the individuals SINGAPORE and HONG KONG both had high values for variables DENS and AGDS. These individuals also had high contributions in the PCA. They were therefore omitted and the best subset selection procedure was run again in order to have a glance on the stability of the results. The only change was the following: the best subset of size r = 2 now becomes {LIT, AGDS}. It was the second best subset of size r = 2 in the preceeding analysis. Subset {DENS, LIT} is now the second best subset of size r=2.

2. Best subset selection in PCAIV

The data of this example can be found in [2]. We wish to explain, as well as possible in the sense of RVREG, two variables relative to homicides in Detroit between 1961 and 1973: HOM and ASR. The X matrix is defined by 12 variables. The best subsets of respective sizes $r = 1,2, \ldots, 5$ are the following. It is seen that the best subset selection procedure yields interesting results.

RVREG	Variables				
0,9291	FTP				
0,9566	NMAN	ACC			
0,9840	ACC	WE	LIC		
0,9909	ACC	WE	LIC	WM	
0,9948	ACC	WE	LIC	NMAN	GOV

IV. CONCLUSION

In this paper, a procedue for selecting the best subset of variables in PCA and PCAIV has been proposed. It is purely descriptive. However it can be very efficient as was seen from the examples. It allows to summarize, in some way, the information contained in a data set. The next step would be to develop statistical tests in order to evaluate the significance of a subset of size $l+1$ with respect to a subset of size l. For this, the exact distribution of RVREG, in a given setup, is needed. Such an approach has been taken in a stepwise, algorithm for the PCAIV model using another measure of association [3].

REFERENCE

[1] Bonifas, I., Escoufier, Y., Gonzalez, P.L. and Sabatier, R.: Choix de variables en analyse en composantes principales, Rev. Stat. Appl., 1984, vol XXXII no.2.

[2] Gunst, R.F. and Mason, R.L.: Regression Analysis and its Applications, 1980, Marcel Dekker Inc., New York

[3] Lazraq, A. et Cléroux, R.: Un algorithme pas à pas de sélection de variables en régression linéaire multivariée, Stat. Anal. des Données, 1988, 13, 39-58.

[4] Robert, P. et Escoufier, Y.: A Unifying Tool for Linear Multivariate Statistical Methods: the RV Coeficient, Appl. Stat., 1976, 25, 257-265.

[5] Stewart, D. and Love, W.: A General Canonical Correlation Index, Psycho. Bull., 1968, 70, 160-163.

Interesting Projections of Multidimensional Data by Means of Generalized Principal Component Analyses

H. Caussinus and A. Ruiz, Toulouse

Summary : Principal Component Analysis can produce several interesting projections of a point cloud if suitable inner products are chosen for measuring the distances between the units. We discuss two examples of such choices. The first one allows us to display outliers, while the second is expected to display clusters. Doing so we introduce a robust estimate of a covariance matrix and we investigate some of its properties.

Key words : Projection Pursuit, Principal Component Analysis, Discriminant Analysis, Clustering, Outliers, Robust Estimation.

1. Introduction

Projection Pursuit aims to produce interesting low-dimensional projections of a p-dimensional point cloud ([2],[6],[7]). Principal Component Analysis (PCA) provides interesting projections from the point of view of total variance accounted for by the projected data. However, several other criteria of interestingness may be considered, among them the criterion of heterogeneity has been given most attention. In fact, it is interesting to find projections which reveal the presence of clusters when the data can be divided into homogeneous groups, or projections displaying outliers, etc. Most techniques which have been formerly proposed to achieve these purposes require extensive (and somewhat non-standard) computation even when one looks for q-dimensional projections with small q, say q=2. We propose alternative techniques which require simple and standard computations for any q. Thus, it is very easy to include them in any statistical software designed for exploratory analysis of multidimensional data.

Compstat 1990

Let us denote the data by X_i, i=1,...,n, each X_i being a p-vector, the sample covariance matrix by V, and choose a reference inner product (see e.g. [1],§7.6), i.e. a symmetric positive definite p×p matrix M. (Generalized) PCA looks for the eigenvectors of VM associated with the q largest eigenvalues ; then the centered data are projected on the subspace spanned by these eigenvectors, the projection being orthogonal with respect to the inner product M. For instance, M is the identity matrix I_p for "ordinary" PCA, M^{-1} is the diagonal matrix of the sample variances for standardized PCA, and M is the inverse of the within groups variance W in the case of Multiple Discriminant Analysis ([9],§5.8). Various interesting projections can be obtained by other suitable choices of M. The main purpose of this paper is to present two examples.

2. Outliers.

Let $$S = \sum_{i=1}^{n} K(\|X_i - X^*\|^2_{V^{-1}})(X_i - X^*)(X_i - X^*)' \Big/ \sum_{i=1}^{n} K(\|X_i - X^*\|^2_{V^{-1}})$$

where $\|X\|^2_M = X'MX$, X^* may be the mean vector $\bar{X}$ or a robust estimate of location, and K is some decreasing function : our choice will be K(u)=exp(-hu).

By letting $M=S^{-1}$, we obtain projections which display possible outliers. This technique is no more than a simplification of a similar one proposed by Yenyukov [10]. Instead of S, this author uses a robust estimate of the covariance matrix which requires iterative computation. Our proposal is simpler and may be even more efficient if there are many outliers (see §5 below). Actually, under convenient assumptions, for n large enough, S^{-1} is close to $\Gamma^{-1} + 2hV^{-1}$ where Γ is the covariance matrix of the uncontaminated data. Hence, VS^{-1} is close to $V\Gamma^{-1} + 2hI_p$ whose eigenvectors are those of $V\Gamma^{-1}$.

3. Clusters.

Let

$$T = \sum_{i=1}^{n-1} \sum_{j=i+1}^{n} K(\|X_i - X_j\|^2_{V^{-1}})(X_i - X_j)(X_i - X_j)' \Big/ \sum_{i=1}^{n-1} \sum_{j=i+1}^{n} K(\|X_i - X_j)\|^2_{V^{-1}})$$

By letting $M=T^{-1}$, we get projections which are likely to reveal clusters with proportional covariance structures if such clusters are present in the data. We point out that the technique is close to Multiple Discriminant Analysis, T being expected to play the role of W <u>when the groupings are unknown</u>. A statistic based on pairwise differences of the observations has been proposed in [3] as a possible robust estimate of a dispersion matrix. Our purpose is somewhat different and, in a sense, more general. In fact, let us assume that there are clusters of equal covariance matrices, say Γ, and that the distances between their centroids with respect to the metric Γ^{-1} are large enough. Then, under suitable assumptions, e.g. the sample is drawn from a mixture of normal distributions, it can be proved that VT^{-1} is close to $1/2\ VW^{-1} + 2hI_p$ when n is large. Thus the eigenanalysis of VT^{-1} is equivalent to the eigenanalysis of VW^{-1}.

Some important features of the proposed technique are the following.

a) The graphical displays are invariant under any affine transformation of the data. For example, the raw data and the sphered data provide the same display.

b) The choice of h is a sensible point which deserves further attention. From our present experience, it seems enough to try a few values, say h=.50,.75,1.

c) Our method is easier to implement than other Projection Pursuit algorithms, but it may be less efficient to reveal clusters with very different covariance structures.

d) Since outliers are small clusters, our method tends to display them as well as groups of greater importance. This may result in masking such groups. To overcome this difficulty we propose two solutions :

(i) look for outliers with the method of section 2 and discard them before using the method of section 3,

(ii) use the method of section 3 with a robust covariance matrix instead of V, for instance $(S^{-1} - 2hV^{-1})^{-1}$.

4. Example

The method of section 2 will be denoted by OUT. The two methods of section 3 will be denoted respectively by CL1 when the analysis rests on VT^{-1} and CL2 when V is replaced by a robust estimate as suggested in (ii) above. The data processed are Lubischew's beetle data as in [7] : the number of cases is n=74 and the number of measurements is p=6. To save space, we present only two-dimensional views corresponding to the two largest eigenvalues. The first display has been obtained by standardized PCA. OUT (h=.1) identifies some outliers ; for example, it can be verified that the two cases with small first and second coordinates are characterized by a very low

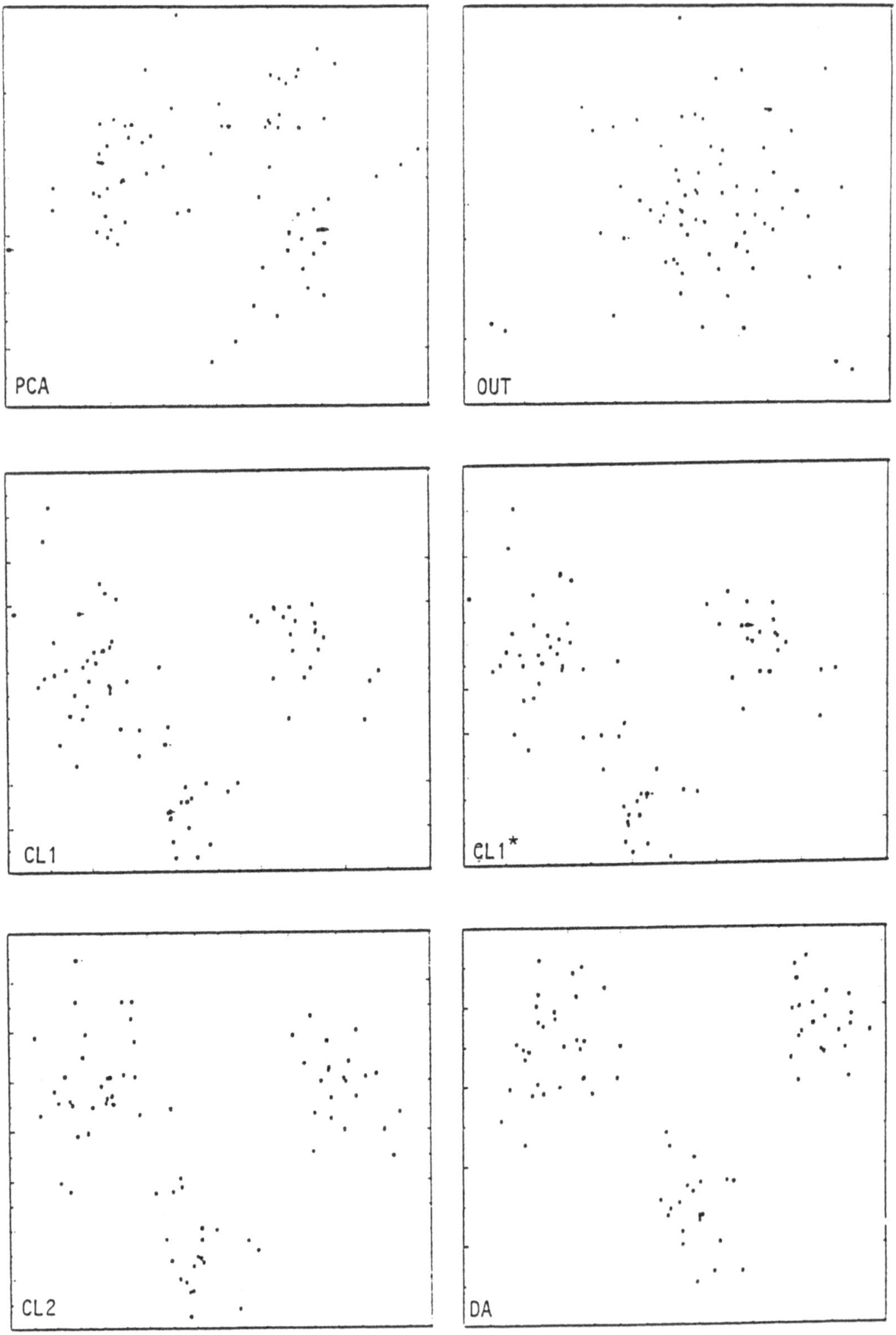

PCA
OUT
CL1
CL1*
CL2
DA

value of the sixth measurement. CL2 (h=1) displays three fairly homogeneous groups, corresponding to the three species while CL1(h=1) and CL1* suggest that one of them might split in two. CL1* has been obtained after deleting the two most outlying cases ; however all cases have been projected. These three views are similar to a large extent but also complementary. Finally, if we make use of the species labelling, Discriminant Analysis leads to DA which is worth comparing with the previous displays. Another interesting comparison is with the results in [7] obtained by another Projection Pursuit method.

5. An alternative robust estimate of a covariance matrix.

Robust estimation of a covariance matrix has been considered by many authors : surveys may be found in [4] or [5]. From S defined in section 2 we can propose the new estimate : $(S^{-1} - 2hV^{-1})^{-1}$. In fact, this estimate is modified as follows to get S^* : X^* stands for the coordinatewise median and V is computed with X^* in place of $\bar{X}$. Now, in many cases, it can be shown that S^* provides a good approximation of Γ, the covariance matrix of the uncontaminated data. Here we leave aside the theoretical aspects of the problem and we give only some simulation results where S^* is compared with a classical M-estimate. The latter, denoted by M, has been computed with the subroutine CYWALG of the program library ROBETH [8] using the Huber's (minimax) weight functions. We have generated samples of size 101 from several mixtures of normal distributions : $\sum_{i=0}^{k-1} \pi_i \; N_p(x_i, I_p)$.

For each case, 100 replications have been performed and the mean (standard-error) have been computed for each element of the estimated matrices. Three examples are presented in the following table with p=3, $x_0 = 0$ and increasing levels of contamination :

A : k=3, $\pi_1 = .02$, $x_1 = (10,5,0)'$
$\pi_2 = .02$, $x_2 = (10,0,15)'$, h=.5

B : k=2, $\pi_1 = .10$, $x_1 = (8,8,0)'$, h=1.5

C : k=3, $\pi_1 = .10$, $x_1 = (10,5,0)'$
$\pi_2 = .10$, $x_2 = (10,0,15)'$, h=1.5

Similar results hold for larger values of p. Since we are

estimating $\Gamma = I_p$, it seems that S^* works rather well in the case of high contamination if a good value of h is chosen. However, this point needs further consideration.

	S^*			M		
A	1.02(.16)	.	.	1.15(.16)	.	.
	-.00(.12)	1.02(.16)	.	.05(.11)	1.01(.13)	.
	-.01(.13)	.02(.11)	1.02(.16)	.06(.12)	.01(.10)	1.08(.15)
B	1.13(.28)	.	.	1.70(.19)	.	.
	-.02(.23)	1.13(.33)	.	.82(.18)	1.72(.23)	.
	.02(.25)	-.01(.24)	1.07(.37)	.00(.11)	-.01(.11)	.93(.13)
C	1.18(.25)	.	.	2.39(.37)	.	.
	.05(.20)	1.12(.28)	.	.73(.19)	1.22(.17)	.
	.04(.16)	.06(.21)	1.08(.21)	.86(.20)	.00(.12)	2.16(.29)

6. References.

[1] DEMPSTER, A.P. (1969) : Continuous multivariate analysis. Addison-Wesley.

[2] FRIEDMAN, J.H. and TUKEY, J.W. (1974) : A projection pursuit algorithm for exploratory data analysis. IEEE Trans. Comp.,C.23,9,881-890.

[3] GNANADESIKAN, R. and KETTENRING, J.R. (1972) : Robust estimates, residuals and outlier detection with multiresponse data. Biometrics,28,81-124.

[4] HAMPEL, F.R., RONCHETTI, E.M., ROUSSEEUW, P.J. and STAHEL, W.A. (1986) : Robust Statistics ; the Approach based on Influence Functions. Wiley, New York.

[5] HUBER, P.J. (1981) : Robust Statistics. Wiley, New York.

[6] HUBER, P.J. (1985) : Projection Pursuit. Ann. Statist.,13, 435-525.

[7] JONES, M.C. and SIBSON, R. (1987) : What is Projection Pursuit ? J.R. Stat. Soc., A,150,1,1-36.

[8] MARAZZI, A. (1985) : Robust affine invariant covariances in ROBETH. ROBETH-85, Doc. N°6. Lausanne, Institut universitaire de médecine sociale et préventive.

[9] SEBER, G.A.F. (1984) : Multivariate observations. Wiley, New York.

[10] YENYUKOV, I.S. (1988) : Detecting structures by means of Projection Pursuit. COMPSTAT88,47-58. Physica-Verlag, Heidelberg.

Maximum Likelihood Estimation of Mixed Linear and Multiplicative Models for Contingency Tables Using Distan

T. Rudas, Budapest

The paper describes a general class of statistical models for the analysis of multidimensional contingency tables, suggests a method of computing maximum likelihood estimates under these models and describes how the computations can be carried out using the DISTAN package. The class of models considered includes log-linear, i.e. multiplicative and linear models including the case when the model is defined by both types of assumptions.

Key words: log-linear model, linear model, contingency table, maximum likelihood, DISTAN.

MIXED LINEAR AND MULTIPLICATIVE MODELS

Let Ω denote a (multidimensional) contingency table. This will be considered as the set of its cells. We shall define certain models regarding the probability distribution $\{p(\omega),\ \omega\in\Omega\}$ on this table.

A linear model is specified by assuming that

$$(1) \qquad \sum_{\omega\in\Omega} c_j(\omega)p(\omega) = d_j ,$$

where d_j, $c_j(\omega)$, $j=1,\dots,l$, $\omega\in\Omega$ are specified numbers, i.e. values of certain contrasts of the cell probabilities are prescribed. Well known examples of this kind are symmetry,

Compstat 1990

marginal homogeneity, known marginal distributions, etc.

The other type of assumptions considered is defined by partitions $\mathcal{P}_1,\ldots,\mathcal{P}_k$ of the cells of the table. These partititons can be represented by functions $f_i(\omega)$, $i=1,\ldots k$. The cells ω and ω' belong to the same subset of partition $\mathcal{P}_i$ if and only if $f_i(\omega) = f_i(\omega')$. Then a general multiplicative model based on the above partitions assumes that

$$(2) \qquad \log(p(\omega)) = \log(q(\omega)) + \sum_{i=1}^{k} g_i(f_i(\omega)) , \qquad \omega\in\Omega ,$$

where q is a specified ditribution on Ω and g_i, $i=1,\ldots,k$ are arbitrary real functions.

The simplest of the models of type (2) are obtained when q is the uniform distribution and every partition is based on a subset of the variables in the following sense: There are as many subsets of cells in each partition as the number of marginal cells defined by the corresponding set of variables. Two cells of the table belong to the same subset if and only if they belong to the same marginal cell. In this case (2) is the well known log-linear model.

When the partitions are defined as above but q is not uniform one gets distributions that preserve certain higher order interactions of q. For a theoretical background see Haberman, 1974, Leimer & Rudas, 1988 and for an application Rudas, 1992.

There are cases when more general definitions of the partititons are necessary. We shall consider such an example later on. See also Rudas, 1990a.

The models considered in this paper are defined by assumptions of both types (1) and (2). These models are called mixed linear and multiplicative models. A simple example of such a model is that of independence and marginal homogeneity of a two dimensional contingency table.

MAXIMUM LIKELIHOOD ESTIMATION

The procedure that is proposed here to find maximum likelihood estimates for models that are defined by (1) and (2) consists of repeated applications of two algorithms. The actual estimates when starting the iterations is the observed distribution.

ALGORITHM 1: Let the new estimate be the maximum likelihood projection of the actual estimate in the linear model (1). Let the new one be the actual estimate. Apply algorithm 2.

ALGORITHM 2: Let the new estimate be the maximum likelihood projection of the actual estimate in the multiplicative model (2) Let the new one be the actual estimate. Apply algorithm 1.

A distribution that remains (relatively) stable after applications of both algorithms can be considered to be the maximum likelihood estimate. The proposed procedure is heuristic for no proof of convergence to the maximum likelihood estimate is known to the author. Our experience is that if there is a distribution that fulfills both (1) and (2), i.e. the assumptions are not contradicting, the above procedure converges to an (at least local) maximum. Note that the procedure described below for algorithm 1 obviously converges to a local maximum in model (1) and the procedure for algorithm 2 yields the maximum likelihood projection in model (2). The same method of obtaining maximum likelihood estimates in a comparable multivariate normal setup and a program to carry out the computations were described in Rudas, 1989a.

Algorithm 1 is a standard application of the gradient method. First (1), as a set of linear equations, is solved. Any solution devides the probabilities $p(\omega)$, $\omega\in\Omega$ into two classes. Those in the first class can be set arbitrarily and then (1) provides the values in the other class. The gradient method is applied to the probabilities in the first class. Such values are looked for that provide positive values for the probabilities in the second class and the probabilities maximize the likelihood function.

Algorithm 2 is a modification of the iterative proportional fitting procedure often used to estimate log-linear models. This, in the traditional case, consists of iterative multiplications of the cell probabilities by such numbers that the respective marginals are the same as in the distribution the maximum likelihood projection of which is sought for. In this procedure the distribution q in (2) is used as starting distribution. In our case the iteration cycles over the partitions and subset totals are adjusted rather than marginal totals. Note that when the distribution q in (2) is not uniform, there are no models with closed form maximum likelihood estimate, even if the partitions are based on a decomposable class of marginals.

A MODEL APPLICABLE IN SOCIAL MOBILITY RESEARCH

Space limitations do not allow for the presentation of the analysis of a real data set but in what follows we shall discuss a mixed linear and multiplicative model that can be applied to certain problems in social mobility research. Suppose the data available are arranged in a 3x3 social mobility table. Every individual (the sample consists of men) is classified according to his own and his father's social status. The categories used are High, Medium and Low. These observations are from a sample. The true present day distribution of the men in the society according to the status categories is known form a census.

The linear part of our model prescribes the above mentioned marginal values. If π_{ij} denotes the probability that someone is in the ith status category and his fater is in the jth then we prescribe that the marginal probability π_{i+} is equal to the known (from the census) proportion of individuals in that category. This part of the model requires two linear contrasts. The coefficients in each of these are given in the tables below. The linear part of the model prescribes that the value of Contrast 1 is $N\pi_{1+}$ and the value of Contrast 2 is $N\pi_{2+}$, where N is the sample size.

Contrast 1

		FATHER H	M	L
S	High	1	1	1
O	Medium	0	0	0
N	Low	0	0	0

Contrast 2

		FATHER H	M	L
S	High	0	0	0
O	Medium	1	1	1
N	Low	0	0	0

The multiplicative part of the model assumes that the distribution of the population in the cells of the mobility table is the result of two factors: One is the distribution of the fathers in their social status categories. The other factor assumed is the social mobility process itself, in the sense that some people remain immobile, some are upward mobile and some are downward mobile. Our model assumes that the these factors are independent. The definitions of the appropriate partitions (with the functions f in (2)) is shown below:

Father factor

		FATHER H	M	L
S	High	1	2	3
O	Medium	1	2	3
N	Low	1	2	3

Mobility factor

		FATHER H	M	L
S	High	1	22	2
O	Medium	3	1	2
N	Low	3	3	1

Note that the multiplicative part of the above model could be described as independence in a Father x Mobility table, that has structural zeros. Models similar to the multiplicative part of the above model were studied by Goodman, 1972. Existence of similar models in a very broad context is studied by Bártfai, Rudas, 1990. Further examples of mixed linear and multiplicative models are given in Rudas, 1990c.

USING DISTAN

DISTAN (short for DIScrete STAtistical ANalysis) is a newly developed package (see Bornemisza & Rudas, 1990, Rudas, 1989b,

1990b, 1990c). DISTAN, beyond having extensive table transforming and printing capabilities, is able to fit a wide variety of statistical models not available elsewhere.

Mixed linear and multiplicative models can be fitted by the MixLinMult module of DISTAN. The coefficients of the linear contrasts in (1) and the functions defining the partitions in (2) can be defined in an interactive way by assigning appropriate values to the cells of various copies of the table using the DISEdit module of DISTAN. MixLinMult allows for arbitrary specification of the distribution q in (2), and reports multiplicative parameters, tests of fit and degrees of freedom.

REFERENCES

Bártfai, P., Rudas, T. (1990) Conditionally independent extension of measures. Sttaisitcs and Decision, to appear.

Bornemisza E., Rudas, T. (1990) The DISTAN package for discrete statistical analysis. Troitzsch (ed.) Proceedings of Computer Aided Sociological Research - CASOR '89, Westdeutscher Verlag, Opladen/Wiesbaden - Akademie Verlag, Berlin, to appear.

Goodman, L. A. (1972) Some multiplicative models for the analysis of cross-classified data. Proc. 6th Berkeley Symposium on Math. Statist. and Probab. Vol 1, 649-696, UCLA Press, Berkeley.

Haberman, S. J. (1974) The Analysis of Frequency Data. University of Chicago Press, Chicago.

Leimer, H.-G., Rudas, T. (1988) Contingency tables with prescribed conditional odds ratios or prescribed log-linear parameters. Berichte zur Stochastik und Verwandten Gebieten, 1988/1, Johannes Gutenberg Universitat, Mainz.

Rudas, T.(1989a) Marginal and partial association models. Preprint 1989/1, Social Science Informatics Center, Budapest.

Rudas, T. (1989b) DISTAN for discrete statistical analysis: an overview. Preprint 1989/2, Social Science Informatics Center, Budapest.

Rudas, T. (1990a) Contribution to the discussion of the papers by Edwards and Lauritzen & Wermuth. J. Roy. Statist. Soc., (Ser. B), Vol. 52, 55-56.

Rudas, T. (1990b) DISTAN for discrete statistical analysis. Faulbaum (ed.) Fortschritte der Statistik, Proceedings of SOFTSTAT'89, Gustav Fischer Verlag, to appear.

Rudas, T. (ed., 1990c) The Discrete Statistical Analysis Handbook Social Science Informatics Center, Budapest, to appear.

Rudas, T. (1992) Prescribed conditional interaction structure models with application to the analysis of mobility tables. Quality and Quantity, to appear.

Alternate Forms of Graphical Modelling - A Comparison

A. Scott and J. Whittaker, Lancaster

SUMMARY: The theoretical ideas underpinning the relatively new technique of graphical modelling have received considerable coverage in recent publications. This paper considers the more practical side of graphical modelling, examining the options open to potential users and the limitations imposed upon them by the currently available software.
KEYWORDS: graphical models, multivariate analysis, directed graphs.

1. INTRODUCTION

Graphical modelling is a relatively new multivariate statistical technique which represents multivariate interactions by means of a graph. Nodes correspond to variables and edges to associations; more precisely the absence of an edge implies that a pair of variables are conditionally independent given the remainder. Distributional assumptions are usually based on the use of the conditional-gaussian distribution.

The technique originated in the work of Darroch, Lauritzen and Speed (1980) who showed how a subset of log-linear models, the graphical models, can be easily interpreted, theoretically and practically, from their associated independence graph. The corresponding theory for continuous variables was examined by Speed and Kiiveri (1986) and more recently the approach of Lauritzen and Wermuth (1989) allowed the technique to be applied in a unified way not only to the separate discrete and continuous situations but also to the mixed case. A further extension to graphs with directed edges (Wermuth and Lauritzen 1990) enables conditional independences with different conditioning sets to be incorporated within the same model. These graphical chain models greatly extend the power and versitility of the technique.

As with other modelling procedures, the application of graphical modelling involves searching for parsimonious models; in this case by dropping or adding edges to the current graph. Edge type, whether directed or undirected, is not alterable by the search procedure but is fixed by the initial specification of a chain structure for the variables. Model comparison is always within a given structure. Sometimes this chain

Compstat 1990

structure is determined by knowledge of the physical situation giving rise to the data and in principle, of course, the distributional assumptions inherent in any chosen structure should be checked for validity. Frequently, however, the wide variety of models made available through the use of chain models means that many structures may fit equally well and choice must be made on other grounds. A common criterion is ease of interpretation, based on the particular research hypotheses that are of interest.

At present the theory of graphical modelling, amply covered in recent publications (Whittaker 1990, Lauritzen 1990, Edwards 1990), has far outstripped its practical implementation. The aim of this paper is to look at the practical side of graphical modelling, to examine the options open to potential users and the limitations imposed upon them. In particular to describe and compare the forms of graphical model available with current software, both theoretically and through their application to a small archaeological data set.

2. CURRENT SOFTWARE

Current software for fitting graphical models can be divided into two categories, standard packages primarily intended for other purposes and programs written specifically for graphical modelling. The class of chain models includes a wide variety of standard techniques such as (multiple) linear regression, anova and manova and the log-linear modelling of contingency tables (Edwards 1990). Hence packages designed to provide these techniques can also be used to fit the corresponding graphical models. The main difficulty with this approach is that in general the range of models made available in this way is not broad enough to allow the full use of model selection techniques. There are two notable exceptions to this. Chain models which contain no continuous variables can be fitted as a sequence of log-linear models, and models in which all edges are directed, and no edge points from a continuous to a discrete variable, can be fitted as a sequence of linear regressions with univariate responses. Thus both of these types can be fitted relatively straightforwardly with packages such as GLIM, GENSTAT, SAS and BMDP.

The second software category contains just three items. The program MIM (Edwards 1987) was designed specifically to fit hierarchical interaction models (Edwards 1990), of which undirected graphical models are a subset. Chain models in which no continuous variable points to a discrete variable can also be fitted as a sequence of undirected models. The program is interactive and command driven and specifies models through the use of model formulae similar to those used by GLIM and GENSTAT. Although the modelling aspects of the package are very effective, the remaining facilities are somewhat limited in scope, lacking for example any data transformation capability and not providing estimates of parameter standard errors.

The program DIGRAM (Kreiner 1989) has been developed to fit a complete range of chain models but for discrete data only. It is an interactive command-driven program which, as well as a model fitting core, provides a considerable range of facilities for displaying, manipulating and analysing graphs. One of its main attractions is that it provides exact significance levels for model testing, in contrast to the other programs which rely on asymptotic results. The central model fitting section is restricted to eight variables but more complex models, containing up to 27 variables, can be built up by combining these smaller ones. The outer shell of the program greatly simplifies this process, although the efficient and effective use of this facility does require a considerable grasp of graphical modelling concepts.

The final item of specialist software is the program ASP, developed at Lancaster to fit restricted forms of the conditional gaussian distribution to incomplete data. The program fits two-block chain models with all discrete variables in the first block and all continuous variables in the second, although, like MIM, more complex chain models can be fitted as a sequence of two-block models. The main disadvantage of ASP is that it is restricted to homogeneous models, i.e. models in which the conditional covariance matrix of the continuous variables does not depend on the discrete variables. Again it is an interactive, command-driven program with slightly more peripheral facilities than MIM or DIGRAM since it provides a basic data transformation capability and also estimates parameter standard errors.

3. EXAMPLE

3.1 Data

The choice of the dataset used here was not made with the intention of testing the software to its limits. Rather it was chosen as a typical example of the type of data arising in fields such as archaeology and is in fact a subset of a much larger and less complete set containing over seventy variables. The data comprise thirteen measurements on a total of 121 adult human skulls from two Egyptian cemeteries: Badari, a predynastic site from Upper Egypt, and Sedment, a IXth Dynasty site from Middle Egypt. All the skulls were sexed for the original publications (Stoessiger 1927, Woo 1930), but 48 were marked as questionable and a more recent study (Hillson 1978, 1985) reclassified some of these skulls. For the purposes of this paper, therefore, these 48 skulls are treated as being of unknown sex.

Table 1 lists the variables used, together with the number of values missing for each. For the continuous variables this ranges from 0% to almost 20% with an average of 5%. Even if sex were always known, over 30% of the cases would be incomplete. When sex is included this proportion rises to almost 60%. Apart from the missing information, however, this can be regarded statistically as a "nice" dataset with no apparent peculiarities.

TABLE 1

Variables used in the study

Variable	Label	Missing values No.	Missing values %
maximum cranial length	L	0	0.0
maximum breadth	B	1	0.8
basibregmatic height	H'	4	3.3
upper facial height	G'H	3	2.4
bimaxillary breadth	GB	7	5.8
basialveolar length	GL	5	4.1
nasal height	NH'	3	2.4
nasal breadth	NB	2	1.7
bidacryonic chord	DC	16	13.2
palatal length	G'1	13	10.7
palatal breadth	G2	22	18.2
orbital breadth	O'1	11	9.1
orbital height	O2	0	0.0
sex	SEX	48	39.7
site	SITE	0	0.0

3.2 Limitations

In attempting to analyse this dataset a number of factors immediately act to limit the possible methodology. The first is the number of cases with missing values. None of the programs except ASP can efficiently cope with incomplete data. If other programs are to be used either the missing values must be imputed in some way or cases and/or variables must be discarded. The second limiting factor is the mixture of discrete and continuous variables involved. This rules out the use of DIGRAM and log-linear modelling packages, unless the continuous variables are discretized. Although this is common practice in some areas of application it is not felt to be appropriate for the present dataset.

A third factor is the number of variables. If a full range of models is to be explored then DIGRAM is effectively limited to 8 variables and so would be unsuitable even if the data were discretized. It could not, for example, fit a saturated model to the data. MIM has a limit of 12 discrete variables and 24 continuous variables and so can be used with the present data. ASP also has size limitations but these are related to storage and not specifically to variable numbers. The use of dynamic storage allocation means that the number of possible variables depends on other factors such as variable type and model complexity and the current dataset falls within these limits. Interestingly enough this factor also limits the use of log-linear modelling packages such as GLIM, not directly through the number of variables but through the size of the corresponding contingency table. The amount of available storage space becomes exhausted for relatively small numbers of discrete variables.

A final limiting factor is the type of model required. The variables in the current dataset naturally fall into two groups, the first containing the discrete variables sex and site, and the second the continuous measurements. A natural model to use is therefore either the two block chain model used by ASP or the undirected model used by MIM. There is no justification for imposing a completely directed model on the data in order to fit models through the use of standard regression packages, and such models would not anyway be of particular interest. For the current data therefore modelling options are limited to the use of either MIM or ASP and to facilitate comparison both programs were used.

3.3 Analysis

Model fitting and selection using ASP were straightfoward. Figure 1 shows the graph of the final fitted model. Although at first sight this appears somewhat confusing, it conveys a considerable amount of information about the data structure. The continuous variables fall into four separate groups, the variables in each group being conditionally independent of those in other groups given sex and site. Furthermore each group is associated with a specific aspect of skull morphology. The first with overall skull size, the second with facial height, and the last with the size of the skull base.

Model fitting via MIM was less easy due to the incompleteness of the data. To produce a dataset of reasonable size the sex variate was completed using the values suggested by Hillson (1978) and the most incomplete continuous variable, palatal breadth, was dropped. Eliminating cases with missing values then left 94 complete cases on which to perform the analysis. Figure 2 shows the graph obtained from the final MIM model. Apart from changes arising from the missing variable and the alteration of model type, the interpretation is the same as for figure 1.

Figure 1

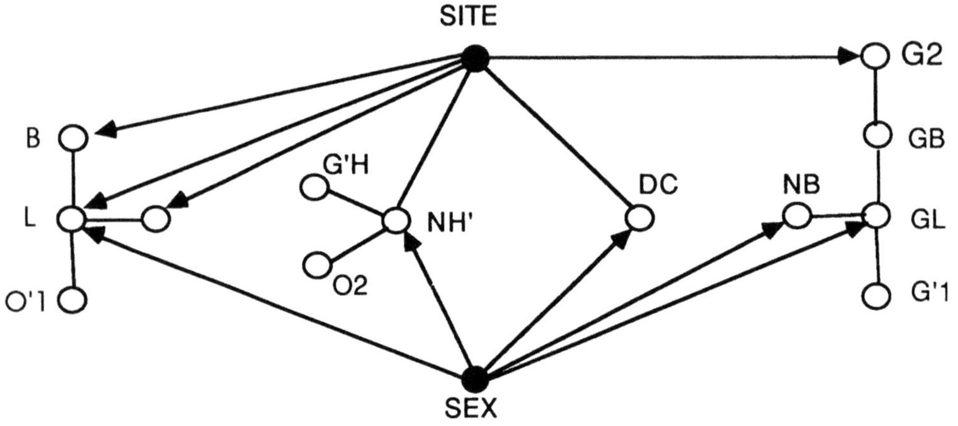

ASP fitted model

Figure 2

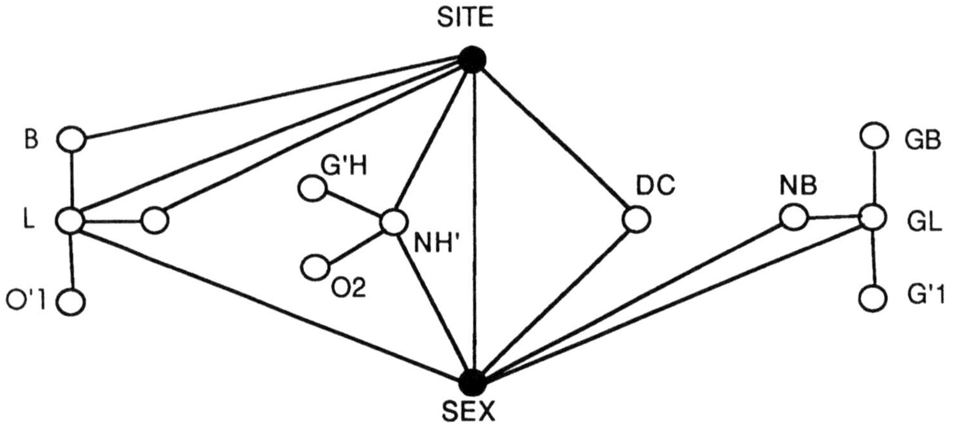

MIM fitted model

4. CONCLUDING REMARKS

Each of the three programs considered here has certain advantages to the potential modeller. MIM fits heterogeneous models, DIGRAM fits complex directed models, and ASP copes with incomplete data. Thus the programs are in many ways complementary. They still fall short, however, of providing the full range of graphical modelling techniques. Apart from combining the special attributes of these programs there remain two main problems to be overcome. At present the only directed models that can be fitted are those in which continuous variables do not preceed discrete variables to which they are connected. Software needs to be produced which can overcome this. The second problem is the limitation on the number of discrete variables which can be included in a model. This arises because of the size of the associated contingency table. DIGRAM shows considerable ingenuity in getting round this problem but the development of fitting techniques which do not require storage proportional to this table would be of considerable value.

5. REFERENCES

Darroch J., Lauritzen S. & Speed T. (1980). Markov fields and log-linear interaction models for contingency tables. *Ann. Stat.* 8: 522-539

Edwards, D.E. (1987). A guide to MIM. Research Report 87/1. Statistical Research Unit, University of Copenhagen.

Edwards, D.E. (1990). Hierarchical interaction models, (with discussion). *J.R. Statist. Soc. B* 52(1): 3-20

Hillson S.W. (1978). Human biological variation in the Nile valley. Ph.D. thesis, London University.

Hillson S.W. (1985). Missing information and collections of skeletal material. In: Fieller N.R.J. et al. Palaeoenvironmental Investigations. Oxford, B.A.R. International Series.

Kreiner, S. (1989). User guide to DIGRAM a program for discrete graphical modelling. Research report 89/10, Statistical Research Unit, University of Copenhagen.

Lauritzen, S.L. and Wermuth, N. (1989). Graphical models for associations between variables, some of which are qualitative and some quantitative. *Ann. Statist.* 17: 31-57.

Speed, T.P. and Kiiveri, H. (1986). Gaussian Markov distributions over finite graphs. *Ann. Statist,* 14(1): 138-150.

Stoessiger B.N. (1927). A study of the Badarian crania recently excavated by the British School of Archaeology in Egypt. *Biometrika* 22: 65-83.

Wermuth, N.. and Lauritzen, S.L. (1990). On substantive research hypotheses, conditional independence graphs and graphical chain models. *J. Roy. Statist. Soc. B* 52(1): 21-72.

Whittaker, J. (1990). Graphical models in applied multivariate statistics. Chichester: Wiley.

Woo T.L. (1930). A study of seventy-one 9th Dynasty skulls from Sedment. *Biometrika* 22: 65-83.

This work was performed under grant GR/E 94951 from the SERC.

Exact Significance Testing by the Method of Control Variates

C. Mehta, N. Patel, Boston and P. Senchaudhuri, Cambridge

1 Introduction

Monte Carlo estimates of exact permutational p-values are an important alternative to both asymptotic and exact inference. For many sparse data sets, asymptotic tests are unreliable, but corresponding exact tests are too difficult to compute, despite the availability of fast numerical algorithms for their execution. Monte Carlo methods are a good compromise between these two extremes. By sampling a predetermined number of times from an appropriate reference set one can obtain an arbitrarily narrow confidence interval for the exact p-value; for instance one could state with 99% confidence that the estimated p-value was accurate to the third decimal place. This level of accuracy is generally sufficient.

Mehta, Patel and Senchaudhuri (*JASA*, 1988) showed how importance sampling may be used in place of crude Monte Carlo sampling for estimating exact p-values. Their importance sampling scheme produced unbiased estimates of permutational p-values with considarably smaller variance than could be obtained by the crude method. In the present communication, these authors present yet another Monte Carlo technique with variance reduction; the method of control variates. This method applies to a broader class of problems than importance sampling. Also, it uses a simpler procedure for sampling elements from the reference set.

2 The Control Variates Algorithm

The main idea of the method of control variates is to compute the exact p-value for an easier problem than the one originally posed, and then "rescue" this possibly inaccurate p-value by Monte Carlo sampling. Let P be the exact p-value for the original problem, and let P_1 be the exact p-value for the simpler problem. We are assuming that it is fairly easy to compute P_1 but not so easy to obtain P. We would like to use P_1 as a surrogate for P but are not sure if this is appropriate since P_1 might differ from P by a large amount. At this stage, we invoke crude Monte Carlo sampling. For each element that is sampled from the reference set define Z to be 1 if that element is in the critical region for the original problem; 0 otherwise. Similarly define Z_1 to be 1 if the sampled element is in the critical region for the simplified problem; 0 otherwise. Now compute

$$P_0 = P_1 - Z_1 + Z.$$

Compstat 1990

It is clear that $E(P_0) = P$, since $E(Z_1) = P_1$ and $E(Z) = P$. Thus repeated sampling and averaging of the sampled P_0 values will yield an unbiased estimate of P. Moreover there is considerable variance reduction relative to simply averaging of the sampled values of Z (which yields the crude Monte Carlo estimate), because Z_1 and Z are positively correlated.

3 Application to Stratified 2×2 Contingency Tables

The data consist of a series of k 2×2 contingency tables. The ith table is of the form

Rows	Col_1	Col_2	Row_Total
Row_1	x_i	y_i	m_i
Row_2	x'_i	y'_i	m'_i
Col_Tot	n_i	$N_i - n_i$	N_i

for $i = 1, 2, \ldots k$. We may regard the two rows of each table as arising from two independent binomial distributions. Specifically, let (x_i, x'_i) represent the number of successes in (m_i, m'_i) Bernoulli trials, with respective success probabilities (π_i, π'_i). The odds ratio for the ith table is defined as

$$\Psi_i = (\frac{\pi_i}{1-\pi_i})/(\frac{\pi'_i}{1-\pi'_i}) \; . \tag{3.1}$$

Although we have specified that the two rows of the 2×2 table represent two independent binomial distributions, this is just a matter of notational convenience. We could equivalently assume that the two columns represent two independent binomial distributions.

Exact Conditional Inference

Let τ denote a generic set of k 2×2 tables of the form displayed above. Let τ_0 denote a specific realization of τ. Exact inference, both for testing that the odds ratio across k 2×2 tables is constant as well as for estimating the common odds ratio, is based on determining how extreme the observed τ_0 is relative to other τ's that could have been observed in some reference set. Different reference sets are used for testing the homogeneity of odds ratios and for estimating the common odds ratio. Define the reference set

$$\Omega = \left\{ \begin{array}{ll} \tau: & x_i + y_i = m_i; \; x'_i + y'_i = m'_i; \\ & x_i + x'_i = n_i; \; y_i + y'_i = N_i - n_i \end{array} \right\} \; . \tag{3.2}$$

Also define the more restricted reference set

$$\Omega_s = \{\tau \in \Omega : x_1 + x_2 + \cdots + x_k = s\} \ . \tag{3.3}$$

Zelen (1971) developed an exact test for the null hypothesis

$$H_0\colon \ \Psi_i = \Psi, i = 1, 2, \ldots k \ .$$

Zelen's test is based on the fact that under H_0 the probability of observing any τ from the conditional reference set Ω_s is a product of hypergeometric probabilities which does not depend on the nuisance parameter Ψ. Specifically, the conditional probability of obtaining any $\tau \in \Omega_s$ is

$$\Pr(\tau|s) = \frac{\prod_{i=1}^{k} \binom{m_i}{x_i}\binom{m_i'}{x_i'} / \binom{N_i}{n_i}}{\sum_{\tau \in \Omega_s} \prod_{i=1}^{k} \binom{m_i}{x_i}\binom{m_i'}{x_i'} / \binom{N_i}{n_i}} \ . \tag{3.4}$$

In addition to its probability interpretation, equation (3.4) may be used to order each $\tau \in \Omega_s$ so as to determine how extreme or discrepant the observed τ_0 is under H_0. Thus, $\Pr(\tau|s)$ is also the test statistic for the homogeneity test and its observed value, $\Pr(\tau_0|s)$, defines the critical region of the exact two sided p-value. Let

$$\Omega_s^* = \{\tau \in \Omega_s\colon \ \Pr(\tau|s) \leq \Pr(\tau_0|s)\} \ . \tag{3.5}$$

The two sided p-value for Zelen's test of homogeneity is

$$p = \sum_{\tau \in \Omega_s^*} \Pr(\tau|s) \ . \tag{3.6}$$

The above p-value is inherently two sided since, if H_0 is rejected, the only conclusion to be drawn is that the Ψ_i's do not have a common value. The direction of the departure from the null hypothesis cannot be known from the value of the test statistic. $\Pr(\tau|s)$ plays a dual role in equation (3.5). It is both the test statistic for identifying the critical region of the reference set and the probability assigned to each τ in the critical region.

We applied the algorithm of Section 2 to the problem of estimating the p-value given by equation (3.6). The control variate p-value, P_1, was obtained by truncating the test statistic, $\Pr(\tau|s)$, on the logarithmic scale, to two significant digits. We tested out our method on the "Alcohol and Oesophageal Cancer" data set of Breslow and Day (1980). The results are given below.

The data below are for six age groups, and relate alcohol to oesophageal cancer.

Age Group	Alcohol Exposure		No Exposure	
	Case	Control	Case	Control
25-34	1	9	0	106
35-44	4	26	5	164
45-54	25	29	21	138
55-64	42	27	34	139
65-74	19	18	36	88
75+	5	0	8	31

We would like to test if the odds ratios across the six age groups are constant. We obtained a p-value estimate of .0981 with a standard error of 0.0009 upon sampling 2000 tables by the method of control variates. Had we sampled 2000 tables by crude Monte Carlo sampling, the standard error of the p-value estimate would have been

$$[(.0981)(1 - .0981)/2000]^{0.5} = 0.0067$$

leading to an efficiency gain of $(.0067/.0009)^2 = 60$. In other words, we needed to sample 60 times fewer tables by the method of control variates than we would have under crude Monte Carlo sampling. At present we are exploring ways to make even larger efficiency gains, and are applying the method to two sample linear rank tests.

References

Breslow NE, Day NE (1980). The analysis of case-control studies. *IARC Scientific Publications No. 32* Lyon, France.

Mehta CR, Patel NR, and Senchaudhuri P. Importance sampling for estimating exact probabilities in permutational inference. *JASA* 88, 404, 999-1005.

Zelen M (1971). The analysis of several 2×2 contingency tables. *Biometrika* 58(1):129-137.

Keywords

Exact inference, nonparametrics, loglinear model, linear rank tests, Monte Carlo sampling, importance sampling, computationally intensive methods.

Testing Collapsibility of Hierarchical Loglinear Models for Contingency Tables

F. M. Malvestuto, Roma

ABSTRACT
A (hierarchical) loglinear model is said to be *collapsible* onto a set of variables if the corresponding marginal totals can be drawn from the tables of marginal totals formed by summing the sufficient marginals of the model over the remaining variables. Collapsibility has important consequences for hypothesis testing and model selection, and can be useful in data reduction. We shall present a procedure for computing marginal totals for arbitrary sets of variables; as a by-product, a simple algorithm for testing collapsibility is obtained.

1. BACKGROUND

Let $X = \{A_1, \ldots, A_t\}$ be a set of classifying factors, henceforth referred to as *variables*, V_i the set of values (or levels) of variable A_i in X and $C = C_1 \times \ldots \times C_t$ the set of all possible combinations (*cells*) of values for variables in X. Let us consider a sample of n objects categorized according to the variables in X and let $n(\mathbf{x})$ denote the number of objects in cell $\mathbf{x}$. If the classifications of the n objects are considered as independent observations, then the sample distribution is multinomial and one can get a maximum-likelihood estimate (MLE) of the probability $p(\mathbf{x})$ of an object falling in cell $\mathbf{x}$ by assuming that $p(\mathbf{x})$ belong to a certain family of probability distributions; typically, loglinear models are employed. A loglinear model over a set of variables X is specified by assigning a *set class* $\mathbf{S} = \{X_1, \ldots, X_s\}$ over X (that is, a *covering* of X: $\cup_k X_k = X$) containing no redundant sets (that is, sets which are subsets of other sets in **S**) and is defined as the family $\mathcal{P}$ of probability distributions $p(\mathbf{x})$ which satisfy the two equivalent conditions:

(i) $p(\mathbf{x})$ can be factorized as a product of functions $\alpha_1(\mathbf{x}_1), \ldots, \alpha_s(\mathbf{x}_s)$;

(ii) $p(\mathbf{x})$ is the maximum-entropy extension of its marginals $p(\mathbf{x}_1), \ldots, p(\mathbf{x}_s)$.

The set class **S** of X is called the *generating class* of $\mathcal{P}$ and the sets in **S** are called the *generators* of $\mathcal{P}$.

Given a set class $\mathbf{S} = \{X_1, \ldots, X_s\}$ over X, the following quadratic-time algorithm, called *reduction algorithm* (RA) [12], reduces **S** to a set class over some subset of X.

Reduction Algorithm

Given a set class $S = \{X_1, \ldots, X_s\}$ over X, apply the two following operations to S repeatedly until neither can be further applied:

(a) delete a variable if it belongs to exactly one set;

(b) remove a redundant set.

After applying RA to **S**, out of the X_k's in **S** some are no longer, some appear in reduced form and, finally, some appear in complete form. The residual sets form a set class, which will be referred to as the *reduction* of **S** and denoted by RED(**S**); their

Compstat 1990

union is called the *support* of RED(S). Moreover, the sets X_k in $\mathbf{S} \cap$ RED(S) are called the *strong elements* of **S** and the sets X_k in **S** - RED(S) are called the *weak elements* of **S**. Finally, the sets removed during the reduction process are called *articulation sets* for **S**; by the *multiplicity* of an articulation set we shall mean the number of times that rule (b) is applied to remove an occurence of it.
A set class **S** is called *fully reducible* [8] (or *acyclic* [7, 12]) if the support of its reduction is an empty set, that is, RED(S) = Ø. A linear-time algorithm to test full reducibility can be found in [12]. A loglinear model is said to be *decomposable* [3, 10] if its generating class is fully reducible [2]. If $\mathcal{P}$ is a decomposable loglinear model and p(**x**) is in $\mathcal{P}$, then

$$p(\mathbf{x}) = \Pi_k\ p(\mathbf{x}_k)\ /\ \Pi_h\ p(\mathbf{z}_h)^{m_h} \qquad (1)$$

where Z_h stands for an articulation set for **S** and m_h for its multiplicity.
Given a contingency table with entries n(**x**), the MLE of the probability of an object falling in cell **x** under the loglinear model $\mathcal{P}$ generated by set class **S** = $\{X_1, ..., X_s\}$ over X, is then given by the unique probability distribution $\pi(\mathbf{x})$ in $\mathcal{P}$, whose marginals $\pi(\mathbf{x}_1), ..., \pi(\mathbf{x}_s)$ equate the marginal sample proportions $n(\mathbf{x}_1)/n, ..., n(\mathbf{x}_s)/n$. The MLE $\pi(\mathbf{x})$ can be calculated by applying the well-known Iterative Proportional Fitting Procedure (IPFP) [2] to the marginal sample proportions. If $\mathcal{P}$ is a decomposable model, then the probability distribution $\pi(\mathbf{x})$ can be computed directly from (1), without resorting to the IPFP. Given a loglinear model $\mathcal{P}$ over X and a nonempty subset Y of X, the covering of Y obtained by removing the redundant sets from $\{Y \cap X_1, ..., Y \cap X_s\}$ will be referred to as the *restriction* of **S** to Y and denoted by RES(S, Y). The sets Y_h in RES(S, Y) can be viewed as the generators of a loglinear model **Q** over Y, called the *restriction* of $\mathcal{P}$ to Y. The MLE $\psi(\mathbf{y})$ of the probability of an object falling in the cell **y** under **Q** is the maximum-entropy extension of the marginal sample proportions $n(\mathbf{y}_h)/n$.

Example 1. Given a sample of n objects classified according to three variables A, B and C, let n(a, b, c) denote the number of objects falling in cell (a, b, c). The loglinear model $\mathcal{P}$ generated by **S** = {AB, AC} is decomposable and $\mathcal{P}$ = {p(a, b, c): p(a, b, c) = p(a, b) p(a, c) / p(a)} . Consequently, the MLE of the probability of an object falling in cell (a, b, c) under $\mathcal{P}$ is $\pi(a, b, c) = (1/n)\ [n(a, b)\ n(a, c)\ /\ n(a)]$, and the MLE of the probability of an object falling in cell (b, c) under $\mathcal{P}$ is $\pi(b, c) = (1/n)\ \Sigma_a\ [n(a, b)\ n(a, c)\ /\ n(a)]$.
Consider now the restriction Q of $\mathcal{P}$ to Y = BC. Since the generating class of Q is RES(S, Y) = {B, C}, then one has Q = {q(b, c): q(b, c) = q(b) q(c)}. Consequently, the MLE of the probability of an object falling in cell (b, c) under Q is $\psi(b, c) = (1/n^2)\ n(b)\ n(c)$. Notice that $\pi(b, c) \neq \psi(b, c)$ for an arbitrary distribution n(a, b, c) of occurrence frequencies.

Definition 1 [1]. A loglinear model $\mathcal{P}$ over a set of variables X is said to be *collapsible* onto a nonempty subset Y of X if for all p(**x**) in $\mathcal{P}$, the marginal p(**y**) of p(**x**) belongs to the restriction of $\mathcal{P}$ to Y.

This definition implies that, if $\mathcal{P}$ is collapsible onto Y, then $\pi(\mathbf{y}) = \psi(\mathbf{y})$, where $\pi(\mathbf{y})$ and $\psi(\mathbf{y})$ are the MLEs of the probability of an object falling in cell **y**, respectively under $\mathcal{P}$

and the restriction of $\mathcal{P}$ to Y.
Of course, every loglinear model is collapsible on any subset of some of its generators. Example 1 presents a case in which the condition of collapsibility is not satisfied.
A test for collapsibility is given by S. Asmussen and D. Edwards [1], based on a suitable graph representation of generating classes.
In this paper, we shall answer the question of collapsibility after solving the general problem of an efficient computation scheme for MLEs and their marginals. We shall solve this problem and derive an algorithm for testing collapsibility as a by-product. Our algorithm is more informative than the test by S. Asmussen and D. Edwards, since it not only answers the collapsibility question, but also provides an efficient procedure for computing marginals of MLEs.

2. COLLAPSIBILITY OF DECOMPOSABLE MODELS

The computation of marginals of the MLE $\pi(\mathbf{x})$ under a decomposable model is an easy task, which can be accomplished by applying a quadratic-time algorithm, called *selective reduction algorithm* (SRA) [12] (or *selective simplification algorithm* [7]):

Selective Reduction Algorithm
Given a set class $\mathbf{S} = \{X_1, \ldots, X_s\}$ over X and a subset Y of X, apply the two following operations to S repeatedly until neither can be further applied:
(a) delete a variable not in Y if it belongs to exactly one set;
(b) remove a set if it is a subset of another.

A linear-time algorithm to selectively reduce fully reducible set classes can be found in [12]. The class of the residual sets $\{U_1, \ldots, U_r\}$ is called the *selective reduction* of $\mathbf{S}$ relative to Y and denoted by RED($\mathbf{S}$, Y). It is well-known that if $\mathbf{S}$ is fully reducible, so is RED($\mathbf{S}$, Y) [6]. If we denote the support of RED($\mathbf{S}$, Y) by U, the articulation sets for RED($\mathbf{S}$, Y) by V_h and their multiplicities by m_h, then we can decompose the marginal $p(\mathbf{u})$ of every $p(\mathbf{x})$ in the loglinear model $\mathcal{P}$ generated by $\mathbf{S}$ according to (1), that is,

$$p(\mathbf{u}) = \Pi_k\ p(\mathbf{u}_k) / \Pi_h\ p(\mathbf{v}_h)^{m_h}.$$

Since Y is definitely a subset of the support U of RED($\mathbf{S}$, Y), the marginal $p(\mathbf{y})$ of $p(\mathbf{x})$ can be taken as

$$p(\mathbf{y}) = \Sigma_{A \in U\text{-}Y}\ \Pi_k\ p(\mathbf{u}_k) / \Pi_h\ p(\mathbf{v}_h)^{m_h} \quad . \qquad (2)$$

Similarly, the MLE $\pi(\mathbf{y})$ of the probability of an object falling in cell $\mathbf{y}$ under $\mathcal{P}$ is

$$\pi(\mathbf{y}) = \Sigma_{A \in U\text{-}Y}\ \pi(\mathbf{u}) = \Sigma_{A \in U\text{-}Y}\ \Pi_k\ [n(\mathbf{u}_k)/n] / \Pi_h\ [n(\mathbf{v}_h)/n]^{m_h}.$$

Example 2. Consider the model $\mathcal{P}$ with generating class $\mathbf{S}$ = {AB, AC, ADE, ADF, AEG}. It proves to be decomposable and every $p(\mathbf{x})$ in $\mathcal{P}$ can be decomposed as follows

$$p(a, b, c, d, e, f, g) = p(a, b)\ p(a, c)\ p(a, d, e)\ p(a, d, f)\ p(a, e, g) / p(a)^2\ p(a, d)\ p(a, e)\ .$$

The selective reductions of **S** relative to the two sets Y = DEF and Z = ADEF are identical: RED(**S**, Y) = RED(**S**, Z) = {ADE, ADF}. The MLEs of the probabilities of an object falling in cell (d, e, f) and in cell (a, d, e, f) under $\mathcal{P}$, can be derived by applying (2) and are found to be respectively

$$\pi(d, e, f) = (1/n) \Sigma_a\ n(a, d, e)\ n(a, d, f)\ /\ n(a, d) \quad \text{and} \quad \pi(a, d, e, f) = (1/n)\ [n(a, d, e)\ n(a, d, f)\ /\ n(a, d)]\ .$$

Let $\mathcal{Q}$ and $\mathcal{R}$ be the restrictions of $\mathcal{P}$ to Y and Z, respectively. The generating classes of $\mathcal{Q}$ and $\mathcal{R}$ are RES(**S**, Y) = {DE, DF} and RES(**S**, Z) = {ADE, ADF}. So, the MLE of the probability of an object falling in cell (d, e, f) under $\mathcal{Q}$ is $\psi(d, e, f) = (1/n)\ [n(d, e)\ n(d, f)\ /\ n(d)]$, and the MLE of the probability of an object falling in cell (a, d, e, f) under $\mathcal{R}$ is $\rho(a, d, e, f) = (1/n)\ [n(a, d, e)\ n(a, d, f)\ /\ n(a, d)]$. Therefore, one can conclude that $\mathcal{P}$ is collapsible onto Z but is not collapsible onto Y.

Theorem 1. A decomposable loglinear model over X is collapsible onto a subset Y of X if and only if the selective reduction of its generating class relative to Y is a covering of Y.

Since the operation of selective reduction is time-linear, Theorem 1 allows us to test decomposable models for collapsibility in linear time.

3. COLLAPSIBILITY OF NONDECOMPOSABLE MODELS

The following example shows that, if one applied the above procedure of selective reduction to a nondecomposable model, he would draw erroneous conclusions about collapsibility.

Example 3. Consider the nondecomposable model generated by **S** = {AB, AC, BC}. Of course, it is collapsible on each of its generators, whereas one can easily see that the support of the selective reduction of **S** relative to each of its generators is ABC, which proves that Theorem 1 is no longer valid for nondecomposable models.

We shall introduce the notion of *connection scheme* for a set of variables in an arbitrary set class (it is implicitly contained in [9]), and show that (1) it reduces to that of selective reduction in the case of fully reducible set classes, and (2) it is sufficient to read "connection scheme" instead of "selective reduction" in Theorem 1 for a general characterization of collapsibility to be stated. The definition of connection scheme passes through that of "decomposable approximation" to an arbitrary loglinear model. The search for decomposable approximations to nondecomposable models has intrigued several authors [4, 5, 7, 10], who have faced the problem of minimizing the computational effort required by the calculation of maximum-entropy distributions. An efficient solution to this problem was recently suggested by the author [8], based on the graph-theoretical notion of "separator" and "nonseparable component" [11].
Given a generating class **S** = $\{X_1, ..., X_s\}$ over X, a generator X_k is a *separator* if RES(**S**, $X-X_k$) (i.e., the restriction of **S** to $X - X_k$) has more connected components than **S** (in the graph-theoretical sense [11]). A subclass (that is, a subset) of **S** is *nonseparable* if it contains no separators; for each ordering of the sets in **S** there is a unique decomposition of **S** into maximal nonseparable subclasses, called the *nonseparable components* of **S** [8, 11], which can be determined in linear time by using a depth-first search scheme [11].

Example 4. Let us consider the generating class $\mathbf{S}$ = {AB, AC, ADF, AEG, DE}. If a lexicographic order of its sets is adopted, the nonseparable components of $\mathbf{S}$ are $\mathbf{S}_1$ = {AB, AC}, $\mathbf{S}_2$ = {AB, ADF} and $\mathbf{S}_3$ = {ADF, AEG, DE}.

At this point, we are ready to state precisely what we mean by the decomposable approximation $\mathcal{P}^*$ to a given loglinear model $\mathcal{P}$ (no matter whether $\mathcal{P}$ is or is not decomposable). After determining the nonseparable components $\mathbf{S}_j$ of $\mathbf{S}$, let us compute their reductions RED($\mathbf{S}_j$).

Definition 2. The *decomposable approximation* $\mathcal{P}^*$ to a given loglinear model $\mathcal{P}$ is the loglinear model generated by the set class $\mathbf{S}^*$, which is obtained by first collecting both the weak elements of the $\mathbf{S}_j$'s and the supports of the nonempty reductions RED($\mathbf{S}_j$) of the $\mathbf{S}_j$'s, and, subsequently, removing redundant sets.

It can be shown that $\mathbf{S}^*$ is fully reducible [8], so that the model $\mathcal{P}^*$ generated by $\mathbf{S}^*$ is decomposable.
A set in $\mathbf{S}^*$ is called *simple* if it is a weak element of some $\mathbf{S}_j$; otherwise - that is, if it is the nonempty support of some RED($\mathbf{S}_j$) -, it is called *compound* and the sets in RED($\mathbf{S}_j$) are called its *components*.

Example 4 (continued). The reductions of the nonseparable components of $\mathbf{S}$ are RED($\mathbf{S}_1$) = RED($\mathbf{S}_2$) = Ø and RED($\mathbf{S}_3$) = {AD, AE, DE}. Therefore, the generating class of the decomposable approximation $\mathcal{P}^*$ to $\mathcal{P}$ is $\mathbf{S}^*$ = {AB, AC, ADF, ADE, AEG} (see Example 2). $\mathbf{S}^*$ contains only one compound set, ADE, which has three components: AD, AE, DE. By (1) the MLE $\pi^*(a, b, c, d, e, f, g)$ of the probability of an object falling in cell (a, b, c, d, e, f, g) under $\mathcal{P}^*$ is

$$\pi^*(a, b, c, d, e, f, g) = \pi^*(a, d, e)\,[n(a, b)\, n(a, c)\, n(a, d, f)\, n(a, e, g)] / [n(a, d)\, n(a, e)\, n(a)^2]$$

where $\pi^*(a, d, e)$ is the maximum-entropy extension of n(a,d)/n, n(a, e)/n and n(d, e)/n.

Let us now consider a subset Y of X and let RED($\mathbf{S}^*$, Y) be the selective reduction of $\mathbf{S}^*$ relative to Y. A set in RED($\mathbf{S}^*$, Y) is called *simplicial* if it is either

(i) the residual part of a simple set in $\mathbf{S}^*$, or
(ii) the residual part of a compound set in $\mathbf{S}^*$ and, simultaneously, a subset of someone of its components.

Definition 3. Let $\mathbf{S}$ be a set class over X and Y a subset of X . The *connection scheme* CON($\mathbf{S}$, Y) for Y in $\mathbf{S}$ is obtained by collecting both the simplicial sets in RED($\mathbf{S}^*$, Y) and the components of the compound sets in $\mathbf{S}^*$ having nonsimplicial residuals in RED($\mathbf{S}^*$, Y).

The computation of CON($\mathbf{S}$, Y) can be carried out by applying a slightly modified SRA, which the author detailed in [9].

Example 4 (continued). Let Y = EF. One has RED($\mathbf{S}^*$, Y) = {ADE, ADF}. ADE is nonsimplicial because it is (the residual part of) a compound set in $\mathbf{S}^*$ and is not a subset of some of its components {AD, AE, DE}; ADF is simplicial because it is (the residual part of) a simple set in $\mathbf{S}^*$. Hence, CON($\mathbf{S}$, Y) = {AD, AE, DE,

ADF}. The MLE $\pi(e, f)$ is given by

$$\pi(e, f) = \Sigma_a \quad \pi(a, d, e)\, n(a, d, f) / n(a, d) \cdot$$

Notice that IPFP is employed only to compute the MLE $\pi(a, d, e)$ corresponding to the nonsimplicial set ADE in RED($\mathbf{S}^*$, Y).
Let Z = ACD. One has RED($\mathbf{S}^*$, Z) = {AC, AD}. Both AC and AD are simplicial; AC is so because it is (the residual part of) a simple set in $\mathbf{S}^*$ and AD is so because it is the residual part of a compound set (ADE) in $\mathbf{S}^*$ and a (subset of a) component of ADE. Hence, CON($\mathbf{S}$, Z) = {AC, AD}.
The MLE $\pi(a, c, d)$ is given by

$$\pi(a, c, d) = (1/n)\, [n(a, c)\, n(a, d) / n(a)] \cdot$$

Theorem 2. A loglinear model over X is collapsible onto a subset Y of X if and only if the connection scheme for Y in its generating class is a covering of Y.

In other terms, Theorem 2 states that $\mathcal{P}$ is collapsible onto Y if and only if (a) $\mathcal{P}^*$ is collapsible onto Y and (ii) each nonsimplicial set in RED($\mathbf{S}^*$, Y) is a subset of Y.
It should be noted that , if $\mathbf{S}$ is fully reducible, then (and only then) each nonseparable component is fully reducible and, therefore, $\mathbf{S}^* = \mathbf{S}$. This implies that, if $\mathcal{P}$ is a decomposable model, then (and only then) the decomposable approximation to $\mathcal{P}$ is nothing more than $\mathcal{P}$ and CON($\mathbf{S}$, Y) = RED ($\mathbf{S}$, Y) and Theorem 2 reduces to Theorem 1.

REFERENCES

1. S. Asmussen and D. Edwards, "Collapsibility and response variables in contingency tables", *Biometrika* 70 (1983), 567-578
2. Y. Bishop, S.E. Fienberg and P.W. Holland, *Discrete multivariate analysis: theory and practice.* MIT PRESS, 1978
3. J.N. Darroch, S.L. Lauritzen and T.S. Speed, "Markov fields and loglinear interaction models for contingency tables", *The Annals of Statistics* 8 (1980), 522-539
4. S.A. Goldman and R.L. Rivest, "Making maximum entropy computations easier by adding extra constraints", *Proc. Workshop on Maximum Entropy and Bayesian Methods in Applied Statistics*, 1986
5. S.L. Lauritzen and D.J. Spiegelhalter, "Local computations with probabilities on graphical structures and their application to expert systems", *J. Royal Statistical Society* B 50 (1988), 157-224
6. D. Maier and J. Ullman, "Connections in acyclic hypergraphs", *Theoretical Computer Science* 32 (1984), 185-199
7. F.M. Malvestuto, "Existence of extensions and product extensions for discrete probability distributions", *Discrete Mathematics* 69 (1988), 61-77
8. F.M. Malvestuto, "Computing the maximum-entropy extension of given discrete probability distributions", *Computational Statistics & Data Analysis* 8 (1989), 299-311
9. F.M. Malvestuto, "A universal table model for categorical databases", *Information Sciences* 49 (1989), 203-223
10. J. Pearl, *Probabilistic Reasoning in Intelligent Systems: Networks for Plausible Inference.* MORGAN-KAUFFMAN, 1988
11. E.M. Reingold et al., *Combinatorial algorithms: theory and practice.* PRENTICE-HALL, 1977
12. R.E. Tarjan and M. Yannakakis, "Simple linear-time algorithms to test chordality of graphs, test acyclicity of hypergraphs, and selectively reduce acyclic hypergraphs", *SIAM J. Computing* 13 (1984), 566-579

The Generalised Biplot: Software Potential

J. C. Gower, Leiden

1.Introduction

The whole area of ordination/multidimensional scaling is one of the most used statistical methodologies in the applied sciences. The very names of *ordination* , coming from ecology, and *multidimensional scaling*, from psychometrics, illustrate the range of applications. Included are well-known methods such as components analysis, biplots, non-metric scaling, correspondence analysis and multiple correspondence analysis. I shall be concerned here with unifying these seemingly diverse techniques and hence providing a solid basis for unified software. Generalisations to samples from several populations, several sets of variables or to so-called multiway techniques (see Coppi and Belasco, 1989) will not be considered here.

I shall be particularly concerned with the multivariate sample expressed as a data-matrix **X** with n rows, the number of samples, and p columns, the number of variables. When all variables are quantitative the classical display uses principal components analysis and its associated biplot technique (Gabriel, 1971); the use of Pythagorean distance to measure the distance between pairs of samples is implicit. When all variables are categorical (qualitive) the classical display uses multiple correspondence analysis (when p=2 formally equivalent to simple correspondence analysis); the use of Chi-square distance is implicit - see Greenacre (1990) for a critique.Principal coordinates analysis / Classical scaling handles any inter-sample distance, preferablly Euclidean, but displays information only on samples. Gower & Harding (1988) showed how to include information on continuous variables in a non-linear biplot, but did not consider categorical variables. Gower & Harding also indicated how the non-linear biplot concept could be embedded in non-metric scaling methodology and this has been developed in detail by Heiser and Meulman (1990) and Underhill (1990). Table 1 summarises the characteristics of all these methods.

Compstat 1990

Some methods concentrate on displaying the samples (e.g. components analysis) and others the variables (e.g. multiple correspondence analysis) although generally this is a matter of historical development and taste, rather than a mathematical necessity. However when metric or non-metric scaling methods operate on a matrix of observed inter-sample distances which has not been derived analytically from information on underlying variables, then there is no possibility of including information on variables in any graphical display. Some methods are tied to particular measures of inter-sample distance and others to particular kinds of variable. The final column of Table 1 indicates that a method termed a Generalised Biplot handles all the situations listed; this method is outlined below.

	PCA	MCA	PCO	Non-linear Biplots	Generalised Biplots
Variables					
Continuous	++		++	++	++
Categorical		++	++		++
Distances					
Pythagorean	++		+-	+-	+-
Chi-Square		++	++	+-	+-
General			++	++	++
Mode					
Units	++	+-	++	++	++
Variables	+-	++		++	++
Criterion					
Metric	++	++	++	++	++
Non-metric			NMS	NMS	NMS

Table 1. ++(main feature), +-(subsiduary feature), NMS(non-metric scaling)

2. A Unified Methodology

All the methods discussed in section 1 are concerned with the same basic problem - the simultaneous display of information on samples and variables. Yet their disparity of detail has required the development of separate computer programs. In the Generalised Biplot, any metric or non-metric scaling method may be used to represent the samples but it is most simple to explain the method first in terms of principal coordinates analysis. It remains therefore only to include information on the variables. For non-linear biplots of quantitative variables information on the k th variable is obtained as follows.

Consider the pseudo-sample with values P(0,0,...,0,r,0,...,0) where the value r occurs for the kth variable and zero represents the mean value of all the other variables (i.e. we are working in deviations from the mean). Given a distance formula of the form:

$$d^2_{ij} = \sum_{k=1}^{p} f(x_{ik}, x_{jk}) \tag{1}$$

then

$$d^2_{ir} = \sum_{k=1}^{p} f(x_{ik}, 0) + f(x_{ik}, r) - f(x_{ik}, 0)$$

represents the squared-distance of P from the ith sample (i=1,2,...,n). Gower (1968) shows how P can then be projected onto any ordination of the samples. The locus (trajectory) of this projected point as r varies gives the non-linear biplot of the kth variable. It is useful to mark unit steps of r on the curvilinear plot; except for the linear trajectory given by Pythagorean distance these marks will not project into equidistant points.

Before proceeding, first note that (1) assumes that each variable contributes independently to squared distance; this is true of many distance and dissimilarity coefficients (see e.g. Gower & Legendre (1986)). However although the assumption aids interpretation, Gower & Harding (1988), it is not essential for constructing a non-linear biplot which requires only that there is some way of evaluating d^2_{ir}. Secondly an additional step is required when the ordination is non-metric. In non-metric scaling there is a mapping between observed distance d_{ij} and plotted distance δ_{ij}. This mapping is chosen so that the monotonic regression of δ_{ij} on d_{ij} is best as judged by some suitable criterion (such as stress) and corresponding to δ_{ij} is a value $\hat{\delta}_{ij}$ fitted by the monotonic regression. Thus calculated values of d_{ir} may be replaced by $\hat{\delta}_{ir}$ and used to extend the non-linear biplot to include mon-metric scaling.This process is not quite so straight foreward as described but Heiser and Meulman (1990) and Underhill (1990) have shown how to resolve certain ambiguities that may arise.

Thus the non-linear biplot satisfactorily handles quantitative variables but cannot handle categorical variables because the concept of mean values and deviations from the mean are not defined. Gower (1990) shows how a bootstrap type of approach can be used to handle both categorical & quantitative variables and in such a way that for quantitative variables the non-linear biplot is subsumed in the methodology. Instead of the pseudo sample (0,0,...,0,r,0,...,0) consider pseudo-samples $(x_{j1},x_{j2},...,x_{j,k-1},r,x_{j,k+1},...,x_{jp})$ for some fixed sample j. This defines n values $d^2_{ir}(i=1,...,n)$ and, as for the non-linear biplot, a point P(j) projected onto the ordination. Now let j vary through all n sample-values and replace the n points by their centroid, P_r. The locus of P_r as r varies, is taken to be the representation of the kth variable. When the kth variable is categorical, r can take only l_k values, the number of categories. When the kth variable is quantitative, r may take a continuum of values, but it is specially interesting to let $r=x_{ik}(i=1,...,n)$ the n values actually occuring in the sample.

In practice the detailed steps of the bootstrap are not necessary as the coordinates of P can be calculated algebraically, at least when the underlying ordination is metric.The required coordinates (Gower 1990) are then given by

$$-\frac{1}{2}(\mathbf{Y}'\mathbf{Y})^{-1}\mathbf{Y}'\{2\mathbf{D}_k\mathbf{1}/n+\mathbf{f}\} \qquad (2)$$

where **Y** is a (nxr) matrix giving the coordinates, referred to their centroid, of the n points in an r-dimentional ordination, the (i,j)th element of $\mathbf{D}_k$ is $f(x_{ik},x_{jk})$ and where **f** is a vector with ith element $f(x_{ik},r)$.

Trajectories for quantitative variables produced in this way have the same shape as those for non-linear biplots, but whereas the latter must be concurrent, the bootstrap approach does not give concurrencey. Gower (1990) shows that a simple translation operation restores the concurrencey without any side-effects. This translation is desirable, as it gives a less cluttered, and hence more readily interpretable,graphical display. Corresponding translations are not defined, and are not needed, for the isolated points corresponding to the levels of categorical variables.

The above has sketched a methodology for extending non-linear biplots to include categorical variables. In so doing it has turned out that continuous variables, categorical variables or mixtures of both kinds are admissible. By choosing the fundamental metric d_{ij} appropriately, various classical methods are subsumed: components analysis and classical biplots with d_{ij} Pythagorean distance, multiple correspondence analysis with d_{ij} Chi-squared distance,principal coordinates and non-linear biplots with general Euclidean distances among quantitative variables. Furthermore other methods of metric scaling and non-metric scaling are admissible. Multiway extensions are easy to recognise but are not explained further here.

3. Software Implications

This unification provides a basis for general-purpose software for all generalised biplots thus obviating the need for the many current special-purpose programs.What then should be the characteristics of this software. The following seem to be basic requirements.

(i) Read the data-matrix **X** allowing for different kinds of variable. Quantitative & categorical variables are essential but it may also be desirable to include dichotomies & their quantitative equivalances in which values corresponding to non-occurence or zero, are handled specially in computation of distance. Missing values too should be accommodated.

(ii) Facilities for the transformation of variables.

(iii)Calculation of **D** according to the information given in (i). Included here should be procedures for calculating the distance between any pair of specified samples and a procedure for calculating $\mathbf{D}_k$.

(iv) Ordination of the samples according to some nominated method. If the chosen method is non-metric then the information required to relate δ_{ij} to d_{ij} should be preserved at this stage.

(v) Calculation of information for the graphical plots of variables:

(a) for trajectories in the continuous case

and (b) for categorical variables.

(vi) Graphics that permit

(a) The plot of the labelled samples in sets of two specified dimensions

(b) The plots of trajectories and labelled categorical points

(c) Facilities for including and excluding scale-points on trajectories

(d) Facilities for deleting unwanted trajectories

(e) Facilities for exhibiting requested interpolants

(f) Facilities for producing hard copy of any graphical display.

The graphical plots in particular are best accommodated in an interactive environment. Much of the above can be achieved within the facilities of a good statistical computing environment such as is given by Genstat (1987); Prof. Michael Greenacre and I are exploring the possibilty of developing free-standing efficient programs. The above list is not intended to be exhaustive; further desirable facilities will occur to many and we plan to expand the system according to needs and experience.

4.References

Coppi, R. & Bulasco, S. (1989) *Multiway Data Analysis*. Amsterdam: North Holland.

Gabriel, K.R. (1971) The biplot graphic display of matrices with applications to principal components. Biometrika **58**, 453-467.

Genstat (1987) *Genstat 5 Reference Manual.* Oxford: Oxford Üniversity Press.

Gower, J.C. (1968) Adding a point to vector diagrams in multivariate analysis. Biometrika **55**, 582-585.

Gower, J.C. (1990) Multiple correspondence analysis and nonlinear biplots: unifications and extensions.(internal report).

Gower, J.C. & Harding, S.A. (1988) Nonlinear biplots. Biometrika **75**, 445-455.

Gower, J.C. & Legendre, P. (1986) Metric and Euclidean properties of dissimilarity coefficients. Journal of Classification **3**, 5-48.

Greenacre, M.J. (1990) The Limitations of multiple correspondence analysis. Computational Statistics Quaterly.(in press).

Heiser, W. & Meulman, J. (1990) Nonlinear biplots for nonlinear mappings. Paper presented at XXII Journees de Statistique, Tours.

Underhill, L. (1990) Untitled typscript-personal communication.

Exploratory Approach and Maximum Likelihood Estimation of Models for Non Symmetrical Analysis of Two-Way Multiple Contingency Tables

R. Siciliano, N. C. Lauro, Napoli and A. Mooijaart, Leiden

Abstract. In non symmetrical analysis of two–way multiple contingency tables we are interested in the dependence between one response variable and two explicative variables. The exploratory approach based on multiple and partial non symmetrical correspondence analysis can be used complementary to asymmetrical association models, logit–linear models and latent budget analysis. In this paper, maximum likelihood estimation of these models is obtained by using alternatingly a multidimensional Newton algorithm. Model parameters are identified by a generalized singular value decomposition with the same metrics as used in non symmetrical correspondence analyses. This provides factorial representations of both the dependence effects among the variables and the residuals from independence.

Key words: non symmetrical correspondence analysis, generalized singular value decomposition, maximum likelihood estimation, association models, latent budget analysis, multidimensional Newton algorithm.

Contents: 1. Introduction. 2. Exploratory methods for dependence analysis. 3. Maximum likelihood estimation of models for dependence analysis. 4. Conclusions. Appendix. References.

Aknowledgements: This research was supported by M.P.I. 60% for the first and the second author.

Compstat 1990

$$p_{i(jk)}/p_{.jk} - p_{i.k}/p_{..k} = \Sigma_m \lambda_m x_{im} y_{jkm} \tag{2}$$

for m=1,...,$M \leq M^*$ ($M^* = \min[I-1, JK-1]$ in multiple NSCA and $M^* = \min[I-1, K(J-1)]$ in partial NSCA). The x_{im} and y_{jkm} are row and column scores associated with the generalized singular values λ_m arranged in descending order. The (1) shows that multiple NSCA represents the departure of column profiles $p_{i(jk)}/p_{.jk}$ from the row margins $p_{i..}$ whereas the (2) shows that partial NSCA represents the departure of column profiles from the row margins in each stratum of the multiple table, i.e. $p_{i.k}/p_{..k}$.
Row and column scores in multiple NSCA satisfy the following centering and orthonormality conditions

$$\Sigma_i x_{im} = 0 \qquad \Sigma_{jk} y_{jkm} p_{.jk} = 0 \tag{3a}$$

$$\Sigma_i x_{im} = \delta^{mm'} \qquad \Sigma_{jk} y_{jkm} y_{jkm'} p_{.jk} = \delta^{mm'} \tag{3b}$$

where $\delta^{mm'}$ is the usual Kronecker delta. In partial NSCA row scores satisfy the same conditions as in (3) whereas column scores satisfy conditions as $\Sigma_j y_{jkm}(p_{.jk}/p_{..k}) = 0$ and $\Sigma_j y_{jkm} y_{jkm'}(p_{.jk}/p_{..k}) = \delta^{mm'}$. Multiple and partial NSCA decompose in M dimensions the total variability of row and column points as measured by two indexes of *dependence* such as the multiple τ_m index and the partial τ_p index both due to Gray and Williams, that is

$$\Sigma_i \Sigma_{jk} [p_{i(jk)}/p_{.jk} - p_{i..}]^2 p_{.jk} = \Sigma_m \lambda_m^2 = \tau_m[1 - \Sigma_i p_{i..}^2] \tag{4}$$

$$\Sigma_k \Sigma_{ij} [p_{i(jk)}/p_{.jk} - p_{i.k}/p_{..k}]^2 p_{.jk} = \Sigma_m \lambda_m^2 = \tau_p[1 - \Sigma_{ik} p_{i.k}^2/p_{..k}] \tag{5}$$

The indexes τ_m and τ_p, which are generalizations of the predictability index τ_s of Goodman and Kruskal for two variables (for references see Lauro and D'Ambra, 1984), can be interpreted as *explained variances* of the response variable that can be (conditionally) accounted for by the explicative variables. The values λ_m^2 in (4) and (5) are proportions of explained variance due to the m–th factorial axis. The sum of the first $M < M^*$ values λ_m^2 gives the percentage of explained variance that is represented in an M–dimensional canonical space. The index τ_m can be decomposed (D'Ambra and Lauro, 1989) to give the following relation

$$\tau_p = (\tau_m - \tau_s)/(1 - \tau_s) \tag{6}$$

The (6) allows to select the two–way multiple table which includes the explicative variables with the highest τ_p, i.e. the variables that contribute more to the explained variance of the response variable (compare Anderson, 1980). This is helpful for table selection and thus for the choice of the explicative variables to include in the models defined in section 3.
In general, NSCA is useful to describe the structure of dependence especially in case of large contingency tables because the GSVD is computationally very efficient and it gives lower–dimensional representations of the table which are *optimal* in a least–squares sense (Greenacre, 1984; Lauro and Siciliano, 1989).

3. Maximum likelihood estimation of models for dependence analysis

Consider now a probability table associated to the observed two–way multiple table in section 2. Models for dependence analysis specify the conditional probability $P_{i(jk)}/P_{.jk}$ (or the logarithmically transformed conditional probability) that an observation falls in the i–th category of variable I given that it falls into the jk–th category of variable JK. We first define asymmetrical versions of association models (compare Goodman, 1986; see also Clogg, 1982; Gilula and Haberman, 1988; Becker, 1989; Becker and Clogg, 1989) and then we consider latent budget analysis (van der Heijden, Mooijaart and de Leeuw, 1989).

The dependence between I and JK and the conditional dependence between I and J given K can be respectively analysed by the following *asymmetrical association models*

$$\log(P_{i(jk)}/P_{.jk}) = \mu_i + \Sigma_m \, \eta_m u_{im} v_{jkm} \tag{7}$$

$$\log(P_{i(jk)}/P_{.jk}) = \mu_{ik} + \Sigma_m \, \eta_m u_{im} v_{jkm} \tag{8}$$

for $m=1,\ldots,M \leq M^*$ ($M^* = \min[I-1, JK-1]$ in model (7) and $M^* = \min[I-1, K(J-1)]$ in model (8)). The positive parameters μ_i and μ_{ik} which sum up to zero pertain to the main effects. In spite of the logit–linear approach (Bishop, Fienberg and Holland, 1975) in model (7) (model (8)) the interaction or *dependence effect* $\Sigma_m \, \eta_m u_{im} v_{jkm}$ is decomposed in M terms allowing factorial representations of row and column points associated to the categories of variables I and JK (J given K). Without loss of generality the parameters u_{im} and v_{jkm} in models (7) and (8) satisfy conditions likewise scores in multiple and partial NSCA respectively. But models (7) and (8) are different from multiple and partial NSCA in (1) and (2) since an additive model is specified for the logarithmically transformed conditional probability. Indeed, the scores decompositions in multiple and partial NSCA, i.e. (1) and (2), approximate in power series expansion the dependence effects in models (7) and (8) respectively when the variables are nearly independent (Lauro and Siciliano, 1989).
Models (7) and (8) are estimated by ML method providing efficient estimates and tests of restricted models if certain assumptions about the observations are fulfilled. Restrictions can be the number M of dimensions, equality between row/column scores, equal spacing between adjacent scores, ordered categories, fixed values of scores (see for instance Gilula and Haberman, 1988).
ML estimates of both expected probabilities and main effects are uniquely determined under centering conditions and eventual restrictions upon scores. For this estimation we propose an iterative algorithm (see Appendix). The goodness of fit can be assessed by the usual likelihood–ratio statistic (Bishop et al., 1975) (the number of degrees of freedom in models (7) and (8) is respectively equal to $[(I-M-1)(JK-M-1)+M(s_I+s_{JK})]$ and to $[(I-M-1)(J-M-1)K+M(s_I+s_{JK})]$ where s_I and s_{JK} are the number of restrictions upon row and column scores). The model parameters are identified by the GSVD of the matrix of elements either $[\log(\underline{P}_{i(jk)}/\underline{P}_{.jk}) - \underline{\mu}_i]$ or $[\log(\underline{P}_{i(jk)}/\underline{P}_{.jk}) - \underline{\mu}_{ik}]$ where the underlying parameters denote ML estimates (compare Becker, 1990). In spite of expected probabilities and model fitting the inference on canonical scores and the geometry of the representations depend on the metrics used in the GSVD (Becker and Clogg, 1989). In order to interpret the structure of dependence among the variables we choose the same metrics as used in multiple and partial NSCA for model (7) and (8) respectively. The resulting identified ML estimates $\underline{\eta}_m$, $\underline{u}_{im}$ and

$\underline{v}_{jkm}$ satisfy orthonormality conditions likewise scores in multiple or partial NSCA. Rows and columns of the centered matrix of logarithmically transformed estimates are represented as points in canonical spaces. The dependence between I and JK or between I and J given K can be interpreted by looking at the geometrical properties of the representations in NSCA. Consider the squared distance between two row points in the JK–dimensional space

$$d^2[i,i'] = \Sigma_{jk} [\log(\underline{P}_{i(jk)}/\underline{P}_{i'(jk)}) - (\underline{\mu}_i - \underline{\mu}_{i'})] p_{.jk} = \Sigma_m \underline{\eta}_m^2 (\underline{u}_{im} - \underline{u}_{i'm})^2$$

It shows that the log–odds linear ratio $\log(\underline{P}_{i(jk)}/\underline{P}_{i'(jk)})$ of being in the jk–th category of variable JK departures from independence $(\underline{\mu}_i - \underline{\mu}_{i'})$ as measured by the difference between $\underline{u}_{im}$ and $\underline{u}_{i'm}$. By testing the model restricted as $u_{im} = u_{i'm}$ we can test such departure. Instead consider the squared distance between two column points in the I–dimensional space

$$d^2[(jk),(jk)'] = \Sigma_i [\log(\underline{P}_{i(jk)}/\underline{P}_{.jk}) - \log(\underline{P}_{i(jk)'}/\underline{P}_{.(jk)'})]^2 = \Sigma_m \underline{\eta}_m^2 (\underline{v}_{jkm} - \underline{v}_{(jk)'m})^2$$

It shows how the difference in dependence effects of categories (jk) and (jk)' of variable JK on variable I is measured by the difference between $\underline{v}_{jkm}$ and $\underline{v}_{(jk)'m}$ that can be tested with the model restricted as $v_{jkm} = v_{(jk)'m}$. In order to analyse separately the dependence effects of J and K on I we can apply an ANOVA decomposition of column score $\underline{v}_{jkm}$, i.e. $\underline{v}_{jkm} = \underline{\mu}_{jm} + \underline{\mu}_{km} + \underline{\mu}_{jkm}$, in the estimated model (7) which yields

$$\log(\underline{P}_{i(jk)}/\underline{P}_{.jk}) - \underline{\mu}_i = \Sigma_m \underline{\eta}_m \underline{u}_{im} (\underline{\mu}_{jm} + \underline{\mu}_{km} + \underline{\mu}_{jkm}) \tag{9}$$

where the μ's parameters taking into account the weights sum up to zero. The (9) also provides factorial representations of first and second order *dependence effects* of J and K on I in logit–linear models (Bishop et al., 1975). If a model with $0 < M < M^*$ dimensions fits to the data then separate displays allow to interpret the set of dependence effects which is more meaningful. Alternatively, if the independent model, i.e. $M = 0$, holds then the (9) can be used to analyse the *residuals from independence* in logit–linear models (compare Lauro and Siciliano, 1989). Analogously, we can apply a similar score decomposition in model (8) with $\Sigma_j \underline{\mu}_{km}(p_{.jk}/p_{..k}) = \underline{\mu}_{km} = 0$ (i.e. the dependence effects between I and K are included in the main effect parameters μ_{ik} which define the independent model).

Another class of models for dependence analysis that is closely related to NSCA is the class of *latent budget models* (van der Heijden, Mooijaart and de Leeuw, 1989) defined as

$$P_{i(jk)}/P_{.jk} = \Sigma_m P_{m/jk} P_{i/m} \tag{10}$$

for $m = 1,\ldots, M \leq \min(I,JK)$, with $\Sigma_m P_{m/jk} = 1$ and $\Sigma_i P_{i/m} = 1$. The $P_{i(jk)}/P_{.jk}$ in column jk constitutes a theoretical budget that is specified by a weighted sum of M latent budgets $P_{i/m}$ with weights given by $P_{m/jk}$. ML estimation of model (10) with eventual restrictions on model parameters is usually obtained by an EM algorithm. The model parameters can be interpreted as conditional probabilities and a latent variable L acts as an *intervening* variable in the dependence relation of I from JK (compare Clogg, 1981). Model (10) is equivalent in terms of ML estimates of theoretical budgets to the following reparametrized model (Siciliano, 1990)

$$P_{i(jk)}/P_{.jk} = P_{i..} + \Sigma_m P_m(P_{i/m}-P_{i..})(P_{m/jk}-P_m)/P_m = P_{i..} + \Sigma_m P_m b_{im} a_{jkm} \quad (11)$$

for $m=1,...,M \leq \min(I-1;JK-1)$ where P_m is the unconditional probability to fall into the m–th category of the latent variable L. Model (11) decomposes the departure from independence for the theoretical budget, i.e. $(P_{i(jk)}/P_{.jk} - P_{i..})$, by a weighted sum of products which terms, i.e. $a_{jkm} = (P_{m/jk} - P_m)/P_m$ and $b_{im} = (P_{i/m} - P_{i..})$, are conditional probabilities departured from their independence. The new parameters b_{im} and a_{jkm} satisfy centering conditions like (3a). Model (11) can be estimated by an iterative algorithm (see Appendix). Without loss of generality we can identify the model parameters with orthonormality conditions like (3b). The GSVD of the matrix of elements $(\underline{P}_{i(jk)}/\underline{P}_{.jk} - \underline{P}_{i..})$ with the same metrics as used in multiple NSCA provides factorial representations of the structure of dependence between JK and L and between L and I.

4. Conclusions

NSCA can be used complementary to models for dependence analysis of two–way multiple contingency tables. ML estimation of both asymmetrical association models and latent budget analysis is obtained by a new algorithm. Restricted models can be also tested. Model parameters are identified by a GSVD with the same metrics as used in NSCA. This provides factorial representations of both the dependence effects among the variables and the residuals from independence.

Appendix

ML estimation of both the asymmetrical association models (7) and (8) and the latent budget model (11) can be obtained by modified versions of the algorithm introduced by Siciliano, Mooijaart and van der Heijden (1990) for ML estimation of NSCA. We now give a brief description of the algorithm for model (7) without restricting the scores parameters. Since the parameters are not identified we can fix $\eta_m = 1$ without loss of generality. Then we can find ML estimates of model $\log(P_{i(jk)}/P_{.jk}) = \mu_i + \Sigma_m u_{im} v_{jkm}$ under conditions $\Sigma_i \mu_i = 0$, $\Sigma_i u_{im} = 0$ and $\Sigma_{jk} v_{jkm} p_{.jk} = 0$ (we assume $P_{i(jk)}/P_{.jk} > 0$). The algorithm minimizes the objective function F = – Log L where Log L is the loglikelihood kernel under any of the usual sampling schemes (i.e. multinomial, product–multinomial, Poisson). An *alternating* use of a multidimensional Newton algorithm provides to estimate the sets of parameters μ_i, u_{im} and v_{jkm}: first the μ_i's are estimated for given u_{im} and v_{jkm}, then the u_{im}'s for each m are estimated for given μ_i and v_{jkm}, whereas the v_{jkm}'s for each m are estimated for given μ_i and u_{im}; the process repeats untill the convergence has been reached (the convergence is guaranteed being positive definite the matrix of second derivatives of each set of parameters).
Let **z** be the column vector of one set of parameters. The parameters should satisfy the side condition $\mathbf{a}'\mathbf{z} = 0$ where the elements of the column vector **a** are equal either to one (for μ_i and for u_{im}) or to sample margins $p_{.jk}$ (for v_{jkm}). In iteration t let $\mathbf{z}^{(t)}$ be the current estimate of **z** for which the condition holds and define the updated estimate as $\mathbf{z}^{(t+1)} = \mathbf{z}^{(t)} + \sigma^{(t)}\mathbf{q}^{(t)}$, where $\sigma^{(t)}$ is the step size parameter (hereby we fix $\sigma^{(t)} = 1$ although it can be reduced in case the function value increases or improper solutions are found) and $\mathbf{q}^{(t)}$ is the search vector to be determined. It results $\mathbf{a}'\mathbf{z}^{(t+1)} = 0$ iff $\mathbf{a}'\mathbf{q}^{(t)} = 0$. By the Taylor expansion the function near the optimum is

$$F(\mathbf{z}^{(t)} + \mathbf{q}^{(t)}) = F(\mathbf{z}^{(t)}) + \mathbf{g}^{(t)\prime}\mathbf{q}^{(t)} + (1/2)\mathbf{q}^{(t)\prime}\mathbf{H}^{(t)}\mathbf{q}^{(t)} \quad (A1)$$

where $\mathbf{g}^{(t)}$ and $\mathbf{H}^{(t)}$ are respectively the gradient vector and the Hessian matrix of F in the point $\mathbf{z}^{(t)}$. The procedure is to find a $\mathbf{q}^{(t)}$ which minimizes (A1) under the side condition $\mathbf{a}'\mathbf{q}^{(t)} = 0$. This leads to minimize the constrained function $\mathbf{g}^{(t)\prime}\mathbf{q}^{(t)} + (1/2)\mathbf{q}^{(t)\prime}\mathbf{H}^{(t)}\mathbf{q}^{(t)} - \varepsilon^{(t)}(\mathbf{a}'\mathbf{q}^{(t)})$ where $\varepsilon^{(t)}$ is the Lagrange multiplier in the step t. In the optimum the first derivatives are equal to zero. As a result the set of equations $\mathbf{H}^{(t)}\mathbf{q}^{(t)} - \varepsilon^{(t)}\mathbf{a} = -\mathbf{g}^{(t)}$ has to be solved together with the equation $\mathbf{a}'\mathbf{q}^{(t)} = 0$ in the parameters $\mathbf{q}^{(t)}$ and $\varepsilon^{(t)}$. These sets of equations can be combined into one set of equations. A solution for $\mathbf{q}^{(t)}$ can be shown to be equal to $\mathbf{q}^{(t)} = -\mathbf{W}^{(t)}\mathbf{H}^{(t)-1}\mathbf{g}^{(t)}$ where $\mathbf{W}^{(t)} = \mathbf{I} - \mathbf{H}^{(t)-1}\mathbf{a}(\mathbf{a}'\mathbf{H}^{(t)-1}\mathbf{a})^{-1}\mathbf{a}'\mathbf{H}^{(t)-1}\mathbf{g}^{(t)}$. Because $\mathbf{H}^{(t)}$ is diagonal the elements of $\mathbf{q}^{(t)}$ can be easily written as $(-g_s^{(t)}/h_s^{(t)} + w_s^{(t)})$ for s = i or j, where $w_s^{(t)}$, $g_s^{(t)}$ and $h_s^{(t)}$ are the general terms of $\mathbf{W}^{(t)}$, $\mathbf{g}^{(t)}$ and $\mathbf{H}^{(t)}$ respectively. The updating formula for the elements $z_s^{(t+1)}$ (s = i or j) of $\mathbf{z}^{(t+1)}$ at the step (t+1) is

$$z_s^{(t+1)} = z_s^{(t)} - g_s^{(t)}/h_s^{(t)} + w_s^{(t)} \qquad \text{(A2)}$$

where $w_s^{(t)}$ is the *adjustment coefficient* that guarantees the side condition $\mathbf{a}'\mathbf{z}^{(t+1)} = 0$ to be satisfied for the step (t+1). The presence of the adjustment coefficient in the updating formula (A2) characterizes the algorithm that we use with respect to the usual algorithm for ML estimation of association models proposed by Goodman (1979) and generalized by Clogg (1982), Becker and Clogg (1989), and Becker (1989; 1990).

References

Anderson, R.J. (1980). Catanova for multidimensional contingency tables in nominal scale response. *Commun. Statist.–Theor. Meth. A9(11)*, 1191–1206.

Becker, M.P. (1989). Models for the analysis of association in multivariate contingency tables. *Journal of the American Statistical Association, 84*, 1014–1019.

Becker, M.P. (1990). Maximum likelihood estimation of RC association models. *Applied Statistics, 39*, 152–167.

Becker, M.P. and Clogg, C.C. (1989). Analysis of sets of two–way contingency tables using association models. *Journal of the American Statistical Association, 84*, 142–151.

Bishop, Y.M.M., Fienberg, S.E. and Holland, P.W. (1975). *Discrete multivariate analysis.* Cambridge: M.I.T. Press.

Clogg, C.C. (1981). Latent structure models of mobility. *American Journal of Sociology, 86*, 836–868.

Clogg, C.C. (1982). Some models for the analysis of association in multiway cross–classifications having ordered categories. *Journal of the American Statistical Association, 77*, 803–815.

D'Ambra, L. and Lauro, N.C. (1989). Non symmetrical analysis of three–way contingency tables. In: R. Coppi and S. Bolasco, *Multiway data analysis*, 301–315. Amsterdam: North Holland.

Gilula, Z. and Haberman, S.J. (1988). The analysis of multivariate contingency tables by restricted canonical and restricted association models. *Journal of the American Statistical Association, 81*, 760–771.

Goodman, L.A. (1979). Simple models for the analysis of association in cross–classifications having ordered categories. *Journal of the American Statistical Association, 70*, 755–768.

Goodman, L.A. (1986). Some useful extensions to the usual correspondence analysis approach and the usual loglinear approach in the analysis of contingency tables (with comments). *International Statistical Review, 54*, 243–309.

Greenacre, M.J. (1984). *Theory and applications of correspondence analysis.* New York: Academic Press.

Lauro, N.C. and D'Ambra, L. (1984). L'analyse non symmetrique des correspondances. In: E. Diday et al. (Eds.), *Data Analysis and Informatics III.* 433–446. Amsterdam: North Holland.

Lauro, N.C. and Siciliano, R. (1989). Exploratory methods and modelling for contingency tables analysis: an integrated approach. *Statistica Applicata. Italian Journal of Applied Statistics, 1*, 5–32.

Siciliano, R. (1990). Relationships between non symmetrical correspondence analysis and latent budget model (in Italian with summary in English, an English version is provided by the author). *Riunione della Societa' Italiana di Statistica, Padova 1990.*

Siciliano, R., Mooijaart A. van der Heijden, P.G.M. (1990). *Non symmetric correspondence analysis by maximum likelihood.* Leiden: Department of Psychometrics and Research Methods, internal report.

van der Heijden, P.G.M., Mooijaart, A. and de Leeuw, J. (1989). Latent budget analysis. In: A. Decarli et al. (Eds.), *Statistical Modelling. Proceedings GLIM, Trento, 1989.* Berlin, Springer Verlag.

An Orthogonal Procrustes Rotation Procedure for Multiple Correspondence Analysis

H. A. L. Kiers and J. M. F. Ten Berge, Groningen

Introduction

Multiple Correspondence Analysis (MCA) is a well-known technique for the exploratory analysis of qualitative variables. For several descriptions of the technique we refer to Tenenhaus and Young (1985). Here we merely give the following rather technical description. Let each qualitative variable be described by a socalled indicator matrix G_j of order $n \times k_j$, $j = 1,...,m$, where n is the number of observation-units, m is the number of variables, and k_j is the number of categories of variable j. An indicator matrix contains one unit element in each row, located in the column corresponding to the category to which the observation unit belongs, and zeroes elsewhere. Let $D_j \equiv G_j'G_j$, the diagonal matrix with category frequencies, then the projector for the space spanned by the columns of G_j is given by $G_jD_j^{-1}G_j$. MCA can be seen as the method that maximizes

$$f(X) = \text{tr } X' \sum_{j=1}^{m} P_j X, \tag{1}$$

over the component scores matrix X, subject to $X'X = I_r$, for some dimensionality r, where P_j is the doubly centered version of the projector for the column-space of G_j. From (1) it is obvious that the solution for X is given by K_r, the matrix with in its columns the first r eigenvectors of $\sum_j P_j$, *or* by any rotation of K_r.

The fact that the component scores are determined up to a rotation is well-known, but this rotational freedom is seldom used. Recently, Kiers (1989b, chapter 8; 1990) has provided techniques for rotations of the component scores such that certain simple structure criteria are optimized. Parallel to simple structure criteria for Principal Components Analysis (PCA), these have been defined in terms of a kind of squared loadings of the components on the variables, called "discrimination measures" (Gifi, 1981). These discrimination measures also play a key role in the method developed below. Let $\mathbf{x}_l$ denote the lth column of X, then $c_{jl} = \mathbf{x}_l'P_j\mathbf{x}_l$ is the discrimination measure for component l and variable j. It is equal to the correlation ratio η^2 based on the variance of component l within categories of variable j. The present paper deals with the problem of rotating X to the effect that the discrimination measures corresponding to the rotated component scores optimally resemble a given set of target discrimination

Compstat 1990

measures, in the least squares sense. Applications of such a rotation procedure are manifold. For instance, if the same data has been analyzed by MCA and a different technique yielding discrimination measures as well, a comparison of the results should take into account rotational freedom. A target rotation procedure can be used to determine how much the MCA discrimination measures can approximate those other discrimination measures. A different application arises where an MCA solution is desired to optimally resemble a set of loadings given by theory or previous (MCA) results, for example in jackknife studies.

An Algorithm for Procrustes Rotation of MCA Solutions

Contrary to what is the case in rotations of the solution of PCA, target rotations for MCA are far from straightforward because a rotation of X does not correspond to a rotation of C, the $m \times r$ matrix with elements c_{jl}. Let the target discrimination measures be given by δ_{jl}, and let T denote an $r \times r$ orthogonal rotation matrix, with columns $\mathbf{t}_1,\ldots,\mathbf{t}_r$. Then the discrimination measure c_{jl} corresponding to the rotated component scores is given by $c_{jl} = \mathbf{t}_l'X'P_jX\mathbf{t}_l$, and a least squares orthogonal target rotation amounts to minimizing

$$\mathrm{f}(T) = \sum_{j=1}^{m} \sum_{l=1}^{r} (\mathbf{t}_l'X'P_jX\mathbf{t}_l - \delta_{jl})^2 = \sum_{j=1}^{m} \| \operatorname{Diag}(T'S_jT) - \Delta_j \|^2 \tag{2}$$

where $S_j \equiv X'P_jX$, of order $r \times r$, and Δ_j is the diagonal matrix with the elements $\delta_{j1},\ldots,\delta_{jr}$ on its diagonal. In order to minimize $\mathrm{f}(T)$ over orthonormal T, we propose to minimize a different function using auxiliary matrices $B_1,\ldots,B_m$ with diagonals constrained to be zero. Let the function g be defined as

$$\begin{aligned} \mathrm{g}(T,B_1,\ldots,B_m) &= \sum_{j=1}^{m} \| T'S_jT - (\Delta_j + B_j) \|^2 \\ &= \mathrm{f}(T) + \sum_{j=1}^{m} \| T'S_jT - \operatorname{Diag}(T'S_jT) - B_j \|^2. \end{aligned} \tag{3}$$

Then, for fixed T, g can be minimized over B_j by setting $B_j = T'S_jT - \operatorname{Diag}(T'S_jT)$, $j = 1,\ldots,m$. Conversely, for fixed $B_1,\ldots,B_m$, g can be minimized over T by a special case of the (majorization) algorithm described by Kiers (1989a). That is, minimizing g over orthonormal T is equivalent to minimizing

$$\mathrm{h}(T) = -\sum_{j=1}^{m} \operatorname{tr} T'S_jT(\Delta_j+B_j) \tag{4}$$

over orthonormal T. Let

$$F \equiv T + (\Sigma_j\alpha_j)^{-1}\Sigma_j S_j T(\Delta_j+B_j), \tag{5}$$

where α_j is the "majorization" parameter chosen as the greatest eigenvalue of $(-S_j \otimes (B_j+\Delta_j))$, with $\otimes$ denoting the Kronecker product. If PDQ' is the singular value decomposition of F (with P and Q orthonormal matrices, and D a diagonal matrix, all of order $r \times r$), then according to Kiers (1989a), the update $T^u = PQ'$ for T decreases (4).

Alternating the procedures for minimizing g over $B_1,...,B_m$ and over T, monotonically decreases the auxiliary function g. Because g is bounded from below, this alternating least squares procedure is guaranteed to converge. Moreover, because each time after updating $B_1,\ldots,B_m$ the second term in (3) vanishes, this procedure also decreases f monotonically, and therefore this method can be used to minimize f(T).

In the above sketched algorithm each iteration requires the computation of m majorization parameters. Since this involves the rather expensive computation of eigenvalues, we modified this algorithm slightly. That is, instead of the F defined in (5) we choose F as

$$F \equiv \Sigma_j S_j T(\Delta_j+B_j). \tag{6}$$

According to Kiers (1989a) using the definition of F in (6) does not jeopardize monotony of the algorithm as long as $\Sigma_j\alpha_j$ is negative, as is often the case in practice. The modified program uses F as defined in (6), but if an increase of g is found the thus found update for T is discarded and replaced by the update based on F defined in (5) so monotony is guaranteed. In practice, this algorithm is typically faster than the algorithm using the majorization parameters. Therefore, we used this modified algorithm in all our analyses.

Some Experiences with the Algorithm

In order to test the performance of our program we constructed 40 data sets with 6 variables, and 60 data sets with 12 variables, with numbers of categories ranging from 3 to 5. These were constructed by combining structured data and data obtained from a random number generator. The data sets were analyzed by MCA, with different dimensionalities: For the data sets with 6 variables we chose dimensionality 2 for the first 20 sets, and dimensionality 3 for the other 20 data sets. For the data sets with 12 variables we chose $r = 2$ for the first 20 sets, $r = 3$ for the next 20 sets, and $r = 4$ for the final 20 data sets. The sets are labelled by their number of variables ($m = 6$, or 12) and by the number of MCA dimensions obtained for them ($r = 2$, 3, or 4) as: (6,2), (6,3), (12,2), (12,3) and (12,4). From the MCA results we computed the S_j matrices to be used in our Procrustes rotation program.

All data sets were subjected to 4 analyses. First a "simple" target was constructed as follows: The first m/r variables were "assigned" to the first component (i.e., had unit discrimination measures for the first component and zeroes elsewhere), the second m/r variables were assigned to the second component, etc. The first two analyses consisted of rotating the MCA solutions to the thus chosen targets, with different starts for T. The first analysis used I_r as a start for T (the "simple" start); the second analysis used a socalled "rational" start for T obtained from ordinary orthogonal Procrustes rotation (Green, 1952) of the matrix of MCA discrimination measures to the target matrix. The next two analyses used the same two starts, but rotated towards a different target, denoted as "perfect" target here. That is, as a target set of discrimination measures we used the discrimination measures resulting from the previous (simple) target rotation. Since there is a matrix T that produced these target discrimination measures, for these analyses a perfect fit can be found. Hence the global minimum is 0, and any function value that is considerably above 0 can be seen as a local minimum. In all analyses iterations were stopped as soon as the function value decreased by less than 0.0001.

The purpose of our study was mainly to check the sensitivity of our procedure to local minima; apart from that we studied the computation times to have a rough indication of the speed of our program on an XT with 8087 processor, and looked for possible differences due to the start for T.

The mean computation times in all 5 groups of 20 data sets are reported in Table 1. Standard deviations are given in brackets. It should be noted that the computation times were not nearly distributed normally, due mainly to some outliers. These outliers also caused some standard deviations to be extremely large. Table 1 shows that computation times increase slowly with m and rapidly with r. This indicates that in practice computation times will not become unreasonably large, since the values for r in our study correspond to the most prevalent in practice.

Table 1. Mean Computation Times (seconds) and Standard Deviations (in brackets) in the 5 Groups of Data Sets, for the 4 Different Analyses.

	Rotation to Simple Target		Rotation to Perfect Target	
	Simple Start	Rational Start	Simple Start	Rational Start
(6,2)	2.6 (.9)	2.2 (.5)	6.5 (3.8)	5.2 (2.7)
(6,3)	12.6 (7.1)	14.6 (22.3)	26.6 (11.6)	21.2 (7.6)
(12,2)	3.2 (.6)	3.1 (.8)	7.4 (4.3)	6.1 (3.8)
(12,3)	14.0 (6.9)	15.2 (12.1)	38.1 (4.6)	37.2 (14.0)
(12,4)	90.6 (46.7)	92.5 (51.5)	149.5 (116.1)	101.1 (39.3)

On the basis of Table 1 we can also compare the performance of the algorithm when started with different starts. These differences are not consistent as far as rotations to simple targets are concerned. In the case of rotations to perfect targets the rational start consistenly produced faster convergence, but differences tend to be small compared to the standard deviations.

More important than computation times is the optimality of the solutions. We checked for local optima by checking the function values obtained from the rotations to perfect targets. In most of the 200 analyses we found function values of approximately 0.001, which seems close enough to zero considering our convergence criterion. We found a function value above 0.006 in only 4 instances. In 3 of these the function value was below 0.06 which is still reasonably close to 0. In only one case we found a value above 0.1, which definitely indicated a local minimum. This local minimum was found when the "simple" start was used. Using the "rational" start for this case yielded a much smaller function value, corresponding to an entirely different rotation matrix. So for all 100 data sets we found only one local minimum, for only one of the two starts. The "near local minima" were found for both simple and rational starts. It is therefore advised to use both starts in practice so a possible local optimum is likely to be detected. Our experience with some empirical data sets confirms that local minima in cases where a perfect fit was possible do not tend to occur.

Results for an Empirical Data Set

The empirical data analyzed here is from Hartigan (1975, p.228), and consists of 24 objects such as screws and nails that are classified according to 6 qualitative variables (whether they have a thread, type of head, head indentation, kind of bottom, whether made of brass, and length (in 5 categories). These data have been analyzed by Kiers (1990) by means of MCA (before and after varimax rotation) and a method called INDOQUAL (which differs from MCA mainly in that it tends to produce more extreme discrimination measures than MCA does). Kiers focused on comparing the varimax rotated MCA solution and the INDOQUAL solution (with $r = 3$), but this comparison does not fully take advantage of the rotational freedom of the MCA solution. Here we fully take advantage of rotational freedom by performing a Procrustes analysis with the INDOQUAL discrimination measures as targets. The original discrimination measures and the discrimination measures after target rotation are reported in Table 2. For comparative purposes the INDOQUAL discrimination measures and those resulting from varimax rotation of the MCA solution are reported in Table 2 as well.

Table 2. Discrimination Measures Before and After Rotation to the INDOQUAL solution, and from the INDOQUAL and Varimax Solutions

	Original MCA Solution			After Rotation to INDOQUAL			INDOQUAL Solution			After Varimax Rotation		
	Dimension			Dimension			Dimension			Dimension		
	1	2	3	1	2	3	1	2	3	1	2	3
Thread	.93	.02	.00	.91	.00	.05	.99	.00	.00	.95	.00	.00
Head	.95	.64	.72	.94	.64	.72	.99	.40	.95	.96	.64	.72
Head ind.	.94	.67	.07	.91	.73	.05	.99	.87	.00	.95	.73	.00
Bottom	.55	.02	.00	.46	.04	.06	.41	.04	.00	.50	.05	.01
Length	.29	.82	.70	.18	.80	.84	.17	.77	.88	.24	.79	.79
Brass	.06	.03	.47	.19	.00	.38	.06	.00	.07	.09	.00	.47

From these results we can see that the Procrustes rotation of the MCA solution to the INDOQUAL solution produces discrimination measures that for the greater part lie in between those from the original MCA solution and from INDOQUAL. Table 2 also shows that the discrimination measures resulting from our target rotation do not resemble the INDOQUAL discrimination measures much better than the varimax discrimination measures do. The greatest differences between the MCA solution and the INDOQUAL solution are found for the variables Head and Brass, and these differences were hardly diminished by our Procrustes rotation procedure. Therefore, it can be concluded that even after rotation to optimal agreement, these variables are represented differently by MCA and INDOQUAL.

References

Gifi, A. (1981). Nonlinear Multivariate Analysis. Leiden: Department of Data Theory.

Green, B.F. (1952) The orthogonal approximation of an oblique structure in factor analysis. Psychometrika, 17, 429–440.

Hartigan, J.A. (1975) Clustering algorithms. New York: Wiley.

Kiers, H.A.L. (1989a). Majorization as a tool for optimizing a class of matrix functions. Psychometrika, in press.

Kiers, H.A.L. (1989b). Three-way methods for the analysis of qualitative and quantitative two-way data. Leiden: DSWO Press.

Kiers, H.A.L. (1990). Simple structure in component analysis techniques for mixtures of qualitative and quantitative variables. Psychometrika, in press.

Tenenhaus, M., & Young, F.W. (1985). An analysis and synthesis of multiple correspondence analysis, optimal scaling, dual scaling, homogeneity analysis and other methods for quantifying categorical multivariate data. Psychometrika, 50, 91–119.

Optimization Techniques and Nonlinear Models

Optimization in Statistics - Recent Trends

T. S. Arthanari, Madras

All statistical procedures are, in the ultimate analysis, solutions, to suitably formulated optimization problems

- C.R.Rao

INTRODUCTION

In the recent past interaction between mathematical programming and statistical problems has grown considerably. Optimization with multi objective functions and global optimization have also found use in solving statistical problems arising in various fields. Espcially in the field of Quality Engineering and On-line Process Control, due to the innovative approaches of Genichi Taguchi [16] some problems have been identified and solved using optimization methods. Since the publication of MATHEMATICAL PROGRAMMING IN STATISTICS [2] there has been a large number of books and paper written, which bring out the connections between Statistical Methods and Optimization.

We find more and more applications talk about computational complexity, which is certainly an outcome of the interaction between Statistics and Mathematical Programming. This paper is a sampler of this recent development, though not exhaustive in that.

2. ACCEPTANCE SAMPLING

Assuring Quality of a lot based on the sample drawn from the lot is the problem of acceptance sampling. In such situations there are risks due to sampling fluctuations. The decision made in the single sampling plan is as follows: If the number of defectives in the sample of size n is less than or equal to the acceptance number c, accept the lot; otherwise reject the lot. The problem is to determine an integer vector (n,c) satisfying the probability of rejecting a 'good' lot, is equal to α, called producer's risk and the proability of accepting a 'bad' lot is equal to β, called the consumer's risk. n,c, integers, satisfying the above requirement may not exist always. so clas ically the problem was solved by relaxing the equality requirement by 'less then equal to' ones, Hald [10]. However with the impact of multiple objective programming, we have goal programming formulations of the problem, which give unique pair (n,c) satisfying the priorities and weightages given to the producer's and consumer's risks respectively Chakraborty [5]. Also use of Fuzzy

Compstat 1990

optimization has found use in this type of problems. Chakraborty [5] gives the following fuzzy goal programming model for the single sampling scheme determination problem:

Obtain non-negative integer pair (n,c) such that the fuzzy goals below are satisfied

(1) probability of rejecting a 'good' lot $\approx \alpha$

(2) probability of accepting a 'bad' lot $\approx \beta$

where $\approx$ is read as "approximately equal to" or "closely around".

Assuming that the decision maker's dissatisfaction of not achieving the specified risks exactly by a sampling plan is a linear function of the deviation from the α , β specified, four equivalent subproblems which are small linear integer programming problems, are solved [5] and the one which maximizes the membership value (a fuzzy set concept) is used to, find the plan. By this approach we have a wider choice of plans than what Hald could provide.

In an attempt to provide single sampling plans that attain a given strengths exactly, Chakraborty [6] obtains a randomized strategy of using atmost three single sampling plans which can be chosen with optimal probabilities, whenever possible.

The following Linear Programming model is used to find the probabilities.

$$\text{minimize} \quad \alpha_1 x_1 + \ldots + \alpha_k x_k$$

$$\text{Subject to} \quad x_1 + \ldots + x_k = 1$$

$$\alpha_1 x_1 + \ldots + \alpha_k x_k = \alpha$$

$$\beta_1 x_1 + \ldots + \beta_k x_k = \beta$$

$$x_1, \ldots, x_k \geq 0$$

Where α_i, β_i are the producer's and consumer's risks respectively corresponding to the ith plan (n_i, c_i). If a feasible solution exists to the above problem,since there are only three constraints in the problem, we observe the fact that atmost three plans are sufficient to be considered.

3. STRATIFICATION PROBLEM

Ramamurthy [14] reviews a recurring problem, which has been encountered in diverse areas for over a period of 65 years. This problem is christened by Ramamurthy as the Basic optimization problem (BOP).

The problem can be stated in a fairly simple manner as follows:

$$\text{Maxinize} \quad \sum p_i(\mu_i - \mu)/\sigma)2$$

$$\text{Subject to} \quad \mu_i = E(X \mid x_{i-1} \leq X \; x_i)$$

$$\text{and} \quad p_i = P(x_{i-1} \leq X \leq x_i), \text{ for } i = 1, \ldots, m+1$$

where $-\infty = x_0 \le x_1 \le \ldots \le x_{m+1} = \infty$

and $X \cap N(\mu,\sigma)$.

Using the notation $\alpha(t) = \sum p_i((\mu_i - \mu)/\sigma)2$

with $t_i = (x_i - \mu)/\sigma$, for i = 1....m

We can restate the problem as

maximize $\alpha(t)$

subject to $t_1 \le t_2 \le \ldots \le t_m$

For properties of the solution to this problem,to this peoble see [14]. It is interesting to note that the same problem is stated and studied in different contexts. Optimal spacing of sample quantiles, optimal grouping for maximum likelihood estimation, stratification for reduction of variability minimizing loss of information due to grouping, selective assembly in Quality Control, minimizing bias by adjustment, optimal grouping of inventoried items to have group reple nishment scheme are a few of the examples where we have the BOP to solve (See [14] for the references).

4. ISOTONIC MEDIAN REGRESSION

Similar in structure to the BOP discussed earlier, we have the isotonic median regression problem, stated below.

$$\text{minimize} \sum \sum |y_{ij} - x_i|$$

$$\text{Subject to } x_1 \le x_2 \ldots \le x_n$$

where Si is a sample drawn from the ith population; $S_i = \{y_{ij} \mid 1 \le j \le m_i\}$, i = 1,...,n; x_i is an estimate of an unknown location parameter of the ith population . In addition, if each x_i should satisfy specified upper and lower bounds we have bounded isotonic median problem. Chakravarati [7] shows that the IMR problem is equavalent to a linear programming problem and also shows that the 'pool adjacent violators' algorithm for solving the corresponding l.p. Problem.

The l.p. problem simplifies the proof of the PAV algorithm as well, further throwing more insight into the problem. Chakravatri also has performed the computational complexity analysis of the PAV algorithm. PAV algorithm is a strongly polynomial algorithm with complexity, O(nlogn). In the case of BIMR, Chakraborty [8] provides a strongly polynomial algorithm with computational compexity O(mn); also since only comparisons are used in the algorithm the algorithm avoids rounding off errors.

5. INTEGRATION OF SURVEYS

The problem of integration of surveys formed a major portion of a chapter in Arthanari and Dodge [2]. Subsequently several papers have appeared in this area. A review of the existing literature appears in this area in Mitra [12]. In Arthanari and Dodge [2] appears a proof of the optimality of the North West Corner rule for the problem.

$$\text{minimize} \quad \sum p_{ij} d_{ij} \qquad (1)$$

$$\text{subject to} \quad \sum p_{ij} = p_{i.},\ 1 \le i \le N \qquad (2)$$

$$\sum p_{ij} = p_{.j},\ 1 \le j \le N \qquad (3)$$

$$p_{ij} \ge 0 \text{ where } d_{ij} = |\, i-j \,|.$$

A problem closely resembling this is the problem of optimally choosing the joint probablities, P_{ij} = Prob ($X = x_i$, $Y = y_j$) of two discrete random variable assuming values x_i, y_j, with probabilities $P_{i.}$ and $P_{.j}$ respectively, $i = 1...n$, $j = 1...n$, so as to maximize the covariance, C(X,Y). Instead of Cov (X,Y) if we consider a more general problem in which we wish to minimize a convex function of (x_i-y_j), the problem can be stated as

$$\text{minimize} \quad \sum f(x_i - x_j)\, p_i$$

subject to (2) and (3)

Mitra and Mohan [13] show that North West Corner rule is optimal for this problem as well. The problem of maximizing overlap between samples chosen for two different surveys from the set of N elements is formulated as a transportation problem in Arthanari and Dodge [2] as follows:

$$\text{Maximizing} \quad \sum\sum d_{ij} P_{ij}$$

$$\text{subject to} \quad \sum P_{ij} = P_{i.},\ i = 1...t$$

$$\sum P_{ij} = P_{.j},\ j = 1...t$$

where d_{ij} = # $S_i \cap S_j$); S_i, S_j are the samples chosen for I, II survey respectively and t is the total number of samples of size n taken from N. Recently Aragon and Pathak [1] have shown that, there exists an optimal solution matrix P to the problem with $P_{ii} = \min(P_{i.}, P_{.i})$, 1,...,t. Mitra and Pathak [11] consider the problem of maximizing the probability of overlaps when sample size equals n and give an algorithm which works for the case of integrating up to three surveys. However this fails in the case of four surveys.

One of the interesting outcome of the interation between statistics and mathematical programming is the complexity analysis of the computation. Integration of surveys problem has a strongly polynomial algorithm when the number of syrveys is fixed, otherwise the problem is NP hard. see [12]

Some applications of transportation theory to statistical problems can be found in Causey, Cox and Ernest [4], who possibly have not read [2].

6. TAGUCHI'S PARAMETRIC DESIGN PROBLEM

Quality Engineering and assurance aim at minimization of deviation of a product's functional characteristics from the target values. Robustness of the product design assures the ability of the product to withstand changes in noise factors, which are both internal and external to the product. While process design assures against manufacturing imperfections, similar activities are necessitated to build robustness of the design. Taguchi advocates parametric design and tolerance design for this purpose, which should be taken up at the R&D stage. The statistical problems associated with this stage of design can be stated as follows:

Let D be the factor space. Let N denote the noise space. A neighbourhood around $d \in D$ is defined by varying levels of noise factors in a specified manner, denoted by n(d). The problem is to find d such that

(1) Expected response at d is as close as possible to the target value (2) the variability at d is as small as possible.

This problem lends itself to be treated as a multi objective programming problem. Arthanari [3] gives a simulated annealing algorithm for solving this problem. Taguchi's parametric design uses orthogonal array designs to arrive at the optimal parameter values. The concept of monotone likelihood principle is applied in justifiying the simulated annealing method to this problem.

In all practical situations, in Quality Engineering, design parameters need to meet several functional performance characteristics of the final product or system. In these cases, methods from nulti objective programming find use. For instance apart from target value being achieved as closely as possible, and variability being kept as small as possible cost also might have to be kept as low as possible to be competitive. Interactive approaches available for Multi objective programming are adoptable in such cases. See Evern [9].

7. CONCLUSION

Recent applications of mathematical programming in acceptance sampling, stratification problems, isotonic median regression, integration of surveys and robust designs are considered in this paper. Even though it is not exaustive, it is certainly exemplary.

References

1. Aragon,J., and Pathak, P.K.,(1990), An algorithm for optimal integration of two surveys, to appear in Sankhya.

2. Arthanari,T.S., and Dodge, Y.,(1981), Mathematical Programming in Statistics, Wiley & Sons ,New York.

3. Arthanari,T.S.,(1988); Parametric design and global optimization, paper presented at the First International Conference on Optimal Design and Analysis of Experiments, Neuchatel, Switzerland.

4. Causey,B.D., Cox,L.H., and Ernst,L.R.,(1985), Applications of transportation theory to ortation theory to statistical problems. J.Amer.Statis.Assoc.,80,903-909.

5. Chakraborty,T.K.,(1988), Single Sampling attribute plan of given strength based on Fuzzy Goal Programming, OPSEARCH, Vol.25,No.4,pp 259-271.

6. Chakraborty,T.K.,(1989), A Group Sampling attribute plan (1989) to attain the given strength, OPSERACH, Vol26, No.2,pp 122-204.

7. Chakravarti,N.,(1987),Isotonic Median Regression; A Linear Programming approach, private communication.

8. Chakravarti,N.,(1987), Bounded Isotonic Median Regression, private communication.

9. Evren,R.,(1987), Interactive compromise programming J.Opl.Res.Soc.,Vol.38,No.2, pp 163-172.

10.Hald,A.,(1981), statistical theory of sampling inspection by attributes, Academic Press, London.

11.Mitra,S.K., and Pathak,P.K.,(1984), Algorithms for optimal integration of two or three surveys, Scand.J.Statist,Vol.11,pp 257-263.

12.Mitra,S.K.,(1988), On the method of overlappling maps in survey sampling, Sankhya, Vol.50,Series B, Pt.I, pp 9-38.

13.Mitra,S.K. and Mohan,S.R.,(1988), On the optimality of North-West Corner solution in some applications of the transportation theory, Tech. Report No.8801, Indian Stat. Inst. Delhi Centre, India.

14.Ramamurthy,K.G.,(1988), A Review of an optimization problem, OPSEARCH, Vol.25, No.1,pp 29-48

15.Robertson,T. and Waltmen,P.(1968), On estimating monotone parameters, Ann.Maths. Stat.,Vol.31,pp 1030-1039.

16.Taguchi,G.,(1985), Quality Engineering in Japan,Commu. Stat. Theo. Meth., Vol.14, No.1,pp 2785-2801.

Fitting Non Linear Models with Two Components of Error

G. J. S. Ross, Harpenden

Summary

The problem of fitting non linear models to a sample of individual data sets is discussed. For each individual a response curve is proposed, with errors of measurement, and the response curve parameters are themselves considered to be samples from a population of curve parameters, with a covariance matrix to be estimated. By studying cases where full information is available by a conventional analysis of each individual data set, it is shown that a direct attempt to fit the full model by maximum likelihood estimation does not always lead to acceptable estimates of the covariance matrix.

Introduction

A 'Regression Model of the Second Kind' (Fisk (1967), Nelder (1968)) is one in which the parameters are themselves random variables. Given a random sample of individuals j=i,...n, and a set of observations(x_{ij},y_{ij}), $i=1,...,n_j$ for each individual, we are interested in fitting a common form of relationship $y=f(x,\theta)$ such that

$$y_{ij} = f(x_{ij},\theta_j)+\epsilon_{ij} \tag{1}$$

where θ_j is distributed with mean θ and covariance matrix Ω, and ϵ_{ij} is Normally and independently distributed with mean zero and variance σ^2.

The linear regression model was first discussed by Rao (1965) in the context of growth curves of individuals, and developed in terms of orthogonal polynomials. The concept of linear models with several error strata is of course familiar in the context of surveys and split plot experimental designs, where the components of variance may be estimated independently of the regression parameters. The parameter σ^2 is used to make inferences about the response curve for an individual, while Ω measures the variability

Compstat 1990

of the response curve between individuals.

The model may be extended in several ways:

1. The function f may be non linear in the parameters; for example, an exponential or logistic curve.
2. The data may be discrete, in which case a Poisson or Binomial error distribution may be more appropriate than the Normal; or the variance of ϵ_{ij} may be non-uniform, or the residuals correlated.
3. Some of the parameters θ_j may be the same for all data sets, so that the corresponding row and column of Ω is zero. This may be more useful for interpreting the results than if all the parameters are allowed to vary, even if the fit is not so precise.

Maximum likelihood estimation

For the linear model, $f(x,\theta_j)=x'\theta_p$ where θ_j is multivariate Normal with expectation θ , the log likelihood function (multiplied by -2) is

$$L(\theta,\Omega,\sigma^2)=\Sigma(y_{ij}-x_{ij}'\theta)'c_{ij}^{-1}(y_{ij}-x_{ij}'\theta)+\Sigma\log(c_{ij}) \tag{2}$$

where $c_{ij} = x_{ij}'\Omega x_{ij}+\sigma^2$.

For the non linear model, it is necessary to use a first order approximation to evaluate the variance of each observations by expanding $f(x,\theta)$ in a Taylor series about its expectation. The results are not completely independent of the parameterisation used because of non linearity, which may make a considerable difference in some circumstances. Asymptotically however we may approximate the quantity c_{ij} by

$$c_{ij}^{*}=(\frac{\partial f}{\partial\theta})'\Omega(\frac{\partial f}{\partial\theta})+\sigma^2 \tag{3}$$

If the amount of data is sufficient it is preferable to fit models to each individual to obtain estimates of θ_j and σ^2 directly, and then to estimate θ and Ω from the matrix of n rows of $\hat{\theta}_j'$.

The likelihood function (2) loses information relating observations on the same individual, and if the x_{ij} are the same across individuals, there may be insufficient distinct observations to estimate all the parameters simultaneously. Furthermore it is not usually possible to estimate σ^2 independently from (2) when there is a constant term in the model, because the corresponding variance in Ω is added to σ^2. In some situations when between-individual variation is large compared with σ^2, it is possible to assume $\sigma^2 = 0$, but otherwise it must be estimated independently.

Direct estimation becomes necessary when there are relatively few observations per individual, or insufficient to estimate the parameters individually. The package NONMEM (Beal and Sheiner, 1980) is designed to cope with this situation. To get good estimates of the parameters of Ω large numbers of individuals are required, as scale parameters are always less easy to estimate than location parameters, and a few aberrant individuals may have considerable influence on the results.

Estimation from individual curve fitting

The data in Table 1 consist of the annual mean yields in tonnes/hectare of spring barley on the Hoosfield continuous barley experiment at Rothamsted, between 1975 and 1989. The values refer to four levels of applied nitrogen (0, 1, 2, 3) in multiples of 48 kg/ha. The exponential regression model

$$y = \alpha + \beta \rho^x \qquad (4)$$

fits the data very well, but the parameters vary from season to season depending on temperature and rainfall and other factors. Estimates of the parameters are given for each year, as follows:

Table 1
Yields of Spring Barley at Rothamsted in t/ha

Year	Levels of nitrogen				Estimated parameters		
	0	1	2	3	$\hat{\alpha}$	$\hat{\beta}$	$\hat{\rho}$
1975	1.54	2.93	3.66	3.78	4.00	- 2.47	.416
1976	2.17	2.54	2.68	2.88	3.12	- 0.95	.649
1977	2.33	3.92	4.61	4.69	4.85	- 2.52	.352
1978	2.25	3.80	4.20	4.36	4.39	- 2.14	.281
1979	1.59	3.29	4.23	4.43	4.76	- 3.18	.445
1980	2.44	3.91	5.02	5.29	5.95	- 3.53	.550
1981	1.87	3.59	4.19	4.12	4.23	- 2.37	.254

Table 1

Yields of Spring Barley at Rothamsted in t/ha

Year	**Levels of nitrogen**				**Estimated parameters**		
	0	**1**	**2**	**3**	$\hat{\alpha}$	$\hat{\beta}$	$\hat{\rho}$
1982	2.17	3.83	4.55	4.44	4.60	- 2.44	.288
1983	0.98	1.88	2.43	2.76	3.27	- 2.30	.607
1984	2.14	3.45	4.13	4.08	4.25	- 2.12	.346
1985	2.43	4.04	4.44	4.52	4.56	- 2.13	.241
1986	1.79	3.05	3.51	3.68	3.78	- 1.99	.366
1987	2.29	3.45	3.84	3.96	4.03	- 1.74	.330
1988	1.47	2.98	3.66	4.15	4.58	- 3.10	.527
1989	1.26	1.99	2.26	2.33	2.38	- 1.12	.345

The pooled estimate of σ^2 (from the single degree of freedom for lack of fit from each year) is 0.111 on 15 d.f.

The means, standard deviations and correlations of the parameters are as follows:

Parameter	**Mean**	**Standard deviation**	**Correlation**		
α	4.18	0.837	1		
β	- 2.27	0.694	- 0.840	1	
ρ	0.400	0.129	- 0.108	- .099	1

If we assume ρ is common to all sets the variance ratio in the analysis of parallelism (Ross, 1984) is 2.2 on 14 and 15 d.f. which is not significant. Then the pooled estimate of σ^2 rises to 0.0170 on 2.9 d.f. and the estimates become

α	4.12	0.824	1		
β	- 2.21	0.625	- .837	1	
ρ	0.388	0	0	0	0

The parameters fitted by least squares to all data, ignoring the year, are (4.11, - 2.20, 0.386) showing that there is a difference between the parameter means over years and the parameters fitted to the mean yields, because the model is non linear.

Estimation of parameters in the Stochastic Model

Superficially it is only necessary to minimise the log likelihood function (2), modified by formula (3) using the first derivatives $z' = (1, \rho^x, \beta x \rho^{x-1})$

with respect to α, β, ρ, the elements of Ω and the variance σ^2. A solution will no doubt be found, with some effort, particularly if for some values Ω is not positive-definite. The parameter σ^2 will be found to be redundant, being not separately estimable from the variance of α in Ω.

To avoid non-positive definite matrices, it is simpler to write Ω as a product of triangular matrices T'T, where

$$T = \begin{matrix} t_{11} & 0 & 0 \\ t_{12} & t_{22} & 0 \\ t_{13} & t_{23} & t_{33} \end{matrix}$$

Then the quantities c_{ij} are obtained for sums of squares of the vector Tz, and by implication Ω is always positive definite, and no value of c_{ij} can be negative. Elements of Ω are then computed from the fitted parameters in T.

In practice it is found that while the curve parameters are easily estimated, the elements of T cannot be uniquely determined, and different solutions are found from different starting values, but always with the same minimised log likelihood.

The reason in the case of the Barley data above is that there are only four different values of x, and therefore only four different values of c_{ij}. Since six parameters in T are fitted, there are two redundant parameters, and it would have sufficed to use the values of c_{ij} directly as parameters. The solution is then rapidly found to be
$(\hat{\alpha}, \hat{\beta}, \hat{\rho}, c_0, c_1, c_2, c_3) = (4.12, -2.20, .389, .194, .436, .625, .579)$ and no inferences about Ω or σ^2 are possible in this case.

Is direct estimation possible?

Beal and Sheiner (1980) recommended the stochastic model in situations where there are too few points per individual to obtain satisfactory individual curves. If there are sufficient points then the procedure is unnecessary because there is loss of information in ignoring the individual relationships. However, as shown above, if there are fewer distinct values of x than elements of Ω direct estimation is not possible. When there are more distinct values of x_{ij} (and hence c_{ij}) than elements of Ω it should be possible to find solutions. But whether these are unique or well-determined, is another question. For example, if the values of x are only slightly different, the equations to solve for T given a set of values of c which correspond to an optimal curve will be ill-conditioned and the solutions will be very unreliable.

A compromise is to reduce the number of parameters in Ω by imposing restraints. The extreme case is to assume no variation in one or more parameters, based on a judgement as to which parameter is most suitably regarded as common to all data sets. In the exponential example above the parameter ρ may be chosen, so that the third row of the matrix T is set to 0. There are then only three extra parameters to estimate and the model may be fitted. A further option is to assume that ρ is uncorrelated with α or β, and to set t_{13} and t_{23} to zero, leaving four parameters to be estimated.

Since the number of parameters to be fitted increases quadratically with the number of parameters in the curve, the difficulty of fitting the model increases rapidly as the curve becomes more complex. The earlier work of Rao (1965) and his followers used only orthogonal polynomials where the matrix Ω could be assumed diagonal, and the complexity was manageable.

Conclusions

The general 'non linear regression model of the second kind' is important as a summary statement of the distribution of curves fitted to individuals in a population. However, attempts to fit it directly by maximum likelihood without taking into account the structure of the data set lead to very poor estimates of the covariance parameters, if they can be estimated at all.

References

Beal, S.L. & Sheiner, L.B. (1980) The NONMEM System. *The American Statistician* **34**, **118-119.**

Fisk, P.R. (1967) Models of tghe Second Kind in regression analysis. *J.Roy. Statist.Soc. B* **29**, 266-281

Nelder, J.A. (1968) Regression, Model building and invariance. *J. Roy. Statist. Soc. A.* **131**, 303-329.

Rao, C.R. (1965) The theory of least squares when the parameters are stochastic. *Biometrika* **52**, 447-458.

Ross, G.J.S. (1984) Parallel Model Analysis. *Compstat 1984* 458-463.

Computing for Robust Statistics

Some Proposals for Fast HBD Regression

P. J. Rousseeuw, Edegem and B. C. van Zomeren, Delft

Abstract

Existing high-breakdown regression estimators need substantial computation time. In this paper we propose a fast estimator for robust regression with a breakdown point of 1/3. This is not the highest value possible, but it is independent of the number of predictor variables. Also a more refined variant of this method is proposed. Both algorithms start by identifying leverage points by computationally cheap robust estimates of location and scale in **x**-space, and then apply L_1 regression to the non-leverage points. Some examples are given, and a small simulation study is presented. It appears that in these cases both new algorithms perform equally well as more time-consuming estimators.

1 Introduction

Robust regression attempts to cope with outliers as well as bad leverage points. Regression outliers are observations deviating from the linear pattern of the majority of the data, whereas leverage points are outlying in the space of the predictor variables.
A non-robust analysis of data containing bad leverage points may lead to erroneous results. When there are no leverage points in the data one could use the L_1 estimator, which minimizes the sum of the absolute residuals. The L_1 method resists regression outliers, but is sensitive to leverage points. To overcome this, many approaches have been proposed with varying degrees of robustness (as measured by their breakdown point) and different computational complexity. Among those are the bounded influence estimators, which can be computed rather easily, but of which the breakdown point decreases when the number of predictor variables (p) increases. On the other hand, the repeated median (Siegel 1982) has a high breakdown point but requires consideration of all possible subsets of p observations.
Rousseeuw (1984) introduced the Least Median of Squares estimator (LMS), which achieves the highest possible breakdown point. The LMS is usually approximated by a resampling method which produces good results, but needs substantial computation time, which is sometimes seen as a disadvantage. Rousseeuw and van Zomeren (1990) introduced a

Compstat 1990

new diagnostic plot, of robust regression residuals versus robust distances in the predictor space. In such a display observations are classified as belonging to one of four categories: regular observations, vertical outliers, well-behaving leverage points and bad leverage points. Obtaining robust distances in **x**-space boils down to constructing robust estimates of location and scatter for the predictor variables. One way to do this is to compute the Minimum Volume Estimator (MVE) which is also computed by means of a resampling algorithm. The MVE and LMS can be combined into a single program to produce the plot.

In this paper the emphasis is on computation time. The idea is to first identify leverage points by means of a fast algorithm which produces distances in the **x**-space. Then the L_1 estimator is applied to the non-leverage points, yielding robust regression coefficients, as well as regression residuals for all observations. This has the advantage that the diagnostic plot can still be obtained.

In section 2 we describe two proposals for fast regression with high breakdown point. In section 3 these methods are applied to real data. We also present some results of a small simulation study. In section 4 the proposals are discussed and possible developments are indicated.

Our framework is the linear regression model $y_i = \theta_0 + \theta_1 x_{i1} + \ldots + \theta_p x_{ip} +$ error for $i = 1, \ldots, n$, which includes an intercept. An observation will be denoted by $(\mathbf{x}_i, y_i)$.

2 Methods

As described in Rousseeuw and van Zomeren (1990) we propose to identify outliers in multidimensional point clouds by computing for every observation $\mathbf{x}_i$ the index

$$u_i = \max_{\mathbf{v}} \frac{|\mathbf{x}_i \mathbf{v}^t - L(\mathbf{x}_1 \mathbf{v}^t, \ldots, \mathbf{x}_n \mathbf{v}^t)|}{S(\mathbf{x}_1 \mathbf{v}^t, \ldots, \mathbf{x}_n \mathbf{v}^t)} \tag{1}$$

where L and S are robust estimators for location and scale in one dimension, such as the midpoint and the length of the shortest half. This is a variant of an idea of Gasko and Donoho (1982) and Stahel (1981) who suggested taking the median and the median absolute deviation for L and S. The directions $\mathbf{v}$ are generated by considering $\mathbf{v}_l = \mathbf{x}_l - M$ where $l = 1, \ldots, n$ and M is the coordinatewise median of the data. The u_i can be computed very fast. They can be thought of as robust "distances", revealing outliers in $\mathbf{R}^p$. The algorithm itself is permutation invariant, but not affine equivariant, because it involves the coordinatewise median and because it does not project in *all* directions. We therefore propose to scale the data first, using some computationally cheap semi-robust estimators of location and scatter, or maybe even the classical mean and covariance matrix.

In the regression situation, we scale the $\mathbf{x}_i$ and then compute robust "distances" u_i by applying the projection method to the $\mathbf{x}_i$. By setting aside 1/3 of the data with the highest u_i we avoid $n/3$ possible leverage points. On the remaining $2n/3$ points we apply L_1 regression, which is resistant against regression outliers and has a breakdown point of 1/2, which is 1/3 of the original dataset. Hence the resulting procedure can cope with up

to $n/3$ bad leverage points and up to $n/3$ vertical outliers, so it has a breakdown point of $1/3$, independent of the dimension p.
Finally, the estimate for the intercept is adjusted by adding the midpoint of the shortest half of the residuals (which is the univariate LMS location estimate). The scale of the residuals is estimated by the length of their shortest half multiplied by a correction factor to obtain consistency in case of normal errors. This leads to the following proposal.

Algorithm PLA:

1. Standardize the $\mathbf{x}$-cloud.
2. Apply the projection method on the $\mathbf{x}_i$ yielding u_i.
3. Select the $2n/3$ observations with the smallest u_i.
4. Compute the L_1 regression on the selected observations.
5. Adjust the intercept term using the midpoint of the shortest half of the residuals, and estimate the scatter of the residuals by the length of the shortest half.

Algorithm PLA has the disadvantage of discarding a fixed number of $n/3$ observations, which need not all be leverage points. It can be refined by attempting to actually identify the outlying $\mathbf{x}_i$. Because the projection method is not affine equivariant it is impossible to find cutoff values for the resulting distances. We therefore propose to use another high-breakdown estimator of multivariate location, the Minimum Volume Ellipsoid estimator (MVE), which was introduced by Rousseeuw (1983). It is defined as the center of the ellipsoid covering half of the observations. This ellipsoid also determines an estimate for the covariance matrix of the data.
In Rousseeuw and van Zomeren (1990) a resampling algorithm is used to approximate the MVE. It is also suggested to calculate robust distances

$$RD_i = \sqrt{(\mathbf{x}_i - T(\mathbf{X}))\mathbf{C}(\mathbf{X})^{-1}(\mathbf{x}_i - T(\mathbf{X}))^t}, \tag{2}$$

where $T(\mathbf{X})$ and $\mathbf{C}(\mathbf{X})$ are the MVE estimates for location and scale. We propose to use the MVE as follows. By drawing subsamples *only* from the $2n/3$ observations selected in the first step, the probability of drawing at least one good subsample is increased drastically, which reduces the necessary number of subsamples. (In our simulation study good results were achieved using only 100 subsamples.) Then for all $\mathbf{x}_i$ we compute robust distances using (2). These are compared to the square root of the 97.5% quantile of the corresponding $\chi^2(p)$ distribution, where p is the number of predictor variables. This turns out to be a better way of selecting the non-leverage points than the first proposal. Finally, L_1 is used to compute the regression coefficients. The intercept is adjusted as in PLA, and residuals are computed. This leads to the following.

Algorithm PRLA:

1. Standardize the $\mathbf{x}$-cloud.
2. Apply the projection method on the $\mathbf{x}_i$ yielding u_i.

Figure 1: **Robust Distances produced by the projection method (upper panel) and by the MVE resampling method (lower panel) for the Phosphorus Content data.**

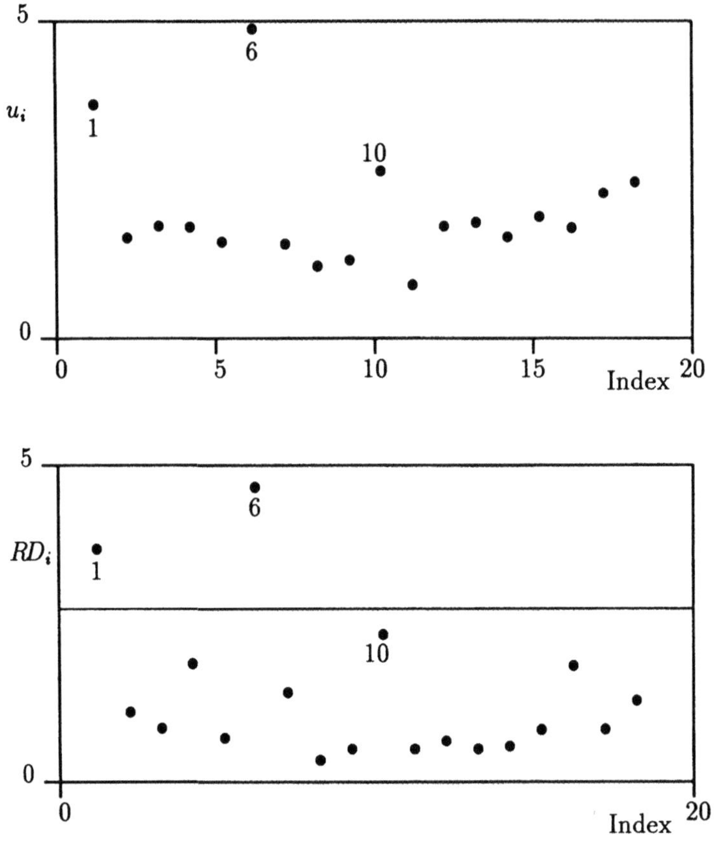

3. Select the 2n/3 observations with smallest u_i.
4. Apply the MVE resampling algorithm to the **x**-part of the selected observations. Draw only 100 subsamples. Evaluate the MVE objective function on *all* observations.
5. Calculate the robust distances of all observations with respect to the MVE. Apply the small sample correction factor proposed by Rousseeuw and van Zomeren (1990).
6. Select those observations of which the distance does not exceed the square root of the 0.975 quantile of the corresponding χ^2_p distribution.
7. Compute the L_1 regression on the selected observations.
8. Adjust the intercept, and estimate the scale of the residuals.

Figure 2: **Diagnostic plot of standardized residuals versus robust distances, produced by proposal PRLA, for the Salinity data.**

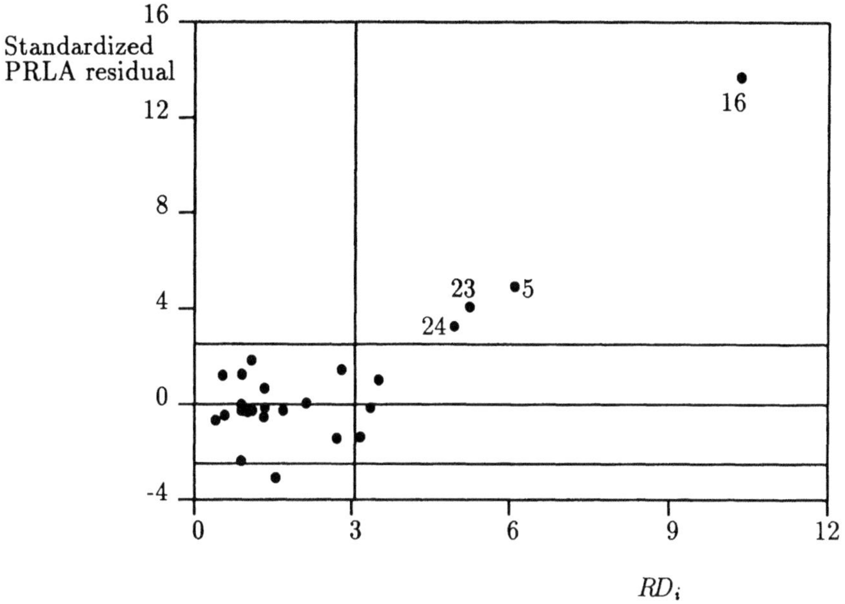

3 Results

To get an impression of the performance of both proposals, several well-known data sets were analyzed, like the salinity data and the phosphorus content data. We also carried out some simulations, similar to a study in Rousseeuw and Leroy (1987).

3.1 Real data

The phosphorus content data (Prescott 1975) consist of 18 observations on 2 predictor variables and a response variable. Observations 6 and 1 are the most outlying in **x**-space, followed by observation 10. These leverage points are identified by the projection method, as is shown in the upper panel of figure 1. This plot also demonstrates that by discarding one third of the data as in PLA, one would also disregard some of the non-leverage points. The lower panel of figure 1 shows the robust distances with respect to the MVE estimates for location and scale, as used in PRLA. The horizontal line gives the value of $\sqrt{\chi^2_{0.975}(2)} \approx 2.72$. Clearly, only the two most outlying observations exceed this cutoff, so several non-leverage points are brought back into the L_1 fit. The final regression exposes the observations 17 and 10 as vertical outliers, because they have very large residuals (both with respect to PLA and PRLA).

Our second example is the salinity data set, which contains 28 measurements on 3 explanatory variables and a response variable (Ruppert and Carroll 1980). Here observation

Table 1: **Mean results of 50 simulated regression datasets with n=40 and p=4 in three different situations.**

	Normal data			Outliers			Leverage points		
Parm	L_1	PLA	PRLA	L_1	PLA	PRLA	L_1	PLA	PRLA
θ_0	0.990	0.980	0.988	1.349	1.050	1.045	-0.227	1.053	0.996
θ_1	0.998	1.005	0.999	0.999	1.009	1.003	0.045	0.991	1.000
θ_2	1.002	1.005	1.002	0.998	1.004	0.996	1.022	1.012	1.007
θ_3	1.008	1.015	1.006	1.007	1.016	1.005	1.001	0.999	1.000
σ	0.741	0.871	0.782	2.517	1.322	1.126	7.211	1.170	1.096

16 is an extreme leverage point, but also 5, 23 and 24 are outlying in **x**-space. These points are also regression outliers and are identified by both algorithms. Figure 2 shows the diagnostic plot of standardized residuals versus robust distances in factor space, computed using PRLA. This plot shows at one glance that there are four regression outliers, which are leverage points as well. The cutoff value on the horizontal axis is $\sqrt{\chi^2_{0.075}(3)} \approx 3.06$. The plot also contains a few points with *slightly* large **x**-distances but small residuals. There is also a non-leverage point with a robust residual outside the [-2.5,2.5] confidence band, but this is another boundary case.

3.2 Simulated data

The behaviour of both new algorithms was investigated in three different situations: first in regression data generated without any outliers, then in data containing 20% regression outliers, and finally using data containing 20% bad leverage points. Rousseeuw and Leroy (1987, section 5.4) presented the results of a simulation study of high breakdown regression estimators. We used a similar setup in order to compare the new algorithms to the existing ones.

We considered 3 batches of 50 datasets of 40 observations and a regression model containing 4 parameters (including intercept). The data were generated using $\boldsymbol{\theta} = (1,1,1,1)^t$. In the normal situation the **x**-variables were generated as $N(0,100)$. The errors were $N(0,1)$. In the regression outlier situation 20% of the errors were generated as $N(10,1)$. In the third batch 20% of the $\mathbf{x}_{i1}$ were replaced by $N(100,100)$ variables, while keeping the original y_i. In every run the four regression coefficients were estimated. An estimate for the scale of the error term was also provided.

Table 1 presents the mean coefficients, averaged over the 50 runs. As a reference the results of the L_1 estimator are also given in the first column for every batch. Table 1 shows that the two new proposals perform as well as the L_1-estimator when applied to outlier-free data. In the case of regression outliers, both new algorithms give good results

Table 2: **Variances of the estimates in 50 simulated regression datasets with n=40 and p=4 in three different situations.**

	Normal data			Outliers			Leverage points		
Parm	L_1	PLA	PRLA	L_1	PLA	PRLA	L_1	PLA	PRLA
θ_0	0.0288	0.1536	0.1236	0.0526	0.3566	0.1274	3.4911	0.1609	0.1423
θ_1	0.0003	0.0023	0.0004	0.0006	0.0025	0.0011	0.0032	0.0009	0.0007
θ_2	0.0006	0.0012	0.0007	0.0012	0.0030	0.0012	0.0601	0.0012	0.0007
θ_3	0.0006	0.0017	0.0007	0.0012	0.0065	0.0016	0.0414	0.0008	0.0007
σ	0.0098	0.0376	0.0225	0.0122	0.2737	0.0424	0.7356	0.0311	0.0339

for the regression coefficients. The estimates for the error scale are slightly large. Note that L_1 gives a wrong intercept, as well as an excessive error scale (which is its mean absolute residual).

In the situation with the bad leverage points the L_1 estimator gives completely erroneous results for $\hat{\theta}_0$, $\hat{\theta}_1$, and, consequently, for $\hat{\sigma}$. The algorithms PLA and PRLA succeed in producing meaningful estimates for all regression parameters, although the error scale is slightly overestimated. Table 1 also shows that the algorithm PRLA comes somewhat closer to the correct values than does PLA.

When comparing our results to table 4 on page 212 of Rousseeuw and Leroy (1987) we see that the new methods, and especially PRLA, perform almost as well as the LMS. It makes sense to compare the new methods to the LMS, because the estimate for θ_0 is in fact the LMS estimate of location in one dimension. Like the LMS, both PLA and PRLA yield coefficients very near the true value, whereas the estimated error scale is somewhat smaller than 1 in the normal case and somewhat larger in the other situations.

Table 2 gives the variances of the estimated parameters over the 50 runs. It shows that the variances of PRLA are smaller than those of PLA. Also in this respect PLA and PRLA behave like the LMS.

4 Discussion

The breakdown point of the PLA method is about 1/3, which is lower than the 50% breakdown point of the Least Median of Squares estimator. However, the resampling method used to approximate the LMS is computationally intensive. For instance, for the salinity data, the program PROGRESS of Rousseeuw and Leroy (1987) would draw 2000 subsamples whereas PRLA uses only 100 subsamples and produces similar results. So because the projection method and the L_1 method are very fast, PRLA will be between 10 to 20 times faster than the LMS for a dataset of this dimension. The ratio will be even more in favor of PRLA when p increases. The algorithm PLA takes even less time

because there no resampling is involved at all.
From the simulation study by Rousseeuw and Leroy it was clear that the Reweighted Least Squares estimator, which uses the residuals from LMS to get 0-1 weights, is much more efficient than LMS itself. We suggest that one could also insert PLA or PRLA instead of LMS. Alternatively the PLA or PRLA could be followed by a one-step M-estimator. One could also think of a multi-step procedure in which the first step is PLA or PRLA, and the second step is LMS using selective resampling.
Finally, we would like to emphasize the usefulness of the diagnostic plot of standardized robust residuals versus robust **x**-distances (see figure 2). The PRLA proposal (which contains a cutoff for the **x**-outliers) produces this plot automatically, which in our opinion is an important advantage.

5 References

Gasko, M. and Donoho, D., *Influential Observation in Data Analysis*, in "American Statistical Assocation Proceedings of the Business and Economic Statistics Section", 104-109 (1982).

Prescott, P., *An approximate test for outliers in linear models*, Technometrics, 17, 129-132 (1975).

Rousseeuw, P. J., *Multivariate Estimation with High Breakdown Point*, paper presented at the Fourth Pannonian Symposium on Mathematical statistics and Probability, Bad Tatzmannsdorf, Austria, 1983. Appeared in "Mathematical Statistics and Applications, Volume B", eds. W. Grossmann, G. Pflug, I. Vincze, and W. Wertz, Dordrecht: Reidel Publishing Company, (1985).

Rousseeuw, P. J., *Least Median of Squares Regression*, J. Amer. Statist. Assoc., 79, 871-880 (1984).

Rousseeuw, P. J. and Leroy, A., "Robust Regression and Outlier Detection", New York: John Wiley (1987).

Rousseeuw, P. J. and van Zomeren, B. C., *Unmasking Multivariate Outliers and Leverage Points*, J. Amer. Statist. Assoc. (1990), to appear.

Ruppert, D., and Carroll, R. J., *Trimmed least squares estimation in the linear model*, J. Amer. Statist. Assoc., 75, 828-838 (1980).

Siegel, A. F., *Robust regression using repeated medians*, Biometrika, 69, 242-244 (1982).

Stahel, W. A., *Robuste Schätzungen: Infinitesimale Optimalität und Schätzungen von Kovarianzmatrizen*, Ph.D. Thesis, ETH Zürich (1981).

Robust Selection of Variables in the Discriminant Analysis Based on MVE and MCD Estimators

V. K. Todorov, N. M. Neykov and P. N. Neytchev, Sofia

Abstract: The paper presents a method for selecting the most discriminative variables in the discriminant analysis, using Wilks' lambda statistic, robustified by means of high breakdown point estimates of multivariate means and covariance matrices. Some examples are given for comparison of the method to the selection of variables based on classical and M-estimates.

Keywords: Discriminant analysis, Wilks' lambda statistic, High breakdown point estimators, Outliers.

Introduction. A common method for selection of the most discriminative variables in the discriminant analysis is based on the Wilks' lambda criterion defined as a ratio of two determinants:

$$\text{Wilks' lambda} = \frac{\det W}{\det T}, \tag{1}$$

where W is the within sums-of-crossproducts matrix and T is the total sums-of-crossproducts matrix. This statistic varies from zero to one and the smaller its value, the better the distinction between groups.

Krusinska and Liebhart (1989) proposed to robustify the Wilks' lambda statistic by substituting the classical estimates of W and T with their M-estimates as defined by Campbell (1980). However a disturbing aspect of the M-estimation approach is that the breakdown point is no more than 1/(p+1) (p is the number of variables) which is disappointingly low especially in the cases of higher dimensions as pointed out Hampel et al (1986). Therefore when a large fraction of outliers is expected we propose to compute the lambda statistic relative to the Minimum Volume Ellipsoid or Minimum Covariance Determinant estimates of location and scatter, introduced by Rousseeuw. For detailed consideration of this estimators see Rousseeuw and Leroy (1987).

Compstat 1990

Definitions and computation. For a data set of n points in p dimensions $\mathbf{X} = (x_1, x_2, \ldots x_n)$ the MVE and MCD estimators are defined as follows:

$M_{MVE}(\mathbf{X})$ = centre of the minimum volume ellipsoid covering (at least) h points of $\mathbf{X}$,

$M_{MCD}(\mathbf{X})$ = mean of the h points of $\mathbf{X}$ for which the determinant of the covariance matrix is minimal.

where h = [n/2] + 1. Both MVE and MCD yield a robust covariance estimate at the same time:

$C_{MVE}(\mathbf{X})$ = the minimum volume ellipsoid itself, multiplied by a constant to obtain consistency at multivariate normal data,

$C_{mcd}(\mathbf{X})$ = classical covariance matrix of the selected h points multiplied by a constant.

In practice approximate MVE and MCD estimates are computed, using the resampling algorithm proposed by Rousseeuw and Zomeren (1989). Random search is performed in order to minimize the objective function - the volume of the ellipsoid, covering at least h points or the determinant of the sample covariance matrix of h points respectively. The obtained estimates are used further as initial solutions on which a one-step improvement is based. The Mahalanobis distance $d_i^2 = (x_i-M(X))C^{-1}(x_i-M(X))$ with the above M and C is computed and a weight

$$w_i(d_i^2) = \begin{cases} 1 & \text{if } d_i^2 \le c \\ 0 & \text{elsewhere,} \end{cases} \tag{2}$$

is assigned to each observation x_i. The cutoff value c is taken equal to $\chi^2_{p,0.975}$.

Another way to improve the estimates is to perform one iteration of M-estimate, taking M and C as initial values. The weights are calculated by the function, discussed by Campbell (1980).

$$w_i(d_i) = \begin{cases} 1 & \text{if } d_i \le d_0 \\ d_0\exp\{-1/2(d_i - d_0)^2/b_2\}/d_i & \text{elsewhere,} \end{cases} \tag{3}$$

where $d_0 = \sqrt{p} + b_1\sqrt{2}$. With $b_1 = 2$ and $b_2 = \infty$ the non-descending form (Huber weights) is obtained and with $b_1 = 2$ and $b_2 = 1.25$ - the redescending one (Hampel weights).

The reweighted estimates are computed as

$$M_1(X) = (\sum_{i=1}^{n} w_i x_i)/(\sum_{i=1}^{n} w_i)$$
$$C_1(X) = (\sum_{i=1}^{n} w_i^2 (x_i - \bar{x})(x_i - \bar{x})^T) \; / \; (\sum_{i=1}^{n} w_i^2 - 1) \qquad (4)$$

where w_i is taken either from (2) or form (3).

Robust mean and covariance matrix are computed for each group, and the obtained weights are used to calculate W and T in the following way:

$$N_k = (\sum_{m=1}^{n_k} w_{km}), \quad \bar{x}_k = (\sum_{m=1}^{n_k} w_{km} x_{km})/N_k, \quad \bar{x} = (\sum_{k=1}^{g} N_k \bar{x}_k)/\sum_{k=1}^{g} N_k$$
$$W = \sum_{k=1}^{g} \sum_{m=1}^{n_k} w_{km}^2 (x_{km} - \bar{x}_k)(x_{km} - \bar{x}_k)^T, \quad B = \sum_{k=1}^{g} N_k (\bar{x}_k - \bar{x})(\bar{x}_k - \bar{x})^T \qquad (5)$$

where g is the number of groups, B is the between-groups sums-of-crossproducts matrix and T = W + B.

When the determinants of W and T in (1) are replaced by their robust analogues described above, the robust version of the Wilks' lambda statistic is obtained.

Examples. The robust variable selection was tested on some artificial data sets generated on the bases of the Fisher's Iris data (3 groups with 50 observations in each in 4 variables). Outliers were introduced by substituting some of the original values. In Table 1 the configurations of outliers for the different examples are presented (the original observations and the ones they are substituted with). The following methods for variable selection were considered:

CLAS: Classical selection;

MHUB: Robust selection with M-estimates (Huber weights). The estimates of means and covariance matrices of the groups are computed iteratively by (4), using the weighting function (3) with b_1 = 2 and $b_2 = \infty$. The convergence criterion is $\|C_{k+1} - C_k\| \leq 0.0001$, where C_k is the covariance matrix at the k-th iteration and $\| C \| = (\Sigma_{ij} c_{ij}^2)^{1/2}$ for a p x p matrix C.

MHAM: Robust selection with M-estimates (Hampel weights). The same as MHUB, but b_1 = 2 and b_2 = 1.25;

MCDO and **MCDM**: Robust selection with MCD estimates, improved by one iteration with weights (2) and (3) respectively;

MVEO and **MVEM**: Robust selection with MVE estimates, improved by one iteration with weights (2) and (3) respectively;

The actual computations are carried out by a FORTRAN 77 program. The results are given in the tables below. For each method the order in which the variables are entered is given together with the value of Wilks' lambda statistic obtained after entering the corresponding variable. The lower part of each table contains the selections with high breakdown point estimates, obtained with fewer (500) iterations (1000 for the main part). The order of entering variables produced by the classical method over the original data (supposed without outliers) serves as a base for comparing the performance of the robust selection methods.

Example 1. Original data. All methods behave well when applied to the original data set, although some information is lost when MCDO and MVEO are used (Table 2).

Example 2. The presence of only one drastic outlier may distort the classical estimate of the Wilks' lambda statistic as is seen in Table 3. All robust methods deal with the case almost equally.

Example 3. Here several outliers are inserted. The classical method and the M-estimate with Huber weights break down but the rest (using variants of MVE and MCD estimates) select the variables properly (Table 4).

Example 4. In the presence of large fraction of "severe" outliers both M-estimates break down and the order of entering variables has nothing to do with their real importance. The selections based on MVE and MCD still retain their properties (Table 5).

Conclusions. The above simple examples show that robustified by means of high breakdown point estimates Wilks' lambda, used over contaminated data yields the same order of entering variables as the classical one used over the original data. The performance of the M-Wilks' lambda is not better than that of the classical when the level of contamination is high. The main drawback of the estimators with high breakdown point however is that they are time consuming.

Table 1. Original data and outlier configurations

Group	Original X1	Original X2	Original X3	Original X4	Substituted with X1	Substituted with X2	Substituted with X3	Substituted with X4
	Example 2							
I. Setosa	5.0	3.3	1.4	0.2	5.0	3.3	8.0	0.2
	Example 3							
I. Setosa	5.0	3.3	1.4	0.2	5.0	3.3	4.4	0.2
.	5.1	3.8	1.6	0.2	1.1	3.8	1.6	0.2
.	5.4	3.4	1.7	0.2	5.4	3.4	3.7	0.2
.	4.9	3.1	1.5	0.2	4.9	3.1	4.5	0.2
II.Versicolor	6.0	3.4	4.5	1.6	6.0	3.4	4.5	0.6
.	5.6	3.0	4.5	1.5	5.6	3.0	2.5	1.5
.	6.0	2.9	4.5	1.5	6.0	2.9	2.5	1.5
.	5.5	2.5	4.0	1.3	5.5	2.5	4.0	0.9
III. Virginica	7.4	2.8	6.1	1.9	7.4	2.8	3.1	1.9
	Example 4							
I. Setosa	5.0	3.3	1.4	0.2	5.0	3.3	14.	0.2
.	5.1	3.7	1.5	0.4	5.1	3.7	15.	0.4
.	5.2	3.5	1.5	0.2	5.2	3.5	15.	0.2
.	5.4	3.9	1.3	0.4	54.	3.9	1.3	0.4
.	5.0	3.4	1.6	0.4	5.0	3.4	16.	0.4
.	5.4	3.9	1.7	0.4	5.4	3.9	17.	0.4

Table 2.

	CLAS	MHUB	MHAM	MCDO	MCDM	MVEO	MVEM
1	X3 .05863	X3 .05839	X3 .05819	X3 .04241	X3 .05356	X3 .05419	X3 .05723
2	X2 .03688	X2 .03654	X2 .03628	X2 .02847	X2 .03386	X2 .02908	X2 .03339
3	X4 .02498	X4 .02471	X4 .02454	X4 .02015	X4 .02278	X4 .01818	X4 .02020
4	X1 .02344	X1 .02324	X1 .02311	X1 .01929	X1 .02146	X1 .01679	X1 .02056
				X3 .04163	X3 .05181	X3 .04376	X3 .05281
				X2 .02526	X2 .02952	X2 .02229	X2 .03089
				X4 .01477	X4 .01817	X4 .01450	X4 .02073
				X1 .01434	X1 .01752	X1 .01398	X1 .01940

Table 3.

	CLAS	MHUB	MHAM	MCDO	MCDM	MVEO	MVEM
1	X4 .07112	X3 .06792	X3 .05889	X3 .04464	X3 .05573	X3 .05426	X3 .05699
2	X2 .03832	X2 .04350	X2 .03670	X2 .02722	X2 .03296	X2 .03002	X2 .03360
3	X1 .03343	X4 .02675	X4 .02482	X4 .01764	X4 .02090	X4 .01884	X4 .02243
4	X3 .03257	X1 .02620	X1 .02337	X1 .01717	X1 .02006	X1 .01746	X1 .02095
				X3 .04232	X3 .05258	X3 .04384	X3 .05285
				X2 .02569	X2 .02991	X2 .02224	X2 .03097
				X4 .01502	X4 .01842	X4 .01422	X4 .02077
				X1 .01458	X1 .01776	X1 .01374	X1 .01951

Table 4.

	CLAS	MHUB	MHAM	MCDO	MCDM	MVEO	MVEM
1	X4 .07773	X4 .07720	X3 .06027	X3 .04369	X3 .05717	X3 .04314	X3 .06126
2	X2 .04523	X2 .04294	X2 .03646	X2 .02515	X2 .03351	X2 .02244	X2 .03495
3	X3 .03689	X3 .03222	X4 .02472	X4 .01155	X4 .02178	X4 .01351	X4 .02065
4	X1 .03640	X1 .03111	X1 .02311	X1 .01469	X1 .02061	X1 .01280	X1 .01975
				X3 .04478	X3 .05698	X3 .05374	X3 .05644
				X2 .02606	X2 .03420	X2 .03040	X2 .03269
				X4 .01660	X4 .02238	X4 .01917	X4 .02235
				X1 .01610	X1 .02125	X1 .01829	X1 .02086

Table 5.

	CLAS	MHUB	MHAM	MCDO	MCDM	MVEO	MVEM
1	X4 .07773	X4 .07720	X4 .06027	X3 .04369	X3 .05717	X3 .04314	X3 .06126
2	X2 .04523	X2 .04294	X2 .03646	X2 .02515	X2 .03351	X2 .02244	X2 .03495
3	X1 .03689	X1 .03222	X1 .02472	X4 .01155	X4 .02178	X4 .01351	X4 .02065
4	X3 .03640	X3 .03111	X3 .02311	X1 .01469	X1 .02061	X1 .01280	X1 .01975
				X3 .03945	X3 .05455	X3 .04667	X3 .05601
				X2 .02553	X2 .03490	X2 .02370	X2 .03305
				X4 .01662	X4 .02296	X4 .01512	X4 .02214
				X1 .01636	X1 .02188	X1 .01456	X1 .02075

REFERENCES:

Campbell,N.A.(1980), Robust Procedures in Multivaria Analysis.I.: Robust Covariance Estimation, App. Statist. **29**, 231-237

Hampel,F.R.,Ronchetti,E.M.,Rousseeuw,P.J. & Stahel,W.A.(1986 Robust statistics: The infinitesimal approach, Jol Wiley&Sons, New York.

Krusinska,E. and Liebhart, J.(1989), Some further Remarks Robust Selection of Variables in Discriminant Analysi Biom. J., **31**, 2, 227-233.

Rousseeuw,P.J. and Leroy,A.(1987), Robust Regression a Outlier Detection, John Wiley&Sons, New York.

Rousseeuw,P.J. and Zomeren,B.C.(1989),Unmasking Multivaria Outliers and Leverage Points by Means of Robu Covariance Matrices, to appear in JASA.

Interactively Computing Robust Covariance Matrices

G. Karnel, Wien

Abstract

Covariance matrices may be seen as the basis of many mutivariate methods, like canonical correlation analysis, principal component analysis, factor analysis, to mention a few. Because of the high sensibility of covariance matrices to outliers it is very important to use robust estimates instead of the classical ones. In this paper we consider two types of estimators: M-estimators and high-breakdown-point-estimators. For the easy handling of these estimators it seems very advantageously to work within an interactive environment which guarantees high flexibility. The programs we refer in this paper are written in APL and have been integrated in an interactive program package called **GRIPS** (**G**eneral and **R**obust **I**nteractive **P**rograms for **S**tatistics, see Karnel, 1989a).

Robust Estimation of Location and Covariance Matrices

One way of robustly estimating multivariate location and covariance matrices which is integrated in the program package **GRIPS** is to use M-estimators. M-estimators have been derived by Maronna (1976) and Huber (1981).

Suppose we have a data set z_{ji} with $j = 1, \ldots, p$ and $i = 1, \ldots, n$, where p denotes the number of variables, n the number of observations and $\mathbf{z}_i$ denotes the i-th column of the matrix Z. We now try to find a transformation $(A, \mathbf{b})$ where A is a $(p \times p)$-matrix and $\mathbf{b}$ is a $(p \times 1)$-vector. $\mathbf{z}_i$ should be transformed to

$$\mathbf{x}_i = A(\mathbf{z}_i - \mathbf{b}) \qquad i = 1, \ldots, n,$$

such that

$$\underset{i}{ave} \; \{w(\|\mathbf{x}_i\|)\mathbf{x}_i\} = \mathbf{0},$$

and

$$\underset{i}{ave} \; \{u(\|\mathbf{x}_i\|)\frac{\mathbf{x}_i\mathbf{x}_i^T}{\|\mathbf{x}_i\|^2} - v(\|\mathbf{x}_i\|)I\} = 0,$$

where u, v and w are proper weight functions. A robust estimate of location is then defined by $\mathbf{b}$ and a pseudo-covariance matrix by

$$C = (A^T A)^{-1}.$$

For a detailed description see Karnel (1989b).

As Stahel (1981) has shown, the breakdown point of all affine equivariant M-estimators is at most $1/(p+1)$, that is, with increasing dimensions the breakdown point ε^* decreases quickly. In order to provide a higher ε^* we use an estimator developed by Rousseeuw (1987), the minimum volume ellipsoid estimator. We try to find the ellipsoid covering $[n/2] + 1$ observations which has the minimum volume. Unfortunately it does not seem feasible to take all possible subsets with size $[n/2] + 1$ even for relatively small n because of the exploding number of subsamples. Therefore Rousseeuw and Leroy (1987) suggest to draw subsamples of size $(p+1)$ and to compute the corresponding covariance matrix. Then, one has to inflate the radius of the corresponding ellipsoid in order to contain exactly $[n/2]+1$ observations and compute the volume of this ellipsoid. This is done for all possible subsets of size $(p+1)$ and thus the minimum volume ellipsoid is found. The center of this ellipsoid is the location estimate and the covariance estimate is given by the ellipsoid itself after applying a correction factor for consistency. The breakdown point of this estimator tends to 0.5 with $n \to \infty$. As one can easily see ε^* does not depend on the dimensionality. Unfortunately, the computing time increases drastically with increasing p and n. A possible way to decrease the computation time is to take enough subsamples to provide a high probability of getting at least one out of m subsamples which consists exclusively of 'good' observations. This probability equals

$$q = 1 - (1 - (1 - \varepsilon)^{p+1})^m$$

with $0 \leq \varepsilon < ([n/2] - p + 1)/n$, and n being large in comparison with p. For small n this probability equals

$$q = 1 - (1 - \frac{\binom{n-n\varepsilon}{p+1}}{\binom{n}{p+1}})^m.$$

Thus, one can easily compute the necessary number of subsamples to obtain a given probability. In order to stay within reasonable computing time it is possible to fix the number of subsamples and to work with a smaller probability of getting at least one subsample consisting exclusively of 'good' observations. In the present version of our program package **GRIPS** m has a maximum value of 3000. Each time the program is called with a fixed m the probability q will be displayed.

Furthermore, the minimum volume estimator may used as initial solution for performing an one-step reweighting to increase its statistical efficiency.

The weights can be found in the following way:

$$w_i = \begin{cases} 1 & \text{if} \quad (\mathbf{x}_i - T(X))^T C(X)^{-1}(\mathbf{x}_i - T(X)) \leq c \\ 0 & \text{otherwise} \end{cases}$$

where c might be taken equal to $\chi^2_{p,0.975}$. Then the estimators look like follows

$$T_w(X) = \frac{\sum_{i=1}^{n} w_i \mathbf{x}_i}{\sum_{i=1}^{n} w_i}$$

$$C_w(X) = \frac{\sum_{i=1}^{n} w_i(\mathbf{x}_i - T_w(X))(\mathbf{x}_i - T_w(X))^T}{\sum_{i=1}^{n} w_i - 1}.$$

That means one can carry out the classical computations for the subset of the data with weight $w_i = 1$.

APL Programs

The methods described above are included in a program package called **GRIPS** (**G**eneral and **R**obust **I**nteractive **P**rograms for **S**tatistics).

The most important characteristics of **GRIPS** are

- Interactivity
- Extensibility
- Portability
- Simpilicity of usage

Several setup- and help-functions facilitate the usage of **GRIPS**. For a detailed description see Karnel (1988, 1989a,b).

The programs are written in APL which provides a kind of natural interactive environment. This makes it possible to store any results for further usage. For example, the problem of canonical correlation analysis may be expressed in terms of covariance matrices. The robustification may be done very comfortably in **GRIPS** by first computing the robust covariance matrix with the help of one of the methods described above and then using it as a basis for the canonical correlation analysis. For a worked-out example see Karnel (1990). In the current version of the package robust covariances may also be used as a basis for correspondence analysis (see Karnel, 1990), or for biplots (see Dutter, 1987).

The programs are available on request from the author.

References

R. Dutter (1987): Robust Statistical Methods in the Analysis of Geochemical Variables. In: Contributions to Stochastics, W. Sendler (ed), Physica Verlag, Heidelberg.

P.J. Huber (1981): *Robust Statistics.* J. Wiley & Sons, New York.

G. Karnel (1988): GPSP: A General Purpose Statistical Package. In: COMPSTAT-88, Short Communications and Posters, Physica Verlag, Vienna.

G. Karnel (1989a): A General Purpose Statistical Package with Emphasis on Robust Methods. Ph.D. Thesis, T.U. Vienna.

G. Karnel (1989b): Interactively Computing Robust Covariance Matrices with GRIPS. Res. rep. TS-89-2.

G. Karnel (1990): Robust Canonical Correlation and Correspondence Analysis. To appear in "The Proceedings of the First International Conference on Statistical Computing", Cesme, Turkey.

R.A. Maronna (1976): Robust M-estimators of Multivariate Location and Scatter, Ann. Stat., **4**, 51-67.

P.J. Rousseeuw, A.M. Leroy (1987): *Robust Regression and Outlier Detection.* J. Wiley & Sons, New York.

W.A. Stahel (1981): Robuste Schätzungen. Infinitesimale Optimalität und Schätzungen von Kovarianzmatrizen. Ph.D. Thesis, ETH Zürich.

Sensitivity Analysis in Factor Analysis: Methods and Software

Y. Tanaka, E. Castaño-Tostado and Y. Odaka, Okayama

1. Introduction

The present paper deals with methods and software of sensitivity analysis in some procedures of exploratory factor analysis. Our aim is to investigate how a small change of data affects the results of analysis. To do this Tanaka and Odaka(1989a,b,c) proposed methods of sensitivity analysis in principal factor analysis(PFA), maximum likelihood factor analysis(MLFA) and least squares factor analysis(LSFA), respectively. The major mathematical tools are influence functions for some functions of eigenvalues and eigenvectors of a real symmetric matrix. We treat the influence of omitting or downweighting one or more individuals on the estimates of the unique and common variance matrices $\widehat{\Delta}$ and $\widehat{T}^*$, the precision of $\widehat{\Delta}$, and the goodness of fit.

2. Influence functions for the unique and common variance matrices

We assume an ordinary factor analysis model: $\mathbf{x} = \mu + L\mathbf{f} + \mathbf{e}$. Then, it is well-known that the relation $\Sigma = LL^T + \Delta$ holds, where Σ is the $p \times p$ variance matrix of the manifest random vector $\mathbf{x}$, L is a $p \times q$ loading matrix($q < p$) of the common factor score vector $\mathbf{f}$, and Δ is a $p \times p$ (diagonal) variance matrix of the unique factor score vector $\mathbf{e}$. We wish to evaluate the influence of a small change of data on the results of factor analysis, in particular, on the values of the unique variance matrix Δ and the common variance matrix $T^* = LL^T$. For this purpose we substitute the perturbed quantities expressed in power series expansions into the determinantal equations of exploratory factor analysis procedures and then compare the coefficients of both sides of the equations. From the coefficients of the first order terms of the perturbation parameter ε we get the quantities $\Delta^{(1)}$ and $T^{*(1)} = (LL^T)^{(1)}$, where the superscript (1) indicates the first derivative. When a perturbation is introduced on the cdf as $F \mapsto \widetilde{F} = (1-\varepsilon)F + \varepsilon\delta_{\mathbf{x}}$ (F: cdf, $\delta_{\mathbf{x}}$: cdf of a unit point of mass at $\mathbf{x}$), the first derivative of a functional of $\widetilde{F}$ with respect to ε gives the so-called influence function (Hampel,1974). In the case of MLFA, the influence functions $\Delta^{(1)}$ and $T^{*(1)}$ are obtained by the following four steps(Tanaka and Odaka, 1989b).

Step 1. Calculate $R^{*(1)}_{jk}$'s for $j \neq k$ by $R^{*(1)}_{jk} = \Sigma^{(1)}_{jk}$.

Step 2. Solve linear simultaneous equations with p unknowns as

$$d_j R^{*(1)}_{jj} - \sum_{i=1}^{p} a_{(j)i} R^{*(1)}_{ii} = \sum_{i=1}^{p} \sum_{\substack{i'=1 \\ i \neq i'}}^{p} b_{(j)ii'} R^{*(1)}_{ii'} + C_{jj}, \quad j = 1, \cdots, p. \tag{1}$$

Compstat 1990

Step 3. Calculate $\Delta_{jj}^{(1)}$'s by $\Delta_{jj}^{(1)} = \Sigma_{jj}^{(1)} - R_{jj}^{*(1)}, j = 1, \cdots, p$.

Step 4. Calculate $T_{jk}^{*(1)}$'s for $j \neq k$ by the following equations.

$$T_{jk}^{*(1)} = \sum_{i=1}^{p} c_{(jk)i} R_{ii}^{*(1)} + \sum_{\substack{i=1 \\ i \neq i'}}^{p} \sum_{i'=1}^{p} d_{(jk)ii'} R_{ii'}^{*(1)}, \quad j, k = 1, \cdots, p; j \neq k. \tag{2}$$

The coefficients in equations (1) and (2), $\{d_j\}, \{a_{(j)i}\}, \{b_{(j)ii'}\}, \{c_{(jk)i}\}$ and $\{d_{(jk)ii'}\}$, are quantities obtained from eigenvalues and eigenvectors of $\Delta^{-1/2}(\Sigma - \Delta)\Delta^{-1/2}$, and C_{jj} is the (j,j)-*th* element of a matrix C which is a linear function of the diagonal part of $\Sigma^{(1)} = (\Sigma_{jk}^{(1)})$ representing the influence function for Σ, which is expressed (Radhakrishnan and Kshirsagar, 1981) as

$$\Sigma^{(1)} = (\mathbf{x} - \mu)(\mathbf{x} - \mu)^T - \Sigma. \tag{3}$$

The above four steps show the procedure to obtain the theoretical influence functions. In actual data analysis, however, we have to use sample versions. Among possible sample versions we use the so-called empirical influence curves*(EIC)*, $\widehat{\Delta}^{(1)}$ and $\widehat{T}^{*(1)}$, which are obtained by the above procedure with the population parameters replaced by their estimates. As shown by Tanaka and Odaka(1989a,c), similar procedures can be derived for other exploratory factor analysis techniques such as PFA and LSFA.

With the procedure explained above we only need to solve a system of linear equations with p unknowns $(n+1)$ times(one for the whole data and n for the data without one individual) instead of applying a factor analysis procedure $(n+1)$ times. Moreover, as discussed by Tanaka and Castaño-Tostado(1990) it is sufficient to solve the system of linear equations $(1+p(p+1)/2)$ times instead of $(n+1)$ times.

The procedures explained so far are based on linear perturbation expansions of some functions of eigenvalues and eigenvectors. Those linear expansions were derived by Tanaka(1988). We can use quadratic expansions to improve the approximation. To do this we must solve additionally another system of linear equations with p unknowns.

3. Measures of influence

After the omission or downweighting of one or more individuals we quantify influence on the estimates $\widehat{\Delta}$ and $\widehat{T}^*$, the precision of $\widehat{\Delta}$ and the goodness of fit.

(a) Influence on the estimates. The influence on the estimates for Δ and T^* can be evaluated by $\widehat{\Delta}^{(1)}$ and $\widehat{T}^{*(1)}$, respectively. As scalar measures we define: 1) $\|\widehat{\Delta}^{(1)}\|$ (Euclidean or Frobenius norm, see Golub and Van Loan, 1990), 2) $D_{us}(\widehat{\Delta}^{(1)}) = \widehat{\Delta}^{(1)T}[\widehat{V}_D(\widehat{\Delta})]^{-1}\widehat{\Delta}^{(1)}$ (univariately standardized squared distance), 3) $D_{ms}(\widehat{\Delta}^{(1)}) = \widehat{\Delta}^{(1)T}[\widehat{V}(\widehat{\Delta})]^{-1}\widehat{\Delta}^{(1)}$ (multivariately standardized squared distance). The matrices $\widehat{V}(\widehat{\Delta})$ and $\widehat{V}_D(\widehat{\Delta})$ are the estimated asymptotic variance matrix of $\widehat{\Delta}$ and a diagonal matrix whose diagonal part consists of the diagonal part of $\widehat{V}(\widehat{\Delta})$, respectively.

(b) Influence on the precision of $\widehat{\Delta}$. Here we consider the case of MLFA. Let ψ_{jl} be the asymptotic covariance between $\widehat{\Delta}_{jj}$ and $\widehat{\Delta}_{ll}$, and $\Psi = (\psi_{jl})$. Then, Ψ is given by

$$\Psi = 2(n-1)^{-1}\Xi^{-1}, \quad \Xi = (\xi_{jl}), \quad \xi_{jl} = \phi_{jl}^2, \quad \Phi = (\phi_{jl}) = \Delta^{-1} - \Delta^{-1/2}V_1V_1^T\Delta^{-1/2}, \quad (4)$$

where the column vectors of $V = (\mathbf{v}_1, \cdots, \mathbf{v}_q)$ are the eigenvectors associated with the largest q eigenvalues $\lambda_1, \cdots, \lambda_q$ of $\Delta^{-1/2}(\Sigma - \Delta)\Delta^{-1/2}$. The first derivative $\Phi^{(1)}$ is easily obtained by using the relation

$$(V_1V_1^T)^{(1)} = \sum_{s=1}^{q}\sum_{r=q+1}^{p}(\lambda_s - \lambda_r)^{-1}(\mathbf{v}_s^T R^{(1)}\mathbf{v}_r)(\mathbf{v}_s\mathbf{v}_r^T + \mathbf{v}_r\mathbf{v}_s^T), \quad (5)$$

where $R^{(1)}$ is expressed as a function of Σ, Δ, $\Sigma^{(1)}$ and $\Delta^{(1)}$. Thus we have a linear approximation

$$\Phi_{(i)} \cong \tilde{\Phi}_{(i)} = \Phi - (n-1)^{-1}\Phi^{(1)}(\mathbf{x}_i), \quad (6)$$

where the subscript (i) indicates the omission of the ith observation and $\Phi^{(1)}(\mathbf{x}_i)$ is the value of $\Phi^{(1)}$ at a specified data point $\mathbf{x}_i$. Note that $\tilde{\Phi}^{(1)}$ is a function of $\Sigma^{(1)}$ and $\Delta^{(1)}$. Based on this an approximate value $\tilde{\Psi}_{(i)}$ is obtained by substituting $\tilde{\Phi}_{(i)}$ into (4). We define a measure of influence on the precision as follows:

$$\widetilde{CVR} = det(\tilde{\Psi}_{(i)})/det(\Psi). \quad (7)$$

(c) Influence on the goodness of fit. The likelihood ratio test statistic for the goodness of fit is given by

$$X^2 = nlog[det(\hat{\Delta} + \hat{T}^*)] - nlog[det(S)], \quad (8)$$

where S is the maximum likelihood estimate for Σ. The statistic X^2 follows asymptotically a chi-squared distribution with $(1/2)[(p-q)^2 - p - q]$ degrees of freedom under the null hypothesis that the model fits to the data(see, e.g. Anderson, 1984). An approximate influence function $(\tilde{X}^2)^{(1)}$ can be defined as

$$(\tilde{X}^2)^{(1)} = -(n-1)(\tilde{X}^2_{(i)} - X^2), \quad (9)$$

where $\tilde{X}^2_{(i)}$ is the quantity defined by (8) with $\hat{\Delta}$, $\hat{T}^*$ and S replaced by $\tilde{\Delta}_{(i)}$, $\tilde{T}^*_{(i)}$ and $S_{(i)}$ respectively, $\tilde{\Delta}_{(i)}$ and $\tilde{T}^*_{(i)}$ indicating linear approximations for $\hat{\Delta}_{(i)}$ and $\hat{T}^*_{(i)}$ based on the first derivatives $\hat{\Delta}^{(1)}$ and $\hat{T}^{*(1)}$, respectively.

4. Statistical software SAF/B

We have developed the first version of statistical software named SAF/B for Sensitivity Analysis in Factor analysis, where B indicates "BASIC", based on the methods proposed by Tanaka and Odaka(1989a,b,c). It was developed on NEC PC-9801 personal computers by using Microsoft Quick Basic. The program consists of five parts : (1)data entry and preliminary analysis, (2)estimation of L and Δ, (3)sensitivity analysis(i), (4)sensitivity analysis(ii) and (5)sensitivity analysis(iii). We shall explain the sensitivity analysis parts precisely.

Sensitivity analysis(i): This part treats the so-called single-case diagnostics. For a specified factor analysis procedure the empirical influence curves $\hat{\Delta}^{(1)}$ and $\hat{T}^{*(1)}$ and the measures of influence are computed by the procedures explained in sections 2 and 3 . The results can be displayed in graphical forms such as index plots and scatter diagrams.

Sensitivity analysis (ii): This part treats the so-called multiple-case diagnostics. The basic idea is to gather observations which have similar influence patterns with large values of influence, because of the fact that the influence of a set of observations is equal to the sum of the influence of each one belonging to the set, so far as the influence function (or equivalently the first order perturbation expansion) is used to measure the influence. For this purpose principal component analysis(PCA) and canonical variate analysis(CVA) provide useful information. PCA is applied to the data set of $(\widehat{\Delta}_{11}^{(1)}, \cdots, \widehat{\Delta}_{pp}^{(1)})$ or $(\widehat{\Delta}_{11}^{(1)}/SE(\widehat{\Delta}_{11}), \cdots, \widehat{\Delta}_{pp}^{(1)}/SE(\widehat{\Delta}_{pp}))$ to reduce the dimension, preserving the sum of the squared Euclidean distances of all possible pairs as far as possible. CVA is also applied to the data set of $(\widehat{\Delta}_{11}^{(1)}, \cdots, \widehat{\Delta}_{pp}^{(1)})$ to reduce the dimension, preserving the sum of the squared distances normalized with $\widehat{V}(\widehat{\Delta})$ in the sense of Mahalanobis distance as far as possible.

Sensitivity analysis(iii): Factor analysis is again applied to the data without some specified observations("influential observations") and the changes of the results from those for the whole data are given in the forms of tables and figures. The validity of the proposed measures can be checked by investigating partly or wholly the relationships with the sample influence curves *SIC*, which are defined as $-(n-1)$ times the differences between the estimates for the sample without the ith observation and those for the whole sample.

5. Example

As an illustration a data set taken from Mardia, Kent and Bibby(1979; the open/closed book data, Table 1.2.1) was analyzed with our software. The set of data consists of 88 individuals(students) and 5 variables(tests). First, the ordinary MLFA was applied to the data set, assuming a two-factor model. The estimated L and Δ are given in Table 1a along with the goodness of fit statistic X^2.

Then, "sensitivity analysis(i)" was performed. Fig.1 provides the index plots of the influence measures $D_{ms}(\widehat{\Delta}^{(1)})$, $\widetilde{CVR}$ and $(\widetilde{X}^2)^{(1)}$.

These index plots show that the sets of individuals $\{54, 56, 75, 82\}$, $\{33, 56, 57, 81, 82\}$ and $\{36, 54, 88\}$ are influential on the estimate $\widehat{\Delta}$, its precision and the goodness of fit statistic X^2, respectively. Table 1b shows the estimated L and Δ for the data set without the 56th individual, which is the most influential on $\widehat{\Delta}$. We can find that the estimates of the unique variances differ considerably from those for the whole data especially for variables 3, 4 and 5. Among the individuals numbered 54, 56, 75 and 82, which are influential on the estimate $\widehat{\Delta}$, $\{54, 56\}$ and $\{75, 82\}$ have different effects on the precision of $\widehat{\Delta}$ as well as the goodness of fit. If 54 or 56 is omitted, the goodness of fit and the precision of $\widehat{\Delta}$ become better. On the other hand, omitting 75 or 82 the goodness of fit and the precision become worse.

In the part of "sensitivity analysis(ii)" CVA was applied to $\widehat{\Delta}^{(1)}$'s. The eigenvalues and their cumulative proportions are, in order of magnitude, 126.92(32.38), 97.40(57.06), 67.32(74.18), 56.96(88.67), 44.55. Fig.2 is the scatter diagram of the first two canonical vari-

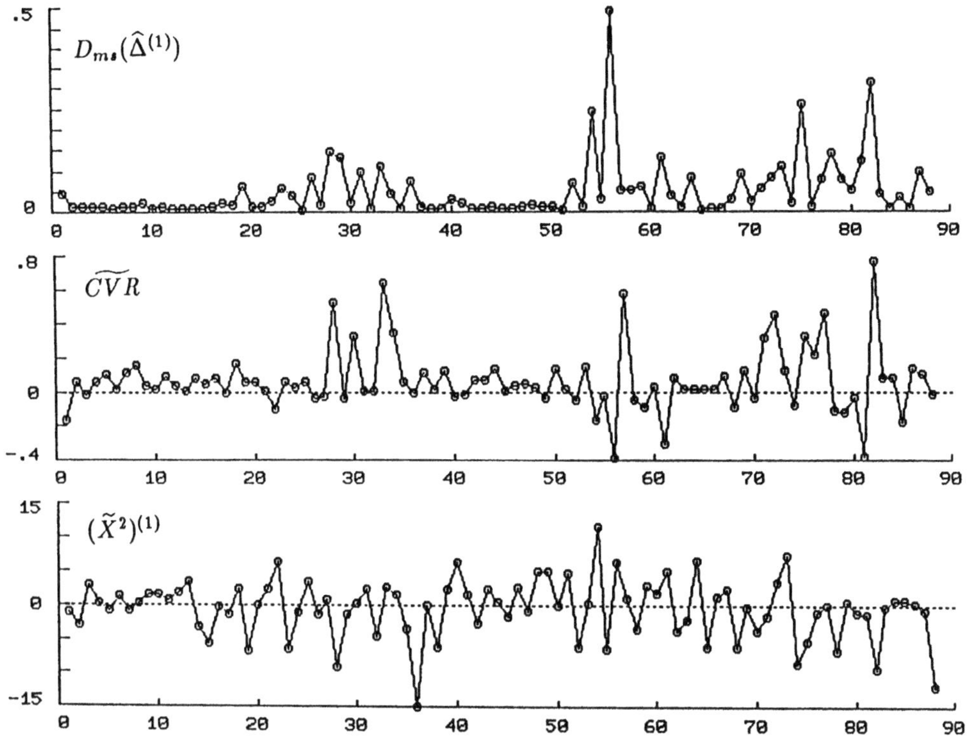

Fig 1. Index plots of $D_{ms}(\widehat{\Delta}^{(1)})$, $\widetilde{CVR}$ and $(\widetilde{X}^2)^{(1)}$.

ates. In this figure clusters $\{75, 82\}$ and $\{54, 56\}$ are clear. So these two sets of individuals will be candidates for influential subsets.

Table 1. Results of the MLFA (The open/closed book data, two-factor model)
(a): the whole data set($n = 88$), (b): the data set without 56 th individual ($n = 87$)

	(a)			(b)		
Var.	Factor loadings		Unique variances(SE)	Factor loadings		Unique variances(SE)
1	0.2700	0.6791	0.4659(0.2006)	0.2680	0.6734	0.4747(0.1694)
2	0.3603	0.6716	0.4190(0.1633)	0.3446	0.6869	0.4095(0.4095)
3	0.7429	0.5094	0.1886(0.0602)	0.7205	0.5258	0.2045(0.0571)
4	0.7403	0.3166	0.3518(0.0869)	0.7662	0.3172	0.3123(0.0860)
5	0.6981	0.2856	0.4310(0.0907)	0.7204	0.2921	0.3957(0.0881)

$X^2 = 0.0791, df = 1$ $\qquad$ $X^2 = 0.0028, df = 1$

Finally in the part of "sensitivity analysis(iii)", to check the validity of the proposed measures, $D(\widehat{\Delta}^{(1)})_{ms}$, $\widetilde{CVR}$ and $(\widetilde{X}^2)^{(1)}$, those quantities were plotted against the corresponding SIC's. The correlations are so high that we can use the proposed influence measures to detect influential observations in practice.

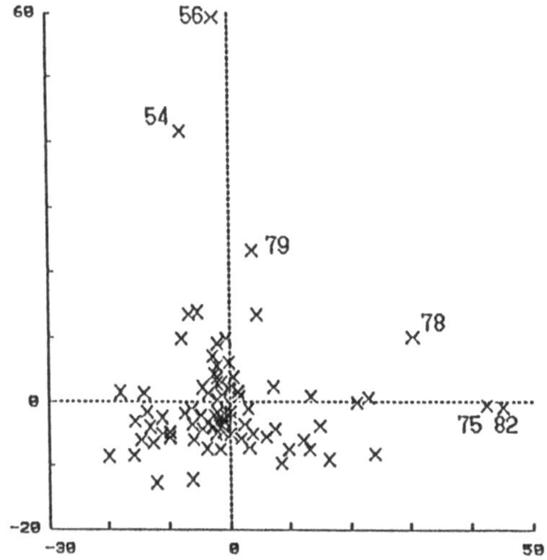

Fig. 2. Scatter diagram of the first two canonical variates.

REFERENCES

Anderson, T.W. (1984). *An Introduction to Multivariate Statistical Analysis.* 2nd Edition. John Wiley.

Golub, G.H. and Van Loan, C.F. (1990). *Matrix Computations.* 2nd edition. The John Hopkins University Press.

Hampel, F. R. (1974). The influence curve and its role in robust estimation. *J. Amer. Statist. Assoc.*, **69**, 383-93.

Mardia, K.V., Kent, J.T. and Bibby, J.M.(1979). *Multivariate Analysis.* Academic Press.

Radhakrishnan, R. and Kshirsagar, A.M. (1981). Influence functions for certain parameters in multivariate analysis. *Comm. Statist.*, **A 10**, 515-29.

Tanaka, Y.(1988). Sensitivity analysis in principal component analysis : Influence on the subspace spanned by principal components. *Comm. Statist.*, **A17**, 3157-3175. (Corrections, **A18** (1989), 4305).

Tanaka, Y. and Castaño-Tostado, E.(1990). Sensitivity analysis in multivariate methods : Decomposition of an arbitrary influence into a finite number of components. *Comm. Statist.*, **A19**, No.4.

Tanaka, Y. and Odaka, Y.(1989a). Influential observations in principal factor analysis. *Psychometrika*, **54**, 475-485.

Tanaka, Y. and Odaka, Y.(1989b). Sensitivity analysis in maximum likelihood factor analysis. *Comm. Statist.*, **A18**, 4067-4084.

Tanaka, Y. and Odaka, Y.(1989c). Sensitivity analysis in least squares factor analysis. *Data Analysis, Learning Symbolic and Numeric Knowledge*(Edited by E. Diday), Nova Science Publishers, 141-148.

Key words: influence function, eigenvalue problem, influence on the unique variance matrix, exploratory factor analysis, maximum likelihood method.

Influence Functions of Eigenvalues and Eigenvectors in Multidimensional Data Analysis

M. Romanazzi, Venice

1. PERTURBATION OF GENERALIZED EIGENPROBLEMS.

Influence functions [2] of the most important parameters in principal component analysis (PCA), e.g., eigenvalues, eigenvectors and projection operators, have been derived by a number of authors [1,3,7] using results from the perturbation theory of the ordinary eigenproblem. In the present paper we show that the perturbation theory of generalized eigenproblems [4] underlies and unifies the treatment of influence functions of eigenvalues and eigenvectors in multidimensional data analysis and we present new applications in canonical variate (CVA) and canonical correlation analysis (CCA).

Let $Q(\varepsilon)$ and $M(\varepsilon)$ be $(p\mathrm{x}p)$ matrices whose entries are real-valued functions of a real variable ε, so that $Q(\varepsilon)$ is symmetric and $M(\varepsilon)$ is symmetric and positive definite. A generalized eigenvalue $\lambda(\varepsilon)$ is defined as a root of $\det[Q(\varepsilon)-tM(\varepsilon)]=0$ and the corresponding generalized eigenvector $a(\varepsilon)$ is a p-vector satisfying $Q(\varepsilon)a(\varepsilon)=\lambda(\varepsilon)M(\varepsilon)a(\varepsilon)$, with $a(\varepsilon)^T M(\varepsilon)a(\varepsilon)=1$. As we shall see in the next section, a number of eigenproblems of multidimensional data analysis fall within this formulation. In particular, the ordinary eigenproblem is obtained by putting $M(\varepsilon)=I_p$.

Suppose $Q(\varepsilon)$ and $M(\varepsilon)$ are twice continuously differentiable on the inner points of a domain D, and let ε_0 be an inner point of D. Then, for $\varepsilon\in D$, we can write the Taylor's expansions:

$$Q(\varepsilon) = Q + Q^{(1)}(\varepsilon-\varepsilon_0) + \tfrac{1}{2}Q^{(2)}(\varepsilon-\varepsilon_0)^2 + O(\varepsilon^3),$$

$$M(\varepsilon) = M + M^{(1)}(\varepsilon-\varepsilon_0) + \tfrac{1}{2}M^{(2)}(\varepsilon-\varepsilon_0)^2 + O(\varepsilon^3),$$

where $Q=Q(\varepsilon_0)$, $M=M(\varepsilon_0)$ and, for $i=1,2$, $Q^{(i)}$ and $M^{(i)}$ are the i-th derivatives of $Q(\varepsilon)$ and $M(\varepsilon)$, evaluated at $\varepsilon=\varepsilon_0$.

Compstat 1990

A major problem in the perturbation theory of linear operators is to establish whether the generalized eigenvalues and eigenvectors of $Q(\varepsilon)$ with respect to $M(\varepsilon)$ also admit a Taylor's expansion. Let $\lambda_1 \geq \ldots \geq \lambda_p$ be the generalized eigenvalues of Q with respect to M and let $a_1, \ldots, a_p$ be the associated eigenvectors normalized in such a way that $a_h^T M a_k = \delta_{hk}$, $h,k \in \{1,\ldots,p\}$. If $\lambda_0 \in \{\lambda_1, \ldots, \lambda_p\}$ is a *simple* generalized eigenvalue and a_0 is the corresponding eigenvector, then the perturbations $\lambda(\varepsilon)$ and $a(\varepsilon)$ of λ_0 and a_0 are real-valued continuous functions of ε in a neighbourhood of ε_0 and are twice differentiable at ε_0, so that we can write the Taylor's expansions:

$$\lambda(\varepsilon) = \lambda_0 + \lambda^{(1)}(\varepsilon-\varepsilon_0) + \tfrac{1}{2}\lambda^{(2)}(\varepsilon-\varepsilon_0)^2 + O(\varepsilon^3),$$
$$a(\varepsilon) = a_0 + a^{(1)}(\varepsilon-\varepsilon_0) + \tfrac{1}{2}a^{(2)}(\varepsilon-\varepsilon_0)^2 + O(\varepsilon^3).$$

Here $\lambda^{(i)}$ and $a^{(i)}$ denote the i-th derivatives of $\lambda(\varepsilon)$ and $a(\varepsilon)$, evaluated at ε_0, $i=1,2$. Putting $C^{(i)}=Q^{(i)}-\lambda_0 M^{(i)}$, $i=1,2$, the following expressions are obtained [4]:

$$\lambda^{(1)} = a_0^T C^{(1)} a_0, \qquad a^{(1)} = -D_0 C^{(1)} a_0 - \tfrac{1}{2}(a_0^T M^{(1)} a_0) a_0, \tag{1}$$
$$\lambda^{(2)} = a_0^T (C^{(2)} - \lambda^{(1)} M^{(1)}) a_0 + 2 a_0^T C^{(1)} a^{(1)}, \tag{2}$$
$$a^{(2)} = -D_0 (C^{(2)} - 2\lambda^{(1)} M^{(1)}) a_0 - 2 D_0 (C^{(1)} - \lambda^{(1)} M) a^{(1)} \tag{3}$$
$$-(2 a_0^T M^{(1)} a^{(1)} + \tfrac{1}{2} a_0^T M^{(2)} a_0 + a^{(1)T} M a^{(1)}) a_0,$$

where $D_0 = M^{-1/2} F_0^+ M^{-1/2}$ and F_0^+ is the Moore-Penrose generalized inverse of $F_0 = M^{-1/2} Q M^{-1/2} - \lambda_0 I_p$. A convenient representation is $D_0 = \Sigma_{(0)} (\lambda_k - \lambda_0)^{-1} a_k a_k^T$, with $\Sigma_{(0)}$ denoting summation of all terms except $\lambda_k = \lambda_0$. Note that, whereas $\lambda^{(1)}$ depends only on λ_0 and a_0, we can write $a^{(1)} = \Sigma_{(0)} (\lambda_0 - \lambda_k)^{-1} (a_k^T C^{(1)} a_0) a_k - \tfrac{1}{2}(a_0^T M^{(1)} a_0) a_0$ and this shows that $a^{(1)}$ is a linear combination of $a_1, \ldots, a_p$.

If λ_0 has multiplicity $m>1$ and A is a $(p\text{x}m)$ matrix satisfying $QA=\lambda_0 MA$, the derivatives of the m eigenvalues which are coincident at ε_0 can be calculated as the eigenvalues of the matrix $A^T C^{(1)} A$ [4].

2. INFLUENCE FUNCTIONS OF EIGENVALUES AND EIGENVECTORS.

Consider a p-dimensional random vector X with cumulative distribution function F and suppose $T(F) \equiv T$ is a parameter expressed as a functional of the distribution function F. Let δ_z be the distribution placing unit mass at $z\in\mathbf{R}^p$. The theoretical influence function $IC[z;T(F)]$ of T at F is defined as the limit as $\varepsilon\text{->}0^+$ of $\{T[(1-\varepsilon)F+\varepsilon\delta_z]-T(F)\}/\varepsilon$, provided this limit exists for all $z\in\mathbf{R}^p$. This implies that the influence function is the ordinary right-hand derivative, evaluated at $\varepsilon_0=0$, of $T[(1-\varepsilon)F+\varepsilon\delta_z]$.

We now obtain the influence functions of eigenvalues and eigenvectors in PCA, CVA and CCA as particular cases of formulae (1). A discussion of their properties and numerical applications may be found in [5,6].

<u>Principal component analysis</u>.

Suppose $\mu(F)\equiv\mu$ and $\Omega(F)\equiv\Omega$ are the mean vector and the covariance matrix of X and let $L=\text{diag }\Omega$. The eigenvalues λ_j of the correlation matrix $R=L^{-1/2}\Omega L^{-1/2}$ are the roots of $\det(\Omega-tL)=0$. Further, if a_j satisfies $\Omega a_j=\lambda_j La_j$ with $a_j^T La_j=1$, then $\beta_j=L^{1/2}a_j$ is a unit-length eigenvector of R associated with λ_j. When F is perturbed to $(1-\varepsilon)F+\varepsilon\delta_z$, $0<\varepsilon<1$, writing w for the vector $z-\mu$, the corresponding perturbations of Ω and L are [1]:

$$\Omega(\varepsilon) = \Omega + (ww^T-\Omega)\varepsilon - ww^T\varepsilon^2, \quad L(\varepsilon) = \text{diag } \Omega(\varepsilon).$$

If λ_j is a simple eigenvalue of R, substituting $\Omega^{(1)}=ww^T-\Omega$, $L^{(1)}=\text{diag }\Omega^{(1)}$ in the first of the equations at (1), it follows:

$$IC(z;\lambda_j) = (\beta_j^T u)^2 - \lambda_j\beta_j^T K\beta_j,$$

where $u=L^{-1/2}(z-\mu)$ and $K=\text{diag } uu^T$. The influence function of β_j is the coefficient of ε in the expansion of $\beta_j(\varepsilon)=L(\varepsilon)^{1/2}a_j(\varepsilon)$. Writing $H=L^{-1/2}(\text{diag } ww^T - L)$ and using the expansion:

$$L(\varepsilon)^{1/2} = L^{1/2} + \tfrac{1}{2}H\varepsilon - \tfrac{1}{2}L^{-1/2}(\text{diag } ww^T + \tfrac{1}{4}H^2)\varepsilon^2 + O(\varepsilon^3), \qquad (4)$$

we obtain:

$$IC(z;\beta_j) = -(\beta_j^T u)R_j^+ u + (\lambda_j R_j^+ + \tfrac{1}{2}I_p)K\beta_j - \tfrac{1}{2}(\beta_j^T K\beta_j)\beta_j,$$

where $R_j=R-\lambda_j I_p$ and $R_j^+=\Sigma_{(j)}(\lambda_k-\lambda_j)^{-1}\beta_k\beta_k^T$. In the same way, substitution of $\Omega^{(2)}=-2ww^T$, $L^{(2)}=-2\text{diag } ww^T$ in (2) and (3), again using (4) and performing the necessary algebra, gives:

$$\lambda_j^{(2)} = -2\lambda_j^{(1)}\beta_j^T K\beta_j - 2(\beta_j^{(1)}-\tfrac{1}{2}K\beta_j)^T R_j(\beta_j^{(1)}-\tfrac{1}{2}K\beta_j),$$

$$\beta_j^{(2)} = \lambda_j^{(1)}R_j^+K\beta_j - [\lambda_j R_j^+ + (3/4)I_p]K^2\beta_j + 2\lambda_j^{(1)}R_j^+\beta_j^{(1)}$$
$$+(2\lambda_j R_j^+ + I_p)K\beta_j^{(1)} + (\beta_j^T Ku - 2\beta_j^{(1)T}u)R_j^+u$$
$$+[(3/4)\beta_j^T K^2\beta_j - \beta_j^T K\beta_j^{(1)} - \beta_j^{(1)T}\beta_j^{(1)}]\beta_j.$$

The following results are easily checked.

1)Choose z in such a way that $u_j=c$, $j=1,...,p$. It follows that $|IC(z;\lambda_j)|=c^2|(\Sigma\beta_j)^2-\lambda_j|$ can reach any positive value and this shows that the usual estimator of $\lambda_j(F)$ is non-robust since it is heavily affected by just one observation with a sufficiently high value of $|c|$.

2)In a neighbourhood of $\varepsilon_0=0$, $\beta_j(\varepsilon)=\beta_j+\beta_j^{(1)}\varepsilon+\frac{1}{2}\beta_j^{(2)}\varepsilon^2+O(\varepsilon^3)$ and the cosine of the angle formed by β_j and $\beta_j(\varepsilon)$ is $\beta_j^T\beta_j(\varepsilon)$ $=1-\frac{1}{2}\varepsilon^2\beta_j^{(1)T}\beta_j^{(1)}+O(\varepsilon^3)$, so it depends only on the Euclidean norm of $\mathrm{IC}(z;\beta_j)$.

3)Let S be a subset of $1<s<p$ elements of the set $\{1,...,p\}$ and let $P_S=\Sigma_{j\in S}\beta_j\beta_j^T$ be the projection operator onto the subspace spanned by the eigenvectors β_j, $j\in S$. If, for $j\in S$, λ_j is a simple eigenvalue, the influence function of P_S is given by:

$$\mathrm{IC}(z;P_S) = \tfrac{1}{2}(KP_S+P_SK) - \Sigma_{j\in S}(\beta_j^TK\beta_j)\beta_j\beta_j^T$$
$$+\Sigma_{j\in S}\Sigma_{(j)}(\lambda_k-\lambda_j)^{-1}(\lambda_j\beta_k^TK\beta_j-\beta_j^Tu\beta_k^Tu)(\beta_k\beta_j^T+\beta_j\beta_k^T).$$

4)Putting $L=I_p$ and substituting $\Omega^{(i)}$ and $L^{(i)}=0_{p,p}$ in (1)-(3), $i=1,2$, gives the first two derivatives of a simple eigenvalue and the associated eigenvector of a covariance matrix [1].

<u>Canonical variate analysis</u>.

Suppose a partition $\{P_g,\ g=1,...,G\}$ of the population P in $G\geq2$ groups is given, and let μ_g and Ω_g be the mean vector and the covariance matrix of X conditional on P_g, $g=1,...,G$. If $\pi_1,...,\pi_G$ ($\pi_g\geq0$, $\Sigma\pi_g=1$) are the mixing proportions of the groups, the mean vector of X in P is $\mu=\Sigma\pi_g\mu_g$ and the covariance matrix is $\Omega=W+B$, where $W=\Sigma\pi_g\Omega_g$ and $B=\Sigma\pi_g(\mu_g-\mu)(\mu_g-\mu)^T$ are the within-group and the between-group covariance matrices, respectively. In CVA, we choose linear combinations of the observed variables which maximize the variation among the groups relative to the variation within the groups. This problem requires the determination of the positive roots λ_j of $\det(B-tW)=0$ and the vectors a_j satisfying $Ba_j=\lambda_jWa_j$, with $a_j^TWa_j=1$. A simple way of perturbing the cumulative distribution function F is to add a small contribution from δ_z to the conditional distribution of only one group, e.g. P_1. The corresponding perturbations of B and W are [5]:

$$B(\varepsilon) = B + (ww^T-uu^T-B)\varepsilon - (ww^T-\pi_1^{-1}uu^T)\varepsilon^2 + O(\varepsilon^3),$$
$$W(\varepsilon) = W + (uu^T-W)\varepsilon - \pi_1^{-1}uu^T\varepsilon^2 + O(\varepsilon^3),$$

where $w=z-\mu$ and $u=z-\mu_1$. If λ_j is a simple root of $\det(B-tW)=0$, substituting $B^{(1)}=ww^T-uu^T-B$, $W^{(1)}=uu^T-W$ in (1), we obtain:

$$\mathrm{IC}(z;\lambda_j) = (a_j^Tw)^2 - (1+\lambda_j)(a_j^Tu)^2,$$
$$\mathrm{IC}(z;a_j) = -(a_j^Tw)D_jw + (1+\lambda_j)(a_j^Tu)D_ju + \tfrac{1}{2}[1-(a_j^Tu)^2]a_j,$$

where $D_j=\Sigma_{(j)}(\lambda_k-\lambda_j)^{-1}a_k a_k^T$. When rank$(B)<p$ the summation in the expression of D_j includes the eigenvectors associated with the (multiple) zero eigenvalue of $W^{-1}B$.

Canonical correlation analysis.

Letting X be partitioned as $X=\begin{pmatrix}X_1\\X_2\end{pmatrix}$ where, for $g=1,2$, X_g has p_g components, the mean vector and the covariance matrix of X are correspondingly partitioned as $\mu=\begin{pmatrix}\mu_1\\\mu_2\end{pmatrix}$, $\Omega=\begin{pmatrix}\Omega_{11} & \Omega_{12}\\\Omega_{12}^T & \Omega_{22}\end{pmatrix}$. The squared canonical correlation coefficients λ_j are the positive roots of $\det(\Omega_{12}\Omega_{22}^{-1}\Omega_{12}^T-t\Omega_{11})=0$ and the canonical vectors a_j and β_j satisfy $\Omega_{12}\Omega_{22}^{-1}\Omega_{12}^T a_j=\lambda_j\Omega_{11}a_j$, $\Omega_{12}^T\Omega_{11}^{-1}\Omega_{12}\beta_j=\lambda_j\Omega_{22}\beta_j$, with $a_j^T\Omega_{11}a_j=\beta_j^T\Omega_{22}\beta_j=1$. For a given $z=\begin{pmatrix}z_1\\z_2\end{pmatrix}$, let $z-\mu=w=\begin{pmatrix}w_1\\w_2\end{pmatrix}$, where $w_g=z_g-\mu_g$, $g=1,2$. When F is perturbed to $(1-\varepsilon)F+\varepsilon\delta_z$, for $0<\varepsilon<1$ and $g,g'\in\{1,2\}$ the corresponding perturbation of $\Omega_{gg'}$ is:

$$\Omega_{gg'}(\varepsilon) = \Omega_{gg'} + (w_g w_{g'}^T-\Omega_{gg'})\varepsilon - w_g w_{g'}^T\varepsilon^2 . \tag{5}$$

For $z\in\mathbf{R}^p$ and $0<\varepsilon<1$, $\Omega_{gg}(\varepsilon)$ is positive definite and a second-order expansion of $\Omega_{gg}(\varepsilon)^{-1}$ is:

$$\Omega_{gg}(\varepsilon)^{-1} = \Omega_{gg}^{-1} + (\Omega_{gg}^{-1}-\Omega_{gg}^{-1}w_g w_g^T\Omega_{gg}^{-1})\varepsilon \tag{6}$$
$$+[\Omega_{gg}^{-1}-(1-\pi_g)\Omega_{gg}^{-1}w_g w_g^T\Omega_{gg}^{-1}]\varepsilon^2 + O(\varepsilon^3),$$

where $\pi_g=w_g^T\Omega_{gg}^{-1}w_g$, $g=1,2$. Let $T_g=\Omega_{g'g}\Omega_{gg}^{-1}\Omega_{g'g}^T$ and let $\tilde{w}_{g'}=\Omega_{g'g}\Omega_{gg}^{-1}w_g$, $g,g'\in\{1,2\}$, $g\neq g'$. Using (5) and (6) it can be shown that the perturbation of T_g is:

$$T_g(\varepsilon) = T_g + (\tilde{w}_{g'}w_{g'}^T+w_{g'}\tilde{w}_{g'}^T-\tilde{w}_{g'}\tilde{w}_{g'}^T-T_g)\varepsilon$$
$$+[(1+\pi_g)(\tilde{w}_{g'}\tilde{w}_{g'}^T-\tilde{w}_{g'}w_{g'}^T-w_{g'}\tilde{w}_{g'}^T)+\pi_g w_{g'}w_{g'}^T]\varepsilon^2 + O(\varepsilon^3).$$

Thus we have two generalized eigenproblems, with $Q(\varepsilon)=T_g(\varepsilon)$, $M(\varepsilon)=\Omega_{g'g'}(\varepsilon)$, $g,g'\in\{1,2\}$, $g\neq g'$. To simplify the notation we put $\Gamma_j = T_2-\lambda_j\Omega_{11}$, $U_j = \Omega_{11}^{-1/2}\Gamma_j^+\Omega_{11}^{-1/2}$ and $\Phi_j = T_1-\lambda_j\Omega_{22}$, $V_j=\Omega_{22}^{-1/2}\Phi_j^+\Omega_{22}^{-1/2}$. Further, $u_j=a_j^T(z_1-\mu_1)$ and $v_j=\beta_j^T(z_2-\mu_2)$ denote the j-th canonical coordinates of z_1 and z_2. If λ_j is a simple root of $\det(T_2-t\Omega_{11})=0$, we obtain [6]:

$$\mathrm{IC}(z;\lambda_j) = -\lambda_j u_j^2 + 2\lambda_j^{1/2}u_j v_j - \lambda_j v_j^2,$$
$$\mathrm{IC}(z;a_j) = -\lambda_j^{1/2}(v_j-\lambda_j^{1/2}u_j)U_j w_1 - (u_j-\lambda_j^{1/2}v_j)U_j\tilde{w}_1$$
$$+\tfrac{1}{2}(1-u_j^2)a_j,$$
$$\mathrm{IC}(z;\beta_j) = -\lambda_j^{1/2}(u_j-\lambda_j^{1/2}v_j)V_j w_2 - (v_j-\lambda_j^{1/2}u_j)V_j\tilde{w}_2$$
$$+\tfrac{1}{2}(1-v_j^2)\beta_j.$$

We note that $\mathrm{IC}(z;\lambda_j)$ is a symmetric function of the canonical coordinates u_j and v_j and, for $0<\lambda_j<1$, the contours of the influence function in the $(u\text{-}v)$ plane are hyperbolas. Thus $IC(z;\lambda_j)$ has the same characteristics as the influence function of the squared linear correlation coefficient.

3. CONCLUDING REMARKS.

The influence functions of the eigenvalues in PCA, CVA and CCA are unbounded and the eigenvectors associated with eigenvalues which are not well separated can be particularly sensitive to perturbations of the data. Thus robust estimators of eigenvalues and eigenvectors are worth investigating and in any case influential observations should be identified by suitable indicators to avoid biased conclusions. Sample versions of the influence functions [1] provide useful diagnostics, but we need criteria to establish when influence is large. The problem of *groups* of influential observations is easily handled by the theory developed in [4], whereas the problem of influential *variables* requires a different approach.

REFERENCES

[1] Critchley,F.(1985). Influence in principal components analysis. *Biometrika* **72**,627-636.

[2] Hampel,F.R.(1974). The influence curve and its role in robust estimation. *J.Am.Statist.Assoc.* **69**, 383-393.

[3] Jolliffe,I.T. *Principal Component Analysis*, Springer-Verlag, Berlin, 1986.

[4] Romanazzi,M.(1989). Derivatives of eigenvalues and eigenvectors in the generalized symmetric eigenproblem dependent on a parameter. Submitted for publication.

[5] Romanazzi,M.(1989). Influence in canonical variates analysis. Accepted for publication in *J.Comput.Stat.Data Anal.*

[6] Romanazzi,M. Influence functions in canonical correlation analysis (in Italian). *Proc. of the XXXV Meeting of the Ital. Statist.Soc.*, CEDAM, Padua, 1990.

[7] Tanaka,Y.(1988). Sensitivity analysis in principal component analysis: influence on the subspace spanned by principal components. *Comm.Statist. A* **17**, 3157-3175.

This research was supported in part by a grant from the Italian National Research Council.

Algorithms for Non-Linear Huber Estimation

H. Ekblom, Luleå and K. Madsen, Lyngby

1. Introduction

In the non-linear least squares problem we minimize

$$\frac{1}{2}\sum_{j=1}^{m} f_j^{\,2}(\mathbf{x}) \tag{1.1}$$

where $f_1,\ldots, f_m$ is a set of non-linear smooth functions $\Re^n \rightarrow \Re$ and $\mathbf{x}$ is an n-vector of "parameters".

During the last decades, there has been much interest in replacing the least squares method by more "robust" criteria, i.e. estimators more resistant to contaminated data. One alternative to the least squares method is to use the Huber function

$$\rho_k(t) = \begin{cases} t^2/2 & \text{if } |t| \le k \\ k|t| - k^2/2 & \text{if } |t| > k \end{cases} \tag{1.2}$$

and instead minimize

$$\sum_{j=1}^{m} \rho_k(f_j(\mathbf{x})/\sigma) \tag{1.3}$$

where σ is a scale parameter.

If we give the constant k of the Huber function a very large value, then we will have the least squares method. If we let k tend to zero, then we will instead approach l_1-approximation. The choice of value of k is connected to the assumed percentage of of "bad points" in the data, as discussed e.g. in [4]. Further discussion on robust estimation is found in [6] and [7].

2. Notations.

To simplify notations, we reformulate (1.2) - (1.3) as follows,

$$\text{minimize } g(\mathbf{x}) \equiv \sum_{j=1}^{m} \rho(f_j(\mathbf{x})/\theta) \tag{2.1}$$

$$\text{where } \rho(t) = \begin{cases} t^2/2 & \text{if } |t| \le 1 \\ |t| - 1/2 & \text{if } |t| > 1 \end{cases} \tag{2.2}$$

and $\theta = k\sigma$ is a constant. f_j, $j = 1,\dots,m$, is a set of (non-linear) continuously differentiable functions, $f_j : \Re^n \to \Re$. We assume that the gradients $\mathbf{f}_j'$, $j = 1,\dots,m$, can be computed.

We shall use the notations

$$\mathbf{f}(\mathbf{x}) \equiv (f_1(\mathbf{x}), \dots, f_m(\mathbf{x}))^T \tag{2.3}$$

and

$$\mathbf{J}(\mathbf{x}) \equiv (\mathbf{f}_1'(\mathbf{x}), \dots, \mathbf{f}_m'(\mathbf{x}))^T \tag{2.4}$$

ρ is twice differentiable almost everywhere. The derivative of ρ at the point $f_j(\mathbf{x})/\theta$ is denoted

$$v_j(\mathbf{x}) \equiv \rho'(f_j(\mathbf{x})/\theta) = \begin{cases} 1 & \text{if } f_j(\mathbf{x}) > \theta \\ f_j(\mathbf{x})/\theta & \text{if } |f_j(\mathbf{x})| \le \theta \\ -1 & \text{if } f_j(\mathbf{x}) < -\theta \end{cases} \qquad j=1,\dots,m \tag{2.5}$$

and

$$\mathbf{v}(\mathbf{x}) \equiv (v_1(\mathbf{x}), \dots, v_m(\mathbf{x}))^T \tag{2.6}$$

The second derivative of ρ at $f_j(\mathbf{x})/\theta$ is denoted $d_{jj}(\mathbf{x})$,

$$d_{jj}(\mathbf{x}) \equiv \bar{\rho}''(f_j(\mathbf{x})/\theta) = \begin{cases} 1 & \text{if } |f_j(\mathbf{x})| \le \theta \\ 0 & \text{otherwise} \end{cases} \qquad j = 1, \dots, m, \tag{2.7}$$

and

$$\mathbf{D}(\mathbf{x}) \equiv \text{diag}\,\{d_{jj}(\mathbf{x})\} \tag{2.8}$$

The gradient of g is

$$\mathbf{g}'(\mathbf{x}) = \sum_{j=1}^{m} v_j(\mathbf{x})\mathbf{f}_j'(\mathbf{x})/\theta = \mathbf{J}(\mathbf{x})^T \mathbf{v}(\mathbf{x})/\theta \tag{2.9}$$

and the Hessian is

$$H(x) = \sum_{j=1}^{m} [d_{jj}(x)f_j'(x)f_j'(x)^T/\theta^2 + v_j(x)f_j''(x)/\theta]$$
$$= J(x)^T D(x)J(x)/\theta^2 + \sum_{j=1}^{m} v_j(x)f_j''(x)/\theta \qquad (2.10)$$

The methods of this paper are of the Gauss-Newton type. In the neighbourhood of the point **x** we consider an approximation $q(\cdot\,; x)$ of g based on linearizations l_j to f_j:

$$l_j(h;x) \equiv f_j(x) + f_j'(x)^T h$$
$$q(h;x) \equiv \sum_{j=1}^{m} \rho(l_j(h;x)/\theta) \qquad (2.11)$$

In the following the notation (**x**) will be left out in case there is no risk of misinterpretation.

3. Algorithms previously proposed.

For linear model problems, where the functions f_j are linear, the Newton iteration is given by

$$\begin{cases} J^T D(x_k)J\ h = -\theta\ J^T v(x_k) \\ x_{k+1} = x_k + h\,\alpha \end{cases} \qquad (3.1)$$

where $\alpha>0$ is a "step length parameter". Here **J** is a constant matrix, whereas **v** and **D** depend on the current value of **x**. They are given by (2.4) - (2.8).

There are modifications to the Newton method, e.g. the IRLS method [7, p.183] and the H-algorithm [7, p.181], where no line search is needed (i.e. $\alpha\equiv1$). When it works, the Newton method is ultimately the fastest. The matrix $J^T DJ$ may become singular, although **J** has full rank. However, this problem can be overcome by a slight modification, see [9].

Each of the methods for the linear case can be extended to non-linear models by letting **J** hold the current value of the Jacobian. For the Newton method, this will lead to a Gauss-Newton type algorithm, as discussed by Dennis [1]. It can be stabilized to the following iteration, which Dennis calls "Generalized Levenberg-Marquardt Method":

$$\begin{cases} [J^T(x_k)D(x_k)J(x_k) + \lambda I]\ h = -\theta\ J^T(x_k)\ v(x_k) \\ x_{k+1} = x_k + h \end{cases} \qquad (3.2)$$

Dutter and Huber [2] studied generalizations of the IRLS- and H-algorithms. However, their sophisticated IRLS-algorithm, guaranteed to converge, was only sligthly better than a standard Davidon-Fletcher-Powell code.

4. New algorithms for the non-linear problem.

The methods we propose for solving (2.1) are similar to those published in [5]. They are of the *trust region* type. We calculate a sequence of points $\mathbf{x}_k$ intended to converge to a local minimum of g. At each point $\mathbf{x}_k$ a linear approximation q_k to g, based on linear approximations l_j to the non-linear functions f_j (j=1,...,m), is considered. q_k is minimized subject to the constraint that the solution should be within a specified neighbourhood N_k of $\mathbf{x}_k$. N_k is intended to reflect the domain in which the approximations l_j of f_j are valid.

The approximation q_k is defined through (2.11), i.e.

$$q_k(\mathbf{h}) \equiv q(\mathbf{h};\mathbf{x}_k) = \sum_{j=1}^{m} \rho\,(l_j(\mathbf{h};\mathbf{x}_k)/\theta) \tag{4.1}$$

The trust region could be defined as follows,

$$N_k \equiv \{\, \mathbf{y} \mid \mathbf{y} = \mathbf{x}_k+\mathbf{h} \,,\, \|\mathbf{h}\| \le \delta_k \} \tag{4.2}$$

where $\delta_k>0$ is given and should reflect the amount of linearity of f_j near $\mathbf{x}_k$. The minimum of (4.1) subject to (4.2) is denoted by $\mathbf{h}_k$.

If $\|\mathbf{h}_k\|<\delta_k$ then $\mathbf{h}_k$ is the unconstrained minimum of q_k. Otherwise the constraint is active and the necessary condition for a minimum implies that $\mathbf{h}=\mathbf{h}_k$ satisfies

$$\begin{aligned} t_h'(\mathbf{h},\lambda) &= \mathbf{0} \\ \mathbf{h}^T\mathbf{h} - \delta_k^2 &= 0 \end{aligned} \tag{4.3}$$

where $\mathbf{t}_h'$ is the gradient, with respect to $\mathbf{h}$, of the Lagrangian t,

$$t(\mathbf{h},\lambda) = q_k(\mathbf{h}) + \lambda(\mathbf{h}^T\mathbf{h}-\delta_k^2) \tag{4.4}$$

The Newton iteration for solving (4.3) can be formulated the following way:

while not STOP **do begin**

solve $$\begin{bmatrix} q_k''(\mathbf{h})+2\lambda\mathbf{I} & 2\mathbf{h} \\ 2\mathbf{h}^T & 0 \end{bmatrix} \begin{bmatrix} \Delta\mathbf{h} \\ \Delta\lambda \end{bmatrix} = - \begin{bmatrix} q_k'(\mathbf{h})+2\lambda\mathbf{h} \\ \mathbf{h}^T\mathbf{h} - \delta_k^2 \end{bmatrix}$$

$$(\mathbf{h},\lambda) := (\mathbf{h},\lambda) + (\Delta\mathbf{h},\Delta\lambda) \tag{4.5}$$

end ;

$(\mathbf{h}_k,\lambda_k) := (\mathbf{h},\lambda)$

Here STOP could be the condition that either the norm of the left hand side of (4.3) is less than ε_1 or $\|(\Delta\mathbf{h},\Delta\lambda)\| < \varepsilon_2 \|(\mathbf{h},\lambda)\|$.

In (4.3) λ and δ_k are dual. If δ_k is given then λ is fixed by (4.3), and vice versa. Therefore, instead of defining a suitable radius δ_k of the trust region to be used in (4.3) one might define suitable value of λ, thus fixing δ_k through (4.3). Then the system (4.3) becomes separable (and thus easier to solve) since the first set of equations is independent of δ_k. Giving $\lambda = \lambda_k$ rather than δ_k is similar to Levenberg-Marquardt type methods, [8],[10], which have been used successfully for solving non-linear least squares. An algorithm of this kind are presented in [4].

The difficulty in minimizing (4.1) subject to (4.2) is that we do not know in advance the distribution between "large" and "small" function values $l_j(\mathbf{h}; \mathbf{x}_k)$ (see the definition (2.1)-(2.2) of the Huber function). If we knew the set $\{ j \mid |l_j(\mathbf{h}_k; \mathbf{x}_k)| \le \theta \}$ in advance then only one iteration of the above Newton algorithms would be necessary. If, however, $\mathbf{x}_k$ is close to the non-linear solution $\mathbf{x}^*$ then the starting value $\mathbf{h}=\mathbf{0}$ will provide a good estimate of this set, and thus (4.5) would take only a few iterations.

We can now specify an algorithm for solving the non-linear Huber problem. We use the usual updating procedures for δ_k based on the ratio

$$r_k = [g(\mathbf{x}_k) - g(\mathbf{x}_k+\mathbf{h}_k)]/[q_k(\mathbf{0}) - q_k(\mathbf{h}_k)] \tag{4.6}$$

i.e. the ratio between the decrease in the non-linear function and the decrease in the local approximation. The new point $\mathbf{x}_k+\mathbf{h}_k$ is accepted if the decrease in the objective function g exceeds a small multiple of the decrease predicted by the local approximation.

Algorithm:

Let $0 < s_1 \ll 0.25$ and $s_2 < 1 < s_3$.

```
given x_0 and δ_0;  k := 0;
while not OUTERSTOP do begin
    find an "approximate" minimum h_k of (4.1) subject to (4.2) using (4.5)
    if r_k ≥ s_1 then   x_{k+1} := x_k+h_k
               else   x_{k+1} := x_k;
     if      r_k ≤ 0.25 then  δ_{k+1} := δ_k * s_2
    else if r_k ≥ 0.75 then  δ_{k+1} := δ_k * s_3
                       else  δ_{k+1} := δ_k
    k := k+1
end
```

In the algorithm OUTERSTOP could be the condition that

$$(\|g'(x_k)\| < \varepsilon_3 \text{ or } \|h_k\| < \varepsilon_4 \|x_k\|) \ .$$

Notice that this algorithm exploits the special structure of the Huber function since they are based on the local approximation function q_k of (4.1). It should be pointed out that the "Generalized Levenberg-Marquardt Method" does not have this property. An iteration of (3.2) corresponds to approximating the Huber function with a smooth approximation function (a quadratic function), and thus the discontinuities in the second order derivatives of g would not be reflected by the local approximation.

It should be noted here that it is not necessary for the convergence of the algorithms that the subproblem (4.1)+(4.2) is solved exactly. In [4] we give a convergence proof which only requires that the decrease in the model function q_k is a small multiple of the best possible decrease in q_k.

6. References.

1. J.E.Dennis, *Non-linear Least Squares and Equations,* In *State of the Art in Numerical Analysis,* (ed. D.A.H. Jacobs), Academic Press, 269-312, 1977.
2 R.Dutter and P.J.Huber, *Numerical Methods for the Nonlinear Robust Regression Problem,* J. Stat. Comp. Sim. (1981).
3. H. Ekblom, *A new algorithm for the Huber estimator in linear models,* BIT 28 (1988), 123-132.
4. H. Ekblom and K. Madsen, *Algorithms for non-linear Huber estimation,* BIT 29 (1989), 60-76.
5. Li Gao and K.Madsen, *Robust Non-linear Parameter Estimation .* In *Numerical analysis 1987,* D.F.Griffiths and G.A.Watson, eds., Pitman Research Notes in Mathematics Series 170, Longman, UK, 176-191, 1988.
6. F.R.Hampel, E.Ronchetti, P.Rousseeuw, W.Stahel, *Robust Statistics: the infinitesimal approach,* John Wiley, New York, 1986.
7. P.Huber, *Robust Statistics,* John Wiley, New York, 1981.
8. K.Levenberg, *A method for the solution of certain problems in least squares,* Quart. Appl. Math. 2. (1944).
9. K.Madsen and H.B.Nielsen, *Finite algorithms for robust linear regression,* DCAMM Report No. 395, Technical Univ. of Denmark, 1989. Accepted for publication in BIT.
10. D.Marquardt, *An algorithm for least-squares estimation of nonlinear parameters,* SIAM J. Appl. Math. (1963).

Statistics and Database Management

Co-Operative Processing - A Challenge for Statistics and Database Management

K. Neumann, Berlin

Keywords: Co-operative Processing; Database Management; Distributed Processing; Knowledge-based Systems; Metainformation; Official Statistical Information System (OSIS); Statistical Analysis

Introduction

According to the essential importance of data for all kinds of statistics the data organisation was and is and will be a very crucial point in statistical work in general and especially in Official Statistical Information Systems (OSIS). This background is reflected in scientific papers and discussions on conferences and workshops over a long time, which however also show a change in emphasis and context in respect to data organisation. Since the end of the sixties the formation of statistical databases has dominated the development of OSIS. With the now arising new conditions, namely the changes in the environment of each OSIS and the evolving of supranational OSIS, both statistics and database management are forced to react. The indispensable new organisation of the division of labour in the whole process of producing and consuming official statistics will

Compstat 1990

require perfecting the up to now dominating distributed processing by a new quality: co-operative processing. But this new quality is not possible without some further development of the management of statistical databases.

Historical Background and First Phase of DBM in OSIS

Data are often called as the assets of each OSIS. The affinity of "data" and "statistics" is obvious, even in common language definitions:

DATA: facts; things known as certain (and from which conclusions may be drawn)

STATISTICS: facts shown in numbers

Nevertheless the term "statistical data" in the meaning of "facts, shown in numbers, from which conclusions may be drawn" is a relative modern one. Its emergence in the statistical literature is correlated with the penetration of the statistical work by electronic data processing activities.
And just this merger between statistics with its long tradition (especially in different kind of computations) and new computing techniques was the starting point of essential changes in the field of data organisation. And it is not difficult to affiliate this child of science to Svein Nordbotten and his paper "Automatic files in statistical systems" /NORD 67/. In an ever more accelerating process both in theory and in practice the creation and management of statistical data bases influenced and changed the Official Statistical Information Systems more and more. At least in the community of the European statistical offices this process was at its beginning dominated by Skandinavians, e.g. Arvas /ARV 68/, /ARV 72/, Langefors /LANG 69/, Bucknell /BUCK 73/, Firth /FIRTH 73/ and Sundgren /SUND 73/. It was also at this time that the outstanding experience of the Canadian Statistical Office in developing, implementing and

using a database-oriented OSIS (called CANSIM) was transferred to Europe via Sweden (SWESSIM). This inspiring influence was at the end of the seventies repeated when on the same route RAPID came from Canada to the wellorganized family of EDP users in official statistics inside the UN-European Commission of Europe (ECE).

In parallel most of the work and the results of CODASYL Data Base Task Group /COD 69/, /COD 71a/, /COD 71b/ influenced the way from a simple file organisation to a sophisticated Data Base Management. Based on these perceptions, on the growing practical experience and on the availability of more powerful hardware and software an increasing number of Statistical Offices reorganized their OSIS during the seventies.
As a main result the former single-survey-oriented data organization was step by step replaced by data bases as a constituent part of data banks. The most important feature of these data bases are the higher accessibility of all the stored data <u>and</u> the maintanance in terms of updating and reordering as a indispensable prerequisit for a multiple use. To ensure this, a permanent management is both a condition and a consequence.
This is i.a. reflected by the widened scope of the discussion about the function and the effect of Data Base Management Systems (DBMS) in OSIS, which started at the turn from the seventies to the eighties.

<u>The Second Phase: DBMS constitute a new quality of OSIS</u>

The widened use of DBMS in OSIS created a very strong impetus in the overall design and/or redesign of such systems in general. We should remember that this took place at a time when
- the demand for more and better statistics not only for

governmental purposes, but also for different public uses was combined with the necessity to improve these statistics in terms of reliability, integrity and accessibility;

- the advance of both applied systems analysis and computer science enables the systems designer to overcome more and more the "single-survey-view" by looking at OSIS as a technological production system having special functions (see especially /ECESEM 77/);
- the progress of the information processing technology strengthened the trend to interactive procedures, thus initiating a broader use of distributed processing techniques based on a growing diversity of hardware and more sophisticated software;
- not only was the OSIS itself undergoing tremendous changes, but also its environment did so, especially by changing the ability of the direct and indirect partner of the statistical production system - the supplier of raw data and the consumer of statistical information - to communicate with the OSIS on a higher technological level.

In this period it was not the question if there was a need for Data Base Management in an OSIS, but again it seemed questionable if there was a specific need for a specific DBMS.

The basic argument for answering this question was logically derived from the unique character of statistical data and therefore of a statistical office producing such data:

> "Modern statistical offices differ from large industrial corporations in many significant ways. The data they produce is in a very real sense their final end product, whereas in an industrial corporation it is a means to an end. Statistical offices tend to build up a large amount of historical data over a long period, and much of this data needs to be preserved indefinitely.

It may be considered as part of the national heritage, and efforts must be made to ensure that it does not become lost due to failure of the materials on which it is stored or because the techniques to record it are no longer available to read it at later date." /JEAS 84/.

These facts which include in addition

- the large quantity and diverse form of statistical data,
- the need for complex analysis of data without detailed determination of such need when the data are collected,
- the need for minimizing the programming burden for statisticians,
- the high priority of efficiency and reliability of storage, maintanance and retrieval,

lead us to the conclusion that no commercial DBMS suits these specific needs of an OSIS.

Therefore, during the eighties different activities in different fields were directed to enlarge the knowledge about statistical databases and their management on the one hand and to perfect the practical use of DBMS in the daily work of OSIS on the other hand. In this paper just three examples should prove, how this was realized in the European family of statistical offices.

1. The Statistical Computing Project (SCP)

Based on "the generally accepted fact that the software industry has not satisfactorily developed software packages for use in statistical offices" /STATJ 84/, and on the extraordinary work of the Working Party on EDP of the Conference of European Statisticians (CES) since 1957 as well as of the "younger" Seminars on Integrated Statistical Information Systems (ISIS), an inter-country project on the use of computers for statistical purposes and the design and development of automated statistical information systems, called Statistical Computing Project (SCP) was estabilished in 1980 by 19 countries and UNDP.

In four Joint Groups the SCP activities were focussed on

- Data editing;
- Relational approach to statistical data base management (RAPID);
- Table Generation; and
- Meta-information system (METIS).

Thanks to the generousity of Statistics Canada, the developer of RAPID, this DBMS was the basis for new software modules to cope with more basic functions. The Base Operators were the most remarkable result of the Joint Group. This relational concept facilitates data stored in RAPID and helps to improve the linking facilities to other program packages. It became obvious already during development that the technology of Base Operators (BOS) is of general significance independently of RAPID. First BOS were available for such functions as "Join", "Select", "Sort", but later the set of BOS was not only enlarged, but brought in to a "pipeline mode" to achieve an internal connection of several BOS. /NERA 85/.

At the end of the eighties the Joint Group finalized the development of the BOS, Version 3, which now includes the Meta Management System (MMS) and the Relational Acces Method (RAM) that can also be used for the development of other software tools /SCP 89/.

The results of the SCP in the DBMS field are an excellent proof not only of the ability to develop real software in international co-operation, but also of the future delopment of DBMS to portable systems. This is precisely an essential condition for the mixed technology approach.

2. The COMPSTAT symposia

The second example for the growth of knowledge about DBMS and exchange of experience about their practical use are the COMPSTAT symposia, especially those of 1984, 1986 and 1988. At COMPSTAT 1984 a section was devoted to

"Integration of statistical data base management and data analysis".

The six papers /NEU 84/, /RAU 84/, /BORK 84/, /KULE 84/,

/RARI 84/ and /JOKR 84/ were focussed at the full spectrum of this topic. If not expressis verbis, in their substance the papers were building a bridge from the technological aspects of DBMS to the influence of data bases on exploratory data analysis.

This coherence became much more obvious at COMPSTAT 1986 with the section

"Statistical Data Base Management Systems",

which comprised two invited papers, one of them dedicated directly to "Data Base Management and Statistical Data Analysis: The need for Integration and for Becoming More Intelligent" /HAJÖ 86/, the second to "Privacy, Informatics and Statistics" /MAR 86/, and another seven papers dealing with different aspects but mainly touching the development of knowledge based tools using DBMS technology: /COMV 86, /BAFB 86/, /JABP 86/, /LEPB 86/, /LESA 86/, /NEU 86/, /SONN 86/.

It is true that COMPSTAT 88 was poorer in its Database section as regards to the number of papers but not in respect to the quality reached by the concentration on

"Statistical Data Bases and Survey Processing".

The first of the three papers was devoted to the new perspectives in computer-assisted survey processing /KEBE 88/, demonstrating the usefulnes of the DBMS concept for all parts of the statistical production process, the second dealt with the aspect of handling missing data /RUB 88/ and the third paper opened the view for using graphical methods in the process of designing data bases /CASA 88/.

It is not only the variety of aspects both in theory and in practice which distinguishes this now already established tradition of DBMS sections at COMPSTAT but also their innovational approach.

3. Statistical and Scientific Database Management (SSDBM)

Writing or talking about DBMS in statistics one must not forget the three workshops and one working conference on statistical and scientific database management, a series initiated in 1981 in California, but since the third SSDBM in Luxembourg and especially with the excellent output of the fourth SSDBM in Rome in 1988 well known also in Europe. It is again not only the variety of topics which is of course in such a specialized conference larger then in a section of e.g. COMPSTAT, but much more the originality of the content both of the panel discussion and of the selected scientific papers which constitutes the high quality of SSDBM.

The output of the open panel

- I: Categorial Numerical Relational Operations for Statistical Database Management
- II: Geographical Database Systems and Statistical Information
- III: Database Management Systems for Statistical and Scientific Applications: Are Commercially Available DBMS Good Enough for SSDBM Applications?

is a condensed expression of the state of art of statistical DBMS and at the same time an example for the similarity of statistical and scientific databases.

This is a proof for the widening of the scope of DBM problems. The main results of the second phase of DBMS use in OSIS as shown on these few examples are

- the recognition of the metainformation and its significance for database management in OSIS;
- the conclusion that the formation and the use of meta-information will lead to an embedding of knowledge-base in future DBMS;
- the emerging necessity to switch from distributed processing to co-operated processing.

Prospect: Database Management and Co-operative Processing

Database Management at first changed very much the mode of production in OSIS. Later the mode of consuming statistics

became also more and more influenced by the extended use of database technology.
This development created an increasing need for more and better, co-operation which had a feedback to the DBMS.
The new quality of this co-operation is mainly based on the presently possible higher level of communication. The new information and communication technologies (ICT) create a necessary pendant to the higher degree of the division of labour. In respect to an OSIS this division of labour is characterized by the increasing use of different modes of computer-aided techniques in the different parts of the statistical production process, e.g.

- design of surveys
- modelling of databases
- data capture
- data processing
- data analysis
- data dissimination.

This results in a new quantity and quality of communication between heterogenous groups of specialistists inside and outside the OSIS.
Meanwhile the use of data base technologies is no longer restricted to the statistical data as such. Especially DBMS have proved to be very usefull tools for managing

- micro data
- macro data
- meta data
- models and methods

and last but not least different types of knowledge.
Co-operation which is always co-operation between human beings is based on the ability of common understanding. In the context of data processing this requires first of all a higher level of standardisation at least of the structure of the different classes of data bases and of the manifold interfaces, especially of those which are user-oriented.
For theoretical and practical reasons we can distinguish

between the following types of co-operative processing:

- co-operation between a single user and a single database;
- co-operation between a single user and n different databases;
- co-operation between m users and a single database;
- co-operation between m users and n databases.

This classification is of special interest in the presently arising technological environment which is more and more characterized not only by a higher complexity of OSIS but also by more and more computer aided workplaces.

References

/ARV 68/ Arvas, C.: Joint Use of Data Banks
Conf. Eur. Stats. /WG. 9/78
/ARV 72/Arvas, C.: Data Structures: Management and Storage
NCBS, Sveden, 1972
/BAFB 86/, Battista, G., Ferranti, G., Batini, C.: A Methodology for Conceptual of Statistical Databases
COMPSTAT 1986, proceedings, p. 431
/BORKO 84/ Borzemski, L., Koszalka, L.: Statistical data analysis using a data base for experiment design and systems identification
COMPSTAT 1984, p. 257
/BUCK 73/ Bucknell, M.: A Plex System for Data Base Organization University of Stockholm and NCBS Sweden, 1973
/CASA 88/ Catarci, T., Santucci, G.: A Complete Graphical Conceptual Language for Definition and Manipulation of Statistical Databases
COMPSTAT 1988, proceedings, p. 401
/COD 69/ CODASYL Systems Committee, A Survey of Generalized Data Base Management Systems, May 1969
/COD 71a/ CODASYL Systems Committee, Report of the CODASYL Data Base Task Group
ACM, April 1971
/COD 71b/ CODASYL Systems Committee, Feature Analysis of Generalized Data Base Mangement Systems
ACM, May 1971
/COMV 86/ Costa, P., Marozza, F., Cinciguerra, F.: Developed with a 4th Genration Language
COMPSTAT 1986, proceedings, p. 425
/ECESEM 77/ ECE-Seminar: Statistical Services in Ten Years Time,
Washington, D.C., 21.-25. March 1977; Pergamon Press
/FIRTH 73/ Firth, J.W.: The Idea of a Data Base
NCBS, Sveden, 1973
/HAJÖ 86/ Haux, R., Jöckel. K.-H.: Database Management and Statistical Data Analysis; The Need for Intergration and for Becoming More Intelligent
COMPSTAT 1986, proceedings, p. 407
/JABP 86/ Jasinski, P.J., Butlewski, B., Paradowski, S.: Toward Natural Integration of Database Management System and Statistical Software
COMPSTAT 1986, proceedings, p. 437
/JEAYS 84/ Jeays, T.M.; Statistical Database Management: The relational approach meets the special needs and basic requirements of a statistical office
ECE/UNDP/SCP/H. 3 - 1984 -
/JORK 84/ Jomier, G., Kezouit, O., Relambondrainy, H.: SILCA-PEPIN: A system integrating data anlysis and relational data base management system
COMPSTAT 1984, p. 263
/KEBE 88/ Keller, W.J., Bethlehem, J.G.: New Perspectives in Computer Assited Survey Processing
COMPSTAT 1988, proceeedings, p. 377

/KULE 84/ Kurzynski, M.W., Lebiediewa, S.: Processing of statistical data for multistage recognition system using data base
COMPSTAT 1984, p. 268
/LANG 69/ Langefors, B.: Data Bases, Goals and Definitions
Stockholm University 1969
/LEPB 86/ Lella, G., Pavan, S., Bucci, P.: Easy-Lind: An Expert Natural Language Interface to a Statistical Data Bank
COMPSTAT 1986, proceedings, p. 442
/LES 86/ Leonard, M., Snella, J.J., Abdeljaoued, A.: Farandole:
A RDBMS for Statistical Data Analysis
COMPSTAT 1986, proceedings, p. 448
/MAR 86/ Marbach, G.: Informatics and Statistics
COMPSTAT 1986, proceedings, p. 415
/NERA 85/ Neumann, K., Rauch, L.: Results of the Statistical Computing Project - An example of succesful international cooperation
Statistical Journal, UN-ECE, Vol. 3, No 4/1985, p. 375 - 379
/NEU 84/ Neumann, K.: Metainformation as a tool for integration
COMPSTAT 1984, proceedings, p. 273
/NEU 86/ Neumann, K.: Model- and Method Banks (MMB) as a Supplement to Statistical Data Banks
COMPSTAT 1986, proceedings, p. 454
/NORD 67/ Nordbotten, S.: Automatic files in statistical systems,
UN-Conf. Eur. Stats. /WG. 9/ Geneva 1967
/RARI 84/ Rafanelli, M., Ricci, F.L.: Statistical database: an interactive language for logical schema definition by means of a model based on graphs
COMPSTAT 1984, p. 279
/RAU 84/ Rauch, L.: Interface problems between software packages used in statistical information system
COMPSTAT 1984, p. 285
/RUB 88/ Rubin, D.B.: Multiple Imputation for Data-Base Construction
COMPSTAT 1988, proceedings, p. 389
/SON 86/ Sonnberger, H.; A Security Mechanism Based of Uniform Rouding to prevent Tracker Techniques
COMPSTAT 1986, proceedings, p. 460
/SUND 73/ Sungren, B.: An infological Approach to Data Bases
Stockholm 1973
/SCP 89/ Recent activities of the Statistical Computing Project
ECE/UNDP/SCP-2/10/1989
/STATJ 84/ Statistical Journal, UN-ECE, UN-ECE, Vol. 2 No 2/1984,
p. 213: The Statistical Computing Project
/WERA 89/ Westlake, A., Rafanelli, M.: Report on the fourth international working conference on statistical and scientific database management, Rome, 21 - 23 June 1989
Statistical Software Newsletter, Vol. 15, No. 1, 1989

A Structured Language for Modelling Statistical Data

T. Catarci, G. D'Angiolini and M. Lenzerini, Roma

In this paper we describe a new language for statistical data modelling, which offers a general framework for the representation of elementary and summary data. Three are the main characteristics of the language: 1) the types of modeling primitives of the language are particularly suited for representing objects from a statistical point of view; 2) the language includes a rich set of structuring mechanisms for both elementary and summary data, which are given a formal semantics by means of logic; 3) the language is equipped with specialized inference procedures, allowing for performing different kinds of checks on the representation.

1 Introduction

In the recent years, several approaches have been proposed for modelling large amounts of data from a statistical point of view [BD88,SW85,Su83,Gh86,OO85,OOM87]. A common characteristic of such approaches is to model elementary data by usual data models, then defining particular representation structures for summary data (i.e. data obtained by grouping elementary data, and by applying statistical operators -- such as total, average, percentage -- to such groupings). It follows that an integrated language for defining both elementary and summary data in a suitable way for statistical applications does not exist. Moreover, the proposed data models are often lacking in formalization, and do not provide the designer with powerful deductive capabilites.

In this paper we present a new language for statistical data modelling, aiming to overcome the above mentioned drawbacks of existing proposals, and to offer a general framework for the representation of elementary and summary data. Three are the main characteristics of our proposal: 1) the types of modeling primitives of the language are particularly suited for representing objects from a statistical point of view; 2) the language includes a rich set of structuring mechanisms for both elementary and summary data, which are given a formal semantics by means of logic; 3) the language is equipped with specialized inference procedures, allowing for performing different kinds of checks on the representation.

In the tradition of the object-centered languages for knowledge representation (see [BS85, NS89]), our language allows both concepts and relationships to be described. A concept is an abstraction of a set of real world entities, whereas a relationship is a binary association between two concepts. Both concepts and relationships can be described by means of suitable constructors, starting from a set of primitive concepts (i.e. concept which are described simply by a name). Due to this property, languages of this kind has been called *concept description languages*. Concept description languages originated with Brachman's KL-ONE [BS85], and grew out of research in semantic networks and frame-based systems.

This work has been supported by the "Progetto Finalizzato Informatica e Calcolo Parallelo" of the Italian National Research Council (CNR).

Using the constructors of the language, the statistician can define a basic collection of concepts, together with derived concepts, which are described by posing some restriction condition on the data in the schema. The language is also equipped with suitable constructors for the definition of summary data, which are not present in the usual concept description languages.

2 Basic Notions for Statistical Data Modelling

We are interested in statistical activities studying phenomena concerning finite populations (e.g. households, persons, enterprises, etc.).

The mean for observing a phenomenon on a finite population is the *statistical survey*. In a statistical survey, the *populations* of interest are defined, and the *phenomenon* is characterized by the values of a set of *variables* observed on each element of the populations. For instance, if the population is *Person*, the variables *Labour Force Status*, *Professional Status*, *Occupation* , *Wage* etc., may be chosen to characterize the phenomenon *Employment*. We use the word *statistical unit* to refer to an abstraction of a set of real world objects, called its instances, which form an enumerable class. The basic property of a statistical unit is the possibility of enumerating its instances and calculating totals on it, obtaining summary data.

The properties of an enumerable class represented by a statistical unit are modeled by means of *variables*. Example of variables describing properties of *Person* are: *Age*, *Sex*, *Personal-income*, *Address*, *Occupation*, etc. We distinguish between qualitative and quantitative variables. The domain of a qualitative variable is a set of states, which can be associated with the elements of the enumerable class. The domain of a quantitative variable is a set of numerical values, representing the observed measures of a measurable characteristic. A summary variable is such that it makes sense to summarize it with respect to a statistical unit. Examples of summary variables are *Personal-Income*, *Expenditure for Vacations*, etc.

Since variables describe properties, they are defined as functions linking their own codomain to a statistical unit. For example *Sex* links *{Male, Female}* to *Person*. Statistical units are linked to each other by *functions* or *relations* (notice that not every function is a variable). For instance, the statistical units *Person* and *Vacation* are linked to each other by the relation *Enjoy*. Links between codomains are realized by 1:n functions. For example, the codomain of *District* is linked to the one of *City*, where each city may be intended as a grouping of several districts.

In a statistical survey, it may be interesting to single out subsets of statistical units, called *statistical sub-units*. A statistical sub-unit S is described starting from a statistical unit U, and specifying the condition that the instances of U must satisfy in order to be instances of S. The condition may be defined either on variables of U, or on statistical units which are linked to U by functions or relations. For example, *Person with age<14* is obtained from the statistical unit *Person* using its variable *Age*; *Household with unemployed head of household* is obtained from *Household* using the variable *Labour-Force-Status* and the statistical unit *Person*. Statistical sub-units can also be obtained by set operations on other sub-units.

The ultimate goal of a statistical survey is to compute summary data concerning the phenomenon of interest. The summary data of a statistical survey are specified by means of statistical goals. A *statistical goal* is the specification of how to compute a value using statistical operators. Two main kinds of goals may be singled out, depending upon the type of the statistical operator.

The first kind of goals refers to either totals computed with respect to summary variables, or count of elements of a statistical unit. These are defined by using statistical units partitioned according to the values of some variables. For example, referring to the statistical unit *Person* and the variables *Sex*, *Age*, and *Personal-income*, we may calculate the goal *Personal-income by Sex and Age*, where *Personal-income* is the summary variable and *Person* is partitioned using the values of the cartesian product of the codomains of *Sex* and *Age*. The second kind of goals refers to computations obtained by using statistical operators such as average, percentage, square mean error, codeviance, etc. Due to the semantics of such operators, the goals of this kind can only defined on previously defined goals. For example, starting from the goals *Personal-income by Sex and Age* and *Number of Persons by Sex and Age*, we may define *Average on Personal-income by Sex and Age.*

3 The Language

In this section we describe the syntax and the semantics of a concept description language for the specification of statistical surveys. The language provides suitable constructors for describing statistical units, variables, statistical sub-units, aggregates and goals.

We assume the existence of two countably infinite sets *P*, *R* of concept and relationship symbols, respectively. The set *R* is partitioned into two subsets, one containing binary relation symbols, called roles, and one containing function symbols. The symbols in *P* and *R* describe primitive concepts and relationships, i.e. elementary concepts and relationships which are simply described by means of a name. We assume that *P* contains two distinguished concept symbols, namely Empty and T, denoting the empty and the universal concept respectively, together with suitable concept symbols denoting the usual domains of values used in programming languages, such as INTEGER, REAL, BOOLEAN, etc..

Concept and relationship descriptions are built out of concept and relationship symbols in *P* and *R* according to the following syntax (P denotes a primitive concept symbol, Q a primitive relationship symbol, $C, D, C_1, \ldots, C_n$ concept descriptions, R,S,T relationship descriptions, $F_1, \ldots, F_m$ function symbols, $v_1, \ldots, v_n$ values and n an integer):

C,D	--->	P | *not* P | C *and* D | C *or* D | *some* R.C |
		all R.C | *atleast* n.R | *atmost* n.R | F.C |
		setof C | $[F_1: C_1, \ldots, F_m: C_m]$ | $(v_1, \ldots, v_n)$
S,T	--->	Q | S ∘ T | S^{-1} | S:C

The set of all the concept descriptions built up from *P* and *R* will be denoted by *L(P,R)*. Notice that, in the above syntax, $v_1, \ldots, v_n$ denote values. In fact, we assume that a symbol is available for each value (of type integer, boolean, real, etc.) of interest. As a notational convenience, we also assume that sets of values which are intervals of an ordered domain can be written in the form: $v_1..v_n$.

The intuitive meaning of the operators *not*, *and* and *or* is simply set complement, set intersection and set union, respectively. Notice that *not* can only be applied to primitive concepts. The operators *setof* is used to describe the set of all the collections of instances of a given concept. The concept $[F_1: C_1, \ldots, F_m: C_m]$ denotes the set of all the (labeled) tuples where each component is labeled by the function symbol F_i and has an instance of C_i as value. The operators *some* and *all* are used to describe concepts on the basis of their linking to roles: intuitively, *some* R.C denotes the set of objects which are linked by the role R to at least one instance of the concept C, whereas *all* R.C denotes the set of objects which are linked by R to objects that are instances of C (analogously for F.C, when dealing with functions). The operators *atleast* and *atmost* are used to constrain the number of links of type R. With regard to relationships, the symbol "∘" denotes composition, the symbol $^{-1}$ is used to denote the inverse of a relationship, and the symbol ":" denotes the restriction of a relationship to those pairs whose second component is an instance of the concept C. More formally, the semantics of the language states that each description in *L(P,R)* is interpreted as a subset of a universe U, according to suitable rules defining the meaning of the operators. The universe U is a set of structured objects, built up from a basic domain D of elementary objects, containing all the values of interest, and defined as the smallest set containing D and such that, if $u_1,...,u_n$ are in U, and $F_1,...,F_n$ are function symbols of *L(P,R)*, then both $<F_1:u_1,...,F_n:u_n>$ (labeled tuple) and $\{u_1,...,u_n\}$ are in U. Let I(·) be a function mapping each concept to a subset of U, and each relationship to a subset of U×U. I is called an interpretation for L(P,R) if the following conditions hold (other obviuos conditions concern complement, intersection and union).

$I(\textit{some } R.C) = \{ x \in U \mid \text{there exists } <x,y> \in I(R) \text{ such that } y \in I(C) \}$

$I(\textit{all } R.C) = \{ x \in U \mid \text{for each } <x,y> \in I(R) \text{ it holds that } y \in I(C) \}$

$I(\textit{atleast } n.R) = \{ x \in U \mid \text{there exist at least n y such that } <x,y> \in I(R) \}$

$I(\textit{atmost } n.R) = \{ x \in U \mid \text{there exist at most n y such that } <x,y> \in I(R) \}$

$I(F.C) = \{ x \in U \mid <x,y> \in I(F) \text{ and } y \in I(C) \}$

$I([F_1: C_1,...,F_m: C_m]) = \{ <F_1: v_1,...,F_m: v_m> \in U \mid v_i \in I(C_i), i=1,...,m \}$

$I(\textit{setof } C) = \{ \{v_1,...,v_k\} \in U \mid v_i \in I(C), i=1,...,m \}$

$I((v_1,...,v_n)) = \{v_1,...,v_n \}$

$I(S \circ T) = \{ <x,y> \in U \times U \mid \text{there exists } z \in U \text{ such that } <x,z> \in I(S) \text{ and } <z,y> \in I(T) \}$

$I(S^{-1}) = \{ <x,y> \in U \times U \mid <y,x> \in I(S) \}$

$I(S:C) = \{ <x,y> \in U \times U \mid <x,y> \in I(S) \text{ and } y \in I(C) \}$

An aggregate is a specification of how to classify all the instances of a given concept, called the target concept of the aggregate, according to their values with respect to a collection of properties. For example, an aggregate can be defined classifying *Persons* according to their *sex* and *age*.

Aggregates are classified into simple and complex. A simple aggregate has the form

$$[| s_1,...,s_n |]$$

where $s_1,...,s_n$ are sets of values, often described as intervals, of the same domain D. In this case D is called the target of the aggregate. A complex aggregate has the form:

$$\textit{agg } C * S_1.B_1 *...* S_n.B_n$$

where C is a concept, $S_1,...,S_n$ are relationships, and $B_1,...,B_n$ are simple aggregates whose target are codomains of variables. In this case, C is the target of the aggregate.

From a semantical viewpoint (see [CDL90]), a simple aggregate is a set of sets, whereas a complex aggregate of the form $\textit{agg } C * S_1.B_1 *...* S_n.B_n$ is defined as a set of labeled tuples. Each tuple corresponds to a combination of values for the specified properties, plus the set of all the instances of the target concept having such values of properties.

A goal is a specification of a meaningful computation to be performed in the context of the statistical survey of interest. In the definition of goals we make use of a set of operators which are usually considered in statistical data bases, such as COUNT, AVERAGE, etc.. Each operator has a set as argument and returns values of a distinguished type.

Goals are classified into simple and composite. A simple goal has the form:

$$(\textit{goal } OP \textit{ on } R \textit{ of } AGG)$$

where OP in an operator, R is a summary variable, and AGG is an aggregate. Intuitively, the above goal is a specification of a collection of values, each one relative to the target component $M=\{c_1,...,c_p\}$ of an instance of AGG; each value is obtained by applying the operator OP to the set of values corresponding to the property R of $c_1,...,c_p$.

A composite goal is simply an application of an operator to a collection of arguments which are goals themselves. The form of a composite goal is

$$(goal\ \mathrm{OP}\ on\ G_1 \ldots G_n).$$

and its semantics is easily obtained from the semantics of both the operator OP and the arguments $G_1,\ldots,G_n$.

A specification is a collection of statements describing all the relevant knowledge about a statistical survey. It is constituted by three parts. In the first part, names are assigned to the various objects involved in the survey, and concepts are classified into statistical units and statistical sub-units, whereas relationships are classified into variables, roles and functions. In the second part, constraints are declared describing the definition of statistical sub-units (in this case the constraint has the form (P *def* α), where P is the name of a sub-unit, and α a concept description), the conditions that units must satisfy (constraints of the form (P *isa* α), where P is the name of a unit, and α a concept description), and the domain and codomain of variables, roles and functions (constraints of the form R: $D_1 \times D_2$ where D_1 and D_2 are concept description). Finally, the third part of a specification is devoted to the definition of both aggregates and statistical goals. An interpretation I is called a model of specification σ, if every statements of σ are true in I.

4 Reasoning About a Specification

The language described in the previous section allows for expressing the knowledge about a statistical survey. Based on the formal definition of the semantics of the language, we can devise specialized techniques for reasoning about a specification, which constitute a valuable support to the designer of statistical data schemas. Notice, first of all, that a specification may suffer from a number of anomalies, due to a wrong usage of the constructs of the language. For example, it may happen that a concept in a specification is defined in such a way that it is invariably empty in all the models of the specification.

In order to formalize the above considerations, we now introduce the notions of consistency and inconsisteny of concepts and specifications. A concept C is said to be inconsistent (consistent otherwise) in a specification σ, if for each model M of σ, it holds that $M(C)=\emptyset$. A specification σ is said to be consistent (inconsistent otherwise) if no concept of σ is inconsistent. Also, we say that the concept C is subsumed by the concept D in a specification σ, if for each model M of σ, M(C) is a subset of M(D).

Consistency can be thought of as the basic property to be checked about a specification. More generally, the designer may want to verify whether a certain property holds in all the models of a specification σ. This can be done by providing the designer with a deductive method that, given a specification σ and an assertion α of one of the forms described for constraints, checks whether $\sigma \models \alpha$, i.e. whether α is satisfied by every model of σ. Notice that, for example, the consistency of the concept C may be checked by testing whether ($\sigma \models$ C *isa* Empty) -- where Empty is a pre-defined concept which is empty in every interpretation -- and subsumption can be checked by testing whether ($\sigma \models$ C *isa* D). For the sake of brevity, we do not delve into the details of such a method for our language. The interested reader is referred to [CDL90].

Other reasoning facilities are provided in our language, concerning aggregates and goals. An aggregate A is said to be derivable from an aggregate B in a specification σ, if there exists a total one-to-many correspondence R from the instances of A and B such that, for each model M of σ, for each x in M(A), $x=y_1\oplus\ldots\oplus y_n$, where $R(x)=\{y_1,\ldots,y_n\}$, and $\oplus$ denotes an operator between tuples, defined as follows:

$$\langle F_1{:}v_1,\ldots,F_n{:}v_n, \mathrm{Target}{:}c\rangle \oplus \langle F_1{:}w_1,\ldots,F_n{:}w_n, \mathrm{Target}{:}d\rangle =$$
$$\langle F_1{:}v_1\cup w_1,\ldots,F_n{:}v_n\cup w_n, \mathrm{Target}{:}c\cup d\rangle.$$

In [CDL90], we describe sufficient conditions for derivability of aggregates. These may be used for verifying what we call consistency of aggregates: an aggregate A is said to be inconsistent in a specification σ, if for each model M of σ, for each tuple x in M(A), the Target component of x is empty. Notice that A=(*agg* C $*$ $S_1.B_1$ $*\ldots*$ $S_n.B_n$) is inconsistent in σ, if and only if (*agg* Empty $*$ $S_1.B_1$ $*\ldots*$ $S_n.B_n$) is derivable from A.

Therefore, we can use derivability for checking the consistency of aggregates. An analogous notion of derivability can be defined for goals, as described in [CDL90].

References

[Br81] Bragg A.W., "Data Manipulation Languages for Statistical Databases - The Statistical Analysis System (SAS)", Proc. of the First LBL Workshop on SDB Management: 147-150, 1981.

[BD88] Batini C., Di Battista G., "Design of Statistical Databases: A Methodology for the Conceptual Step", *Information System*, 13(4): 407-422,1988.

[BS85] Brachman R.J., Schmolze J.G., "An Overview of the KL-ONE Knowledge Representation System", *Cognitive Science*, 9(2): 171-216, 1985.

[CDL90] Catarci T., D'Angiolini G., Lenzerini M., "Concept Description Language for Statistical Data Modelling", Technical Report, Dipartimento di Informatica e Sistemistica, Università di Roma "La Sapienza", Italia, 1990.

[Gh86] Ghosh S.P., "Statistical Relational Tables for Statistical Database Management", *IEEE Transactions on Software Engineering*, SE-12(12): 1106-1116, 1986.

[NS89] Nebel B., Smolka G., "Representation and Reasoning with Attributive Descriptions", IWBS Report 81, IBM Deutschland, Stuttgart, West Germany, 1989.

[OO85] Ozsoyoglu G., Ozsoyoglu Z.M., "Statistical Database Query Languages", *IEEE Transactions on Software Engineering*, SE-11(10): 1071-1081, 1985.

[OOM87] Ozsoyoglu G., Ozsoyoglu Z.M., Matos V., "Extending Relational Algebra and Relational Calculus with Set-Valued Attributes and Aggregate Functions", *ACM Transactions on Database Systems*, 12(4): 566-592, 1987.

[SW85] Shoshani A., Wong H.K.T., "Statistical and Scientific Database Issues", *IEEE Transactions on Software Engineering*, 11(10), 1985.

[Su83] Su S.Y.W., "SAM*: A Semantic Association Model for Corporate and Scientific Statistical Database", *Information Science*, 29: 151-199, 1983.

Time Dependent Models

Spectral Analysis of Non-Stationary Time Series

I. G. Žhurbenko, Moscow

The last decade has seen a great popularity of application of the methods of the correlation and spectral analyses of time series [1-9]. The spectral approach to researches had a necessary condition: a given model of the series must be stationary. But in practice in most researches this condition is possible only within some limited time interval or impossible at all. At the same time a statistical stady of the spectra of these series is drowing more and more attention [4-12] despite the qualitativs and technical complexity of models. Spectral estimates obtained by time shift (they were systematically studied in [8,9]) suggested a new approach which is simple and convenient in most situations. Below is the study of a time series model which not only substantiates investigation of spectra changing in time but also brings us to the problems of the multivariable statistical analysis of these spectra: pattern recognition in the spectral domain, spectral desorders, spectral investigation when there are many strong non-stationary phenomena, and others. The size of the paper and some methodical considerations prevent demonstration of application of this approach to various problems (its simplicity makes these problems quite easy and particularly logical from the researcher's viewpoint. At present on the basis of the enterprize "Spectrum" a group of

Compstat 1990

scientists (I.A.Kozhevnikova, I.G.Niddeker and the author of this paper) are working on a poket of programms RSA on the time series analysis including the above-mentiond approach. The pocket with illustrations is intended for a wide group of researches who have a good intuition and educational background in the field of their researches but superficial knowledge of time series analysis.

The main model

Let X(t,u) be a time series which is stationary in a wide sense with respect to the descrete parameter t and dependent on the continuous real parameter u. Besides, we shall consider the following conditions satisfied:

I. There exist uniformly bounded absolute moments up to the fourth order including for the entire set of random quantities X(t,u):

$$E|X(t,u)|^k \leqslant E|X|^k, \quad k=1,2,3,4. \tag{1}$$

With respekt to the parameter u, the random quantities X(t,u) are continuous in the mean square and

$$E|X(t,u+\Delta u) - X(t,u)|^2 \leqslant \rho^2 |\Delta u| E|X|^2 \tag{2}$$

holds. The coefficient ρ indicates the speed of deviation from stationarity along the parameter u. In some cases we shall require that the fourth cumulants [1,9] of the random quantities X(t,u) should be continuous with respect to the parameter u as well:

$$|S_4(X(t,u+\Delta u)) - S_4(X(t,u))| \leqslant \rho |\Delta u|^2 E X^4 \tag{3}$$

II. The stationary set of random quantities along the parameter t will be considered to have only weak dependence conditions which are sufficiently general and uniform with respect to u. For example, let the Ibragimov weak dependence condition be satisfied:

$$\sup_t \left| p\,(A/B) - p\,(A) \right| = \beta_4(n) \leqslant \beta(n)$$

$$A \in \mathfrak{A}_{-\infty}^{t}(u)$$

$$B \in \mathfrak{A}_{t+n}^{+\infty}(u)$$

where $\mathfrak{A}_a^b(u)$ is a σ-algebra generated by the random quantities X(t,u), $t \in [a,b]$.

The conditions of group I are the general conditions of existence and continuous dependence with respect to the parameter u of the moments of the random quantities X(t,u). Thus, we suppose that there may be weak changes in random quantities and their characteristics in dependence on the parameter u. The conditions on the fourth moments (1), (3) will be used only to derive the equality (7). Along the parameter of stationarity t the supposed conditions of weak dependence are standard for such situations, they can be weakened to the Rosenblatt mixing conditions [8,9].

Suppose we obserbe a sample trajectory of a stationary process X(n,n), n=1,2,..., N. The sample estimates of the covariance function and the spectrum of the time series X(t,u) u=const will be constructed from the sample X(n,n), n=1,...,N which will be called a series with slowly changing covariance function and spectrum.

The process X(t,u) is considered stationary along the

time parameter t. Observing here a short part of the process, we try to notice all of its high-frequency changes. It is also natural to consider the variable u a time parameter, but along this parameter we try to notice the slow changes in the spectrum itself. This natural separation of stationary and non-stationary elements is in conformity wiith our intuitive idea of the physics of non-stationary processes and helps us with a mathematically exact formulation of the problem.

How exactly can we estimate the true spectral density $f(\lambda,u)$ of $X(t,u)$ for u=const from the sample $\{X(n,n)\}$; within what limits with respect to N is this estimate effective; under what conditions can we consider a spectrum invariable; from what moment do we have to mark a meaningful change in it? These and other similar problems will be posed on the basis of the model of interest.

In the conditions (I) for the series $X(t,u)$ the covariance function

$$R(n,u) = \mathrm{Cov}(\,X(t,u),\, X(t+n,u))$$

exists and is bounded, it is independent of the parameter t becaus of stationarity. In the case of convergence, in the terms of condi tion II of the series

$$\sum_{n=-\infty}^{\infty} \beta^{1/2}(n) = \beta$$

the spectral density

$$f(\lambda,u) = \frac{1}{2\pi} \sum_{n=-\infty}^{\infty} e^{-in\lambda} R(n,u)$$

exists and is bounded. With conditions I,II satisfied, the covariance function and spectral density of the stationary series $X(t,u)$ turn out to be weakly dependent on the parameter u in

$$
\begin{aligned}
&\left|R(n,u+\Delta u) - R(n,u)\right| \leqslant 4\rho(\beta(n)\,|\Delta u|)^{1/2} E X^2 \\
&\left|f(\lambda,u+\Delta u) - f(\lambda,u)\right| \leqslant \rho|\Delta u|^{1/2}\beta E X^2
\end{aligned}
\qquad (4)
$$

In practice it is more natural to define the changeability coefficient of a spectrum by formulae (4) because of some a priori considerations. For example, in a speech signal research the change in the spectrum can be considered to be equal to zero in the limits $|\Delta u| = 10^{-1}$ sec and different from zero only for $|\Delta u| > 1$ sec. In many situations the character of the dependence of a real spectrum on u can be defined by the physics of the phenomenon [4,5], in some applications this dependence is evident, in the problem of spectral disorders it is a jumplike function, etc. In these cases it is natural to insert the a priori information into formulae (4).

B i b l i o g r a p h y

I. D.Brillinger, Time Series Data Analysis and Theory, Holt,Rinehart and Winston Inc., New York,1975,500 p.

2. A.M.Yaglom, Correlation Theory Stationary Statistic Functions, L., Hydrometeorological Publishing House, 1981,279 p.,(russian).

3. R.K.Otnes, L.Enochson, Applied Time Series Analisis, John Wiley and Sons, New York,1978,...p.

4. S.M.Ratov, Y.A.Kavtsov, V.I.Tatarsky, Introduction into Statistical Radiophysics and Stochastic Processes, M.,Science,1976,5.. p.,(russian).

5. M.Rosenblatt, Stationary Sequences and Random Fields, Boston, Birkhauser,1985,258 p.

6. Handbook of Statistics,3,Time Series in the Frequency Domain, Ed. by D.Brillinger, P.R.Krishnaiah, Amsterdam, North Holland.1983,485 p.

7. Handbook of Statistics,5,Time Series in the Time Domain, Ed. E.Nannan, P.R.Krishnaiah, M.Rao, Amsterdam, North Holland,1985.490p.

8. I.G.Zurbenko, The Spectral Analisis of Time Series, Amsterdam, North Holland,1986,247p.

9. I.G.Zurbenko, Analisis Stationary and Uniform Stohastic Systems, M.,Moscow University publishing house,1987, 240 p.,(russian).

10. W.Fuller, Non-Stationary Autoregressive Time Series, 1-24, Handbook of Statistic, 5, Amsterdam, North Holland,1985.

11. M.B.Priestley, Spectral Analysis and Time Series, Vol. 1,2,New York, Academic Press,1981,890 p.

12. C.Bhagavan, On Non-Stationary Time Series, 311-320, Handbook of Statistic,5, Amsterdam, North Holland, 1985.

ARMA Simulator for Testing Methods of Process Diagnostic

O. Lessi, Potenza, L. Olivi and P. Parisi, Ispra

1. Introduction

In the management of automatic or semi-automatic plants it is important to check that the system dynamics do not undergo variations beyond limits considered acceptable, or, if this happens, the variations which have occurred must be quantified. The presence of a large number of variables introduces serious obstacles in system management because of: the consistency of behaviour among the variables, the computation time for their control and the difficulties in diagnosis.

A methodology for globally monitoring and on-line diagnosing plants and process operations has been developed. It consists of a set of mathematical and statistical procedures enabling the process under study to be modelled and the actual and model predicted behaviours to be contrasted.

This study refers to physical systems, or parts of them, suitable to be represented by matrix differential equations relating input and output variables, which can be approximated by dynamic stochastic models. These models are characterised by parameters, typical of the phenomena being studied.

The whole study has been subdivided into three sections: identification, analysis, simulation.

i) Identification. In the processes considered [6], the possible stationary states can be identified and, therefore recognized. Under some favourable physical conditions, e.g. soft changes, also operational transients can be enclosed.

ii) Statistical Rupture Analysis. Suitable output functions of the process and its model can be contrasted and analysed on line by statistical tests. Control card-like devices can be used to monitor system operations and report when and where the operations modelled no longer describe a healthy system. The set of these procedures has been compacted in the "Rupture Analyser" programme [2,3].

Compstat 1990

iii) Simulation. A major theme of the work, as described in this paper, is to ascertain the practicality and sensitivity of the statistical rupture tests and their calibration in the cases under study. To this purpose an ARMA-generator simulating the foreseeable situations has been designed and implemented.

2. Phisical systems and their models: Dynamic systems.

A quite general representation of dynamic systems is given by differential equations of the type:

$$\dot{X}(t) = f[X(t)] + g[U(t),X(t)] \tag{1}$$

where $X(t)$ is the k-dimensional vector of output variables from the system at time t; $\dot{X}(t)$, the vector of the derivatives with respect to t; $U(t)$ is the m-dimensional vector of input variations; f and g are continuous differentiable functions.

Observation of the variables $U(t)$ and $X(t)$, at discrete times $t_1,\ldots t_n$, gives rise to an approximating system of finite difference equations, which, in several cases, can be solved by numerical techniques. On the other hand, the analytical representation is often unpractical, because either the properties of stability and observability of the system are not easily derivable, or non-negligible stochastic sources of disturbance are present. For these reasons, a black box approximation, by suitable stochastic models, is malleable and flexible for adaptation to dynamic systems.

3. Stochastic representation of dynamic systems

A process X is said to be in k-variate Auto Regressive Moving Average form (k-ARMA(p, q)), when satisfies the linear equation:

$$X_t = \sum_{j=1}^{p} \Phi_j \cdot X_{t-j} + \sum_{j=1}^{q} \Theta_j \cdot \xi_{t-1} + \xi_t \tag{2}$$

here, X is a stationary k-variate process, the output, ξ_t the white noise process, the input, with variance-covariance matrix Σ, t the time index and $\Phi_1,\ldots,\Phi_p$, and $\Theta_1,\ldots,\Theta_q$ are coefficient matrices with p and q the order of complexity of the model. By introducing the backshift operator z, the (2) becomes:

$$\Phi(z^{-1})\cdot X_t = \Theta(z^{-1})\cdot \xi_t \tag{3}$$

The term $\Theta(z^{-1})\ \xi_t$ summarises and absorbs various types of random error: modelling and stochastic approximation errors, variable measurement errors and stochastic sources of disturbance. Modelling, perfomed by a k-ARMA(p,q) is suitable to fit time dependent stochastic systems, when linearity or quasi-linearity can be invoked: i.e. the input-output relationship can be represented as a linear function between the output and input vectors at time t and their values at the preceeding times.

4. The ARMA generator

The (2) has been taken as the reference model. If the polynomial $Det[\Phi(z^{-1})]$ has no roots inside the unit circle, the process (2) is stationary. The condition is the same as requiring that the eigenvalues of the companion matrix are inside the unit circle [5]. Analogously, if the polynomial $Det[\Theta(z^{-1})]$ has no roots inside the unit circle, the process is invertible. The ARMA generator, implemented according to the algorithm suggested by Barone [1], has been organized in the following steps:

i) Generation of the companion matrices' eigenvalues λ_i, and μ_j, $i = 1,\dots, p$, $j = 1,\dots, q$, such that $|\lambda_i| < 1$ and $|\mu_j| < 1$, for the autoregressive and moving average parts of the model. An automatic procedure, the "rake algorithm" [4] enables the eigenvalues to be encompassed and uniformly distributed without overlapping overall their assigned ranges $r(\lambda)$ and $r(\mu)$. The allowed simultaneous maximum values, confirmed by several exepriments, are $k = 5$, $p, q = 10$, [3,4].

ii) Numerical solutions of the 2k linear equation systems:

$$A_h\cdot \mathbf{v}_h = \alpha_h, \qquad B_h\cdot \mathbf{w}_h = \beta_h,$$

$$(A_h)_{i,j} = [\lambda_{(h-1)p+i}]^{h-j}, \qquad (\alpha_h)_i = [\lambda_{(h-1)p+i}]^{p},$$

$$(B_h)_{r,s} = [\mu_{(h-1)q+r}]^{q-s}, \qquad (\beta_h)_r = [\mu_{(h-1)q+r}]^{q},$$

$$h = 1,\dots, k, \quad i,\ j = 1,\dots, p, \quad r,\ s = 1,\dots, q.$$

A and B are Vandermonde matrices; their determinants even if always non-zero can be very small, leading to numerical instability. In spite of that, the solutions $\mathbf{v}_h$ and $\mathbf{w}_h$, being the ratios between very small numbers of the same order, as the Vandermonde cofactors and the Vandermonde determinant itself, can be easily calculated by a specially taylored algorithm [3].

iii) Provided by a suitable algorithm [1,5] the arbitrary non singular matrices H, the coefficient matrices Φ_i and Θ_j are:

$$\Phi_i = H^{-1} \cdot \mathrm{diag}[(v_1)_i, (v_2)_i, \ldots, (v_k)_i] \cdot H,$$
$$\Theta_j = H^{-1} \cdot \mathrm{diag}[(w_1)_j, (w_2)_j, \ldots, (w_k)_j] \cdot H.$$

iv) Generation of the white noise ξ_t by a classic IMSL library routine, μ and Σ being assigned.

v) By introducing the computed matrices Φ_i and Θ_j and the sequence $\{\xi_t\}$ in (2), an ARMA trajectory is obtained.

It is worth to remark that, as far as the determination of Φ_i and Θ_j is concerned, the matrix H does not play the same role of the vectors $\mathbf{v}_h$ and $\mathbf{w}_h$, because:

$$\mathrm{Det}[\Phi_i] = \mathrm{Det}\{H^{-1} \cdot \mathrm{diag}[(v_1)_i, (v_2)_i, \ldots, (v_k)_i] \cdot H\} =$$
$$= \mathrm{Det}\{\mathrm{diag}[(v_1)_i, (v_2)_i, \ldots, (v_k)_i]\}.$$

Therefore, different choices of H, for the same sets of vectors $\mathbf{v}_h$ and $\mathbf{w}_h$, provide matrix coefficients Φ_i and Θ_j with the same spectra of eigenvalues. That emphasizes the importance of a suitable choice of the covariance matrix Σ to determine the mutual correlation among the k components of $\{X_t\}$.

5. Simulation tests

A simulation campaign for several dynamic systems of different complexity has been performed [2,3]. Ruptures have been simulated by changes in the model parameters, as the mean, the variance-covariance and coefficient matrices (e.g. fig. 1), at assigned times t^*. The resulting models define globally non stationary processes, each one composed by two different ARMA structures, describing the behaviours of the systems before and afterwards the rupture time t^*. The "Rupture Analyser" programme [2,3] explores all the process paths through a window, of assigned width, sliding along all the time axis. Data sampled are processed and analysed according to three statistical tests sensitive to changes of the mean, variance

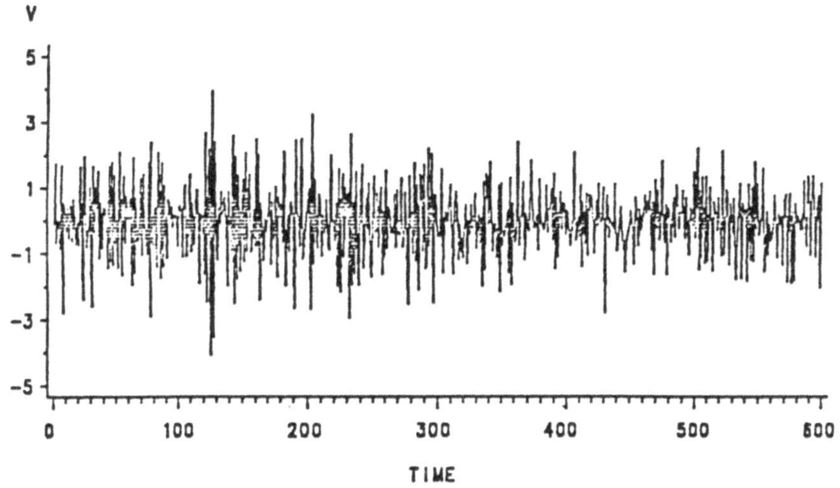

Fig. 1: First output path with coefficient rupture.

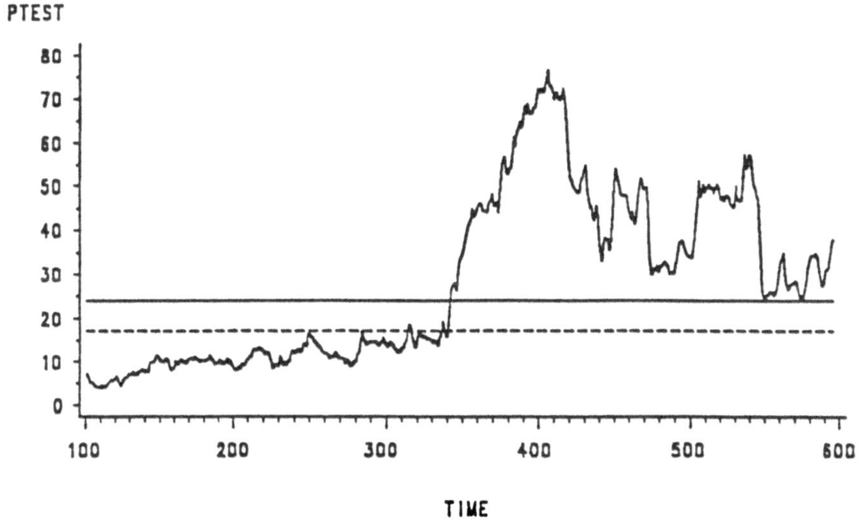

Fig. 2: Portmanteau test with coefficient rupture.

and coefficients, as in the example of fig. 2 .

The figure of merit to measure the effectiveness and performance of such tests and, hence, of all the proposed methodology, is the response time, i.e. the mean delay detection time with respect to the real occurence (in this case the simulated occurence) time of the rupture. The response time depends upon both the rupture detection rapidity of the test and the window width, so that it is a reference value for an optimal use of the "Rupture Analyser" in the cases under study. A selection of the results is given in [2], it shows that, even if in a in vitro experiment, the "Rupture Analyser" can be exploited as an automatic diagnostician in parallel to a human operator, capable to announce possible deviations from healty conditions. In this framework, the ARMA generator presented here, when simulates preidentified real systems, plays the roles of both the calibrator for the automatic diagnostician and the pre-explorer of the possible non standard conditions and, therefore, its exploitation can be of interest for decision support system and ergonomy areas [6].

References:

[1] Barone, P.: A Method for Generating Indenpendent Realizations of a Multivariate Normal Stationary and Invertible ARMA (p, q) Process. Journal of Time Series Analysis Vol. 8, No 2, 125-130, 1987.

[2] Fasso',A., Lessi, O., Olivi, L., Parisi, P.: Statistical Rupture Analysis as Diagnosis Methodology in System Reliability. 7th International Conference on Reliability and Maintainability, 18th-22nd of June 1990, Brest, France.

/3/ Fasso',A. Lessi, O., Olivi, L., Parisi, P.: Methods of Process Diagnostic as Decision Support Systems, EUR report in preparation.

[4] Fasso', A., Lessi, O., Olivi, L., Parisi, P.: Methods of Simulation and Diagnostic of ARMA Modelled Processes, in preparation.

[5] Hannan, E.J.: Multiple Time Series, Wiley, N.Y. 1970.

[6] Lessi, O., Olivi, L., Parisi, P., Varfis, A.: Statistical Aids in Man-Machine Interaction Problems, in Analysis, Design and Evaluation of Man-Machine Systems, Ed. J. Ranta, Oulu 14th-16th of June 1988.

Keywords: Automatic Diagnostician, Decision Support Systems, Ergonomy, Identification, Rupture Analysis.

Statistical Inference in an Extremal Markovian Model

M. I. Gomes, Lisboa

Abstract: The markovian sequence $X_i = k \max(X_{i-1}, Y_i)$, $i\geq 1$, $0<k<1$, X_0 a random variable with distribution function H_0, and $\{Y_i\}_{i\geq 1}$ a sequence of independent, identically distributed random variables, independent of X_0, with d.f. F, is considered in this paper, as the genesis of a model for which statistical inference is developed, under stationarity conditions.

Keywords: Statistical Extremes; Markovian Sequences; Extremal Index; Statistical Inference; Simulation Techniques.

1. Introduction and preliminaries.

The markovian sequence

$$X_i = k \max(X_{i-1}, Y_i),\ i\geq 1,\ 0<k<1, \tag{1.1}$$

X_0 a random variable (r.v.) with distribution function (d.f.) H_0, and $\{Y_i\}_{i\geq 1}$ a sequence of independent, identically distributed (i.i.d.) r.v.'s, independent of X_0, with d.f. F, is considered in this paper, as the genesis of a model for which statistical inference is developed. Related markovian sequences have often been proposed in literature, to model several physical systems, having been studied mainly from a probabilistic point of view . Alpuim (1989) has proved that such a markovian sequence possesses a stationary distribution H(x) if and only if there exists $x>0$, where $F(x/k)>0$ and $0<\sum_{j\geq 1} (1-F(x/k^j)) <\infty$; consequently the class of stationary d.f.'s for the sequence $\{X_i\}_{i\geq 1}$ is a very huge class: for instance all distributions $H(x)=H(x/k)F(x/k)$, the stationarity equation, with support in $\mathbb{R}^+$, and such that $\log H(\exp(x))$ is concave whenever $0<H(\exp(x))<1$, belong to such a class. The *Generalized Pareto* distribution

$$P_\alpha(x) = 1 - (1-\alpha x)^{1/\alpha},\ 1-\alpha x>0,\ x>0,\ \alpha\in\mathbb{R}, \tag{1.2}$$

and the *Generalized Extreme Value* distribution

$$G_\alpha(x) = \exp(-(1-\alpha x)^{1/\alpha}),\ 1-\alpha x>0,\ x\in\mathbb{R},\ \alpha\in\mathbb{R}, \tag{1.3}$$

Compstat 1990

important d.f.'s in *Statistical Theory of Extremes*, are in such class for all α in (1.2) and for $\alpha<0$ in (1.3).

The aim of this paper is to deal with statistical inference on such markovian models, assuming that a random sample $(X_1,X_2,\ldots,X_m)$ has been observed, under stationarity conditions. In §2 of this paper we derive some probabilistic results that help us to estimate k, and to identify models F and H, and in §3, the estimation of the *extremal index* of the underlying stationary sequence, and the influence of a nonparametric empirical rule on such estimation is considered. Notice that the extremal index θ is a key parameter in *Extreme Value Theory*, and may roughly be defined by the approximation $P(\max(X_1,\ldots,X_n)\le x) \approx H^{n\theta}(x)$, H the marginal distribution of the stationary sequence (Leadbetter, Lindgren & Rootzén(1983)). The sequences under study are strong-mixing sequences, that do not verify $D'(u_n)$ condition, but verify $D''(u_n)$ condition (Leadbetter & Nandagopalan (1989)), and consequently they possess extremal index $\theta<1$ in a great variety of situations, referred in more detail in §3 of this paper. See Galambos (1987) for basic concepts on Asymptotic Theory of Extremes.

2. Some probabilistic results and identification of the underlying models.

Let us consider the sequence (1.1) under stationarity, i.e., the d.f.'s F(.) and H(.) are such that

$$H(kx)/H(x) = F(x) \tag{2.1}$$

We have obviously $X_i/X_{i-1}\ge k$, for all $i\ge1$, and consequently, given a sample $(X_1,X_2,\ldots,X_m)$,

$$\hat{k}_m = \min_{1\le i\le m-1} (X_{i+1}/X_i), \tag{2.2}$$

provides an obvious nonparametric estimator of k.

Notice that there is a positive probability of having $X_{i+1}/X_i=k$, once $P(X_{i+1}/X_i=k)=P(X>Y)$, X and Y independent r.v.'s with d.f.'s H(.) and F(.) respectively, i.e.

$$P(X_{i+1}/X_i=k) = \int_0^1 F(H^{-1}(t))dt = \int_0^1 \frac{H(kH^{-1}(t))}{t}dt > 0. \tag{2.3}$$

Consequently, the smaller the value of k, the smaller is also the number of X_{i+1}/X_i equal to k. For k close to 1 we have $P(X_{i+1}/X_i=k)$ also close to 1. On the other hand if we think on the event $\{X_{i+1}>X_i\}$, which implies $\{X_{i+1}=kY_{i+1}\}$, we see that identification of the model F(.) may then be done on the basis of the new sample X_{i+1}/k for all $X_{i+1}>X_i$. The obvious question is whether the size of such a sample is adequate for modelling. We have

$$P(X_{i+1}>X_i)=P(Y>X/k) = 1-\int_0^1 F(H^{-1}(t)/k)dt = 1-\int_0^1 \frac{t}{H(H^{-1}(t)/k)}dt >0, \qquad (2.4)$$

and consequently for k close to zero we identify in average approximately 50% of the sample values as coming from the generating Y , but for k close to 1 the identification of F(.) on the basis of the subsample X_{i+1}/k for all $X_{i+1}>X_i$ becomes extremely difficult.

Given $(X_1,X_2,...,X_m)$, the simplest graphical procedure to make a preliminary informal validation of the markovian model is the identification of a linear horizontal structure for the downpeaks of the plotting points $(i, X_{i+1}/X_i)$, i=1,2,...,m-1. After an informal graphical validation of the model, the previously mentioned results lead us to propose the following method of models identification: a) identify H(.) on the basis of a trivial preliminary data analysis associated to the original sample $(X_1,X_2,...,X_m)$; b) estimate k through $\hat{k}_m = \min_{1\leq i\leq m-1}(X_{i+1}/X_i)$; c) test, on the basis of a batery of goodness-of-fit tests, whether the values $X_{i+1}/k|X_{i+1}>X_i$ may be regarded as observed values from a population F(x) =H(kx)/H(x). We have applied this methodology to several generated markovian samples from (1.1), and to a few real hydrological sequences, and we were successful on validation and identification of the model for all generated samples, and for some of the real samples.

3. Estimation of the extremal index.

We have already roughly defined the *extremal index* θ by the approximation $P(\max(X_1,...,X_n)\leq x) \approx H^{n\theta}(x)$, H(.) the marginal distribution of the stationary sequence. When we are dealing with the stationary markovian sequences $\{X_i\}_{i\geq1}$ in (1.1), it is known that they possess an extremal index if, for some $\alpha\in\mathbb{R}$, H(x), the stationary distribution, and consequently the d.f. of X_i, i≥1, whenever $H_0(x)=H(x)$, belongs to the *domain of attraction for maximum values* of G_α, defined by (1.3). Such sequences may have any extremal index in (0,1], $\theta=1$ if H(x) is in the domain of attraction of G_α, $\alpha\geq0$, and $0<\theta<1$ if H(x) is the domain of attraction of G_α, $\alpha<0$. Notice that the extremal index θ is always between 0 and 1, being equal to 1 if there is no clustering of exceedances of high levels, and being $1/\theta$, in a great variety of situations, the limiting mean cluster size. Once chosen a high threshold, which from a theoretical point of view should be a value u_n, such that $n(1-H(u_n)) \to \tau$, as $n\to\infty$, $\tau>0$, and after identifying the clusters of exceedances, we may estimate θ by the reciprocal of the mean cluster size. Equivalently, since $D''(u_n)$ condition of Leadbetter is valid for these markovian sequences, the validity of the relation $\theta = \lim_{n\to\infty} \frac{P(X_1>u_n,X_2\leq u_n)}{P(X_1>u_n)}$, provides a quick nonparametric estimator of θ, based on a random sample $(X_1,X_2,...X_m)$; once chosen a suitable threshold u, we put

$$\hat{\theta}_m = \min \{1, \frac{\#\{X_i>u \text{ and } X_{i+1}\le u \text{ for } 1\le i\le m-1\}}{\#\{X_i>u \text{ for} 1\le i\le m\}} \frac{m}{m-1}\}. \tag{3.1}$$

We have here used this nonparametric estimator of θ and we have empirically studied the influence of u in the estimation of θ. Unfortunately, the estimation of θ is not so insensitive to the value u as we expected , and as is suggested by Davison & Smith (1990). For instance for the particular Fréchet case $F(x) = \exp(-x^{-\alpha})$, $\alpha=0.5,1.0$, and $k=0.1,0.5,0.9$, we get the following simulated mean values and standard deviations for $\hat{\theta}_m$, m=500, based on 500 replications of markovian samples, and considering thresholds $X_{m-\tau:m}$, $\tau=10,20,30,40$.

k		0.10	0.50	0.90
$\alpha=0.5$	θ	0.68	0.29	0.05
	$\tau=10$	0.70,0.14	0.37,0.14	0.15,0.07
	$\tau=20$	0.67,0.10	0.32,0.10	0.10,0.05
	$\tau=30$	0.66,0.08	0.30,0.08	0.08,0.04
	$\tau=40$	0.64,0.07	0.29,0.07	0.07,0.03
$\alpha=1.0$	θ	0.90	0.50	0.10
	$\tau=10$	0.90,0.10	0.54,0.15	0.19,0.09
	$\tau=20$	0.87,0.07	0.51,0.11	0.14,0.07
	$\tau=30$	0.85,0.06	0.49,0.09	0.13,0.05
	$\tau=40$	0.82,0.05	0.48,0.07	0.12,0.04

Table 1. Mean Values and Standard deviations of $\hat{\theta}_m$, m=500, for thresholds $X_{m-\tau:m}$, $\tau=10,20,30,40$, and markovian structures (1.1), k=0.1,0.5.0.9, with generating Fréchet (α) Y, α=0.5,1.0.

For this important situation, $F(x)=\exp(-x^{-\alpha})$, we have $H(x)=\exp(-x^{-\alpha}/(k^{-\alpha}-1))$, and consequently, $P(X_{i+1}/X_i=k)=k^{\alpha}$, $P(X_{i+1}>X_i) = \frac{1-k^{\alpha}}{1+(1-k^{\alpha})}$. The extremal index of such a sequence is given by $\theta = \lim_{x\to\infty} \frac{P(X_i>x,X_{i+1}\le x)}{P(X_i>x)} = 1 - \lim_{x\to\infty} \frac{1 - H(x/k)}{1-H(x)} = 1- k^{\alpha}$. As was expected, the higher the extremal index (and then the dependent structure behaves in a context of extreme values more likely the independent structure), or the smaller the value of k, the smaller also is the number of X_{i+1}/X_i equal to k [in this particular case, estimation of the extremal index could be done on the basis of $1 - \hat{P}(X_{i+1}/X_i=k)$!]. Figure 1 enhances the difference between sample paths associated to the structure (1.1) for Fréchet(1) generating Y, and for k=0.1 and k=0.9 respectively (notice the clusters of exceedances in the figure of the right, where the limiting mean cluster size should be 10).

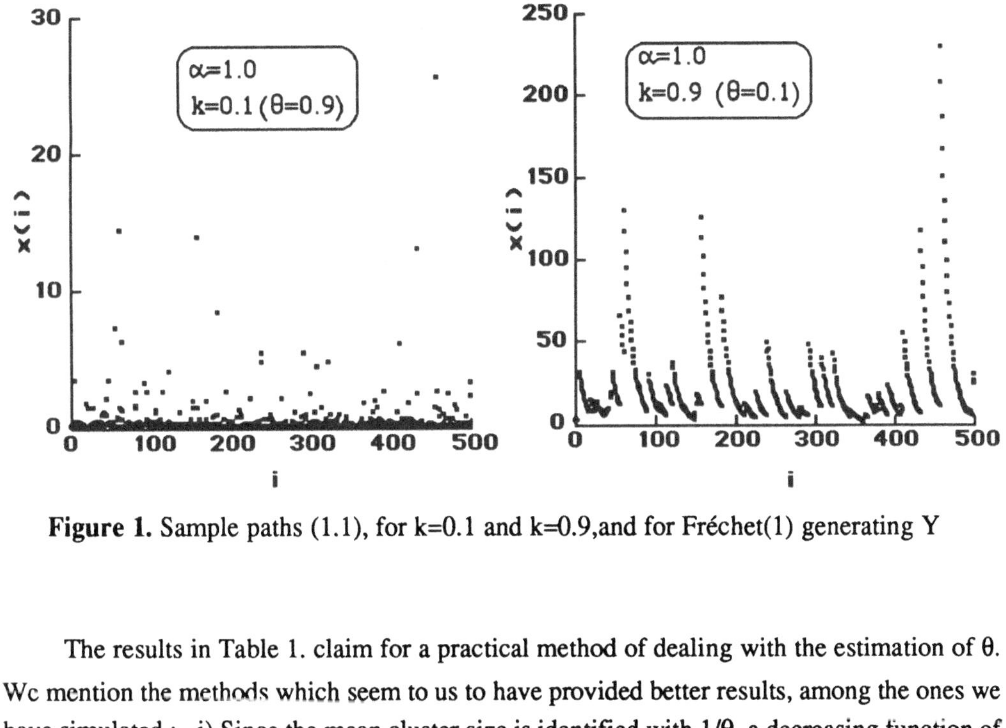

Figure 1. Sample paths (1.1), for k=0.1 and k=0.9,and for Fréchet(1) generating Y

The results in Table 1. claim for a practical method of dealing with the estimation of θ. Wc mention the methods which seem to us to have provided better results, among the ones we have simulated : i) Since the mean cluster size is identified with 1/θ, a decreasing function of θ, and we noticed in different simulations that the ideal percentage of high values chosen was directly proportional to 1/θ, we have implemented an iterative procedure to estimate θ, beginning with an arbitrary value, for instance $\tau=m/5$, chosing the level $u=X_{m-\tau:m}$, and according to the estimate θ_0 of θ obtained, we consider recursively another value $\tau= \min(\frac{m}{2}, \frac{m(0.075-0.05m/1000)}{\theta_0})$ if $m\leq 1000$, and $\tau= \min(\frac{m}{2}, \frac{25}{\theta_0})$ if $m>1000$; ii) the splitting of the sample of size m in k sub-samples of size r, and for a fixed $\tau>0$, obtained as in i) on the basis of t he first sample of size r, the consideration of an increasing threshold $X_{jr-\tau:jr}$, for $(j-1)r+1\leq i <jr$, $1\leq j\leq k$, also provides good practical results; iii) finally, a non-realistic method, suitable only in this situation of a generated sequence with known extremal index $1-k^{\alpha}$ is the

following: since from a theoretical point of view we should choose a threshold $u_m=\left(\frac{m}{\tau(k^{-\alpha}-1)}\right)^{1/\alpha}$ [$m(1-H(u_m)$ should go to τ as $m\to\infty$], we have chosen τ and a threshold $X_{m-\tau:m}$ (really exceeded τ times in the sample $(X_1,X_2,\ldots,X_m)$)] such that $X_{m-\tau:m}-u_m(\tau)$ is minimum.

Since the results obtained by i), ii) or iii) are equivalent, we present the simulated mean values and standard deviations for $\hat{\theta}_m$, m=100,500,1000, obtained through method i), based on 500 replications of markovian samples (1.1), k=0.1,0.5,0.9 and for Fréchet(α) generating Y, α=0.5,1.0.

k		0.10	0.50	0.90
	θ	0.68	0.29	0.05
α=0.5	m=100	0.65,0.15	0.28,0.10	0.05,0.02
	m=500	0.66,0.08	0.28,0.04	0.05,0.01
	m=1000	0.67,0.08	0.28,0.04	0.05,0.01
	θ	0.90	0.50	0.10
α=1.0	m=100	0.84,0.12	0.48,0.14	0.08,0.03
	m=500	0.85,0.06	0.48,0.07	0.08,0.01
	m=1000	0.88,0.05	0.49,0.07	0.09,0.01

Table 2. Mean Values and Standard deviations of $\hat{\theta}_m$, m=100, 500, 1000 obtained from approach i), and for markovian structures (1.1), k=0.1,0.5.0.9, and generating Fréchet (α) Y, α=0.5,1.0.

Similar results have been obtained for other values of k and α, for other models in (1.1), and for other dependent structures either simulated or real.

References:

ALPUIM, M.T. (1989). An extremal markovian sequence. *J. Appl. Probab.* **26**, 219-232.

DAVISON A.C., & SMITH, R.L. (1990). Models for exceedances over High Thresholds. To be published in *J. Royal Statist. Soc.*

GALAMBOS, J. (1987). *The Asymptotic Theory of Extreme Order Statistics.* Krieger, Melbourne.

LEADBETTER, M.R. and NANDAGOPALAN, S. (1989).On exceedance point processes for stationary sequences under mild oscillation restrictions. In Hüsler, J. & R.-D. Reiss (eds.), *Extreme Value Theory*, Springer-Verlag, 69-80.

LEADBETTER, M.R., LINDGREN, G. and ROOTZÉN, H. (1983). *Extremes and Related Properties of Random Sequences and Series.* Springer-Verlag, New York.

Interpretation of Spectral Estimation Results for Nonstationary Signals

S. J. Pliskin and V. A. Karpenko, Moscow

ABSTRACT

We consider application of spectral estimation methods(MSE) to the problem of identification of parameters of nonstationary signals. The main result of our work is obtaining interpretation conditions, that take into account frequency resolution and nonstationarity of different components of the signal at the same time. These conditions establish the connection between the length of stationarity intervals and a priori information about signal parameters.

INTRODUCTION

The identification of parameters of signals with time-varying amplitude and frequency characteristics appear when different wave processes are studied. The considered signals can be presented in the form

$$u(t)=\sum_{j=1}^{N} b_j(t)\exp\{ig(t,\bar{a}_j)\}+n(t), \qquad (1)$$

Compstat 1990

where $b_j(t)$ are unknown amplitude terms, being determined functions or lowfrequency narrow band random processes, $g(t,\bar{a}_j)$ - phase functions, depending on unknown $\bar{a}_j$. It is necessary for given realizations $u(t)$ to obtain estimates $\{ b_j(t), \bar{a}_j \}$. The class of signals (1) describes processes, explorated in different applied fields such as hydrolocation, hydroacoustics, radiolocation, etc.

The analysis of applied problems permits to mark out the common property inherent to every model (1): functions $b_j(t)$ and $f(t,\bar{a}_j)=g'(t,a\ )/2\pi$ are slowly varying time functions. It allows to use the traditional stationarity methods [1]. The signal (1) is approximated by function (2) on consequent intervals $I_k=(t_k-h_k/2, t_k+h_k/2)$

$$u_1(t_k+s)=\sum_{j=1}^{N} b_j(t_k)\exp\{i(g(t_k,\bar{a}_j)+g'(t_k,\bar{a}_j)s)\}, \qquad (2)$$

and we can use power spectra of $u(t)$ on I_k for definition of its parameters. Every spectral peak corresponds to each component, location of spectral maximums allows to estimate instantaneous frequencies of different components, intensities of peaks define $b_j(t_k)$. Estimates of $\bar{a}_j$ are resulted from totality of instantaneous frequensy evaluations in consequent time moments t_k.

When we use the spectral approach, the first step is to chose MSE. Questions of such choice with application to our problems are investigated in detail in [2]. When MSE has choosen, the most important problem is to define processing parameters and to receive information about estimating parameters from spectras. The main of such processing parameters is stationarity interval length. From one hand increasing of I_k improves resolution , but at the same time it leads to spoiling of estimates, that is connected with nonstationarity within I_k. Traditionally the problem of h_k choice is solved euristically and its length is chosen within realization as a constance. It leads to losing or spoiling of information. We solve the problem of formalization of the choice procedure in dependence on a priori information about the considered signal.

We investigate the errors of frequency estimates arising as a result of amplitude and frequensy varying within I_k. Let us use $S_k^h(t)$ as estimate of power spectra

$$S_k^h(f)=\left|\int_{t_k-h/2}^{t_k+h/2} u(s)r(s)\exp\{-2\pi ifs\}ds\right|^2/h \tag{3}$$

where r(s) is window function, satisfying common conditions [3]. For simplicity we consider the case when N=1, $n(t)\equiv 0$.

Let $f_1=\operatorname{argmax} S_k^h(f)$, $f_0=g'(t_k,\bar{a})/2\pi$. Then we prove the next theorems(see the proofs in [2]).

THEOREM 1. IF $b'(t)=0$ on $I_k (h\le 1)$, THEN

$$|f_0-f_1|\le \max_{I_k}|2g''(t,\bar{a})/\pi|^{1/2} \tag{4}$$

THEOREM 2. IF $g''(t,a)=0$ on I_k and $h\le 2*\min_{I_k}|b(t,a)/b'(t,a)|$, THEN

$$|f_1-f_0|\le\{2*\max|b''(t,\bar{a})|/b(t_k,\bar{a})\}^{1/2}. \tag{5}$$

We have found the main factors, defining errors of the estimates of instantaneous frequencies. These characteristics can change rather strongly. The choice of stationarity intervals should made in accordance with a priori information about parameters $\bar{a}$. Further we shall consider that region A_j of possible $\bar{a}_j$ localization is known.

Under interpretation we mean obtaining reliable estimates of signal parameters from spectras. Interpretation conditions involve:

1) concordance of the polyharmonic model (2) with the signal (1) (CC);

2) resolution of spectral peak, which is used for estimation j-th component parameters, from other peaks of spectra (RC).

The first one defines precision demands for presentation (1) on I_k by model (2). Let us formulate CC both for amplitude and frequency characteristics. Frequency and amplitude CC of spectras with precisions Δf and Δb accordingly can be written in the form

$$|f(t_k,\bar{a}_j)-\hat{f}(t_k,\bar{a}_j)|<\Delta f*f(t_k,\bar{a}_j), \tag{6}$$

$$|b_j(t_k)-\{S_k^h(\hat{f}(t_k,\bar{a}_j))\}^{1/2}|<\Delta b*b_j(t_k) \tag{7}$$

where $\hat{f}(t_k,\bar{a}_j)$ is the spectral peak frequency. The level of Δf and Δb is defined by demands of an applied problem.

Now let us introduce frequency deviation function

$$W(t_k,\bar{a},h)=\max_{I_k} f(t,\bar{a})-\min_{I_k} f(t,\bar{a}).$$

(6) and (7) are implemented for that t_k, $\bar{a}$ and h for which

$$W(t_k,\bar{a},h)<C/h, \tag{8}$$

is true, where C - coefficient, depending only on Δf, Δb and on chosen MSE and practically does not depend on $f(t,\bar{a})$. When h is increasing at first Δb is increasing and then Δf is. When $\Delta b>0.3$ spliting of spectral peak takes place. Table 1 reflects values of C when $\Delta f\leq 0.01$ for different Δb and various time windows when $f(t,\bar{a})$ is monotony.

TABLE 1

Δb	0.1	0.2	0.3
rectangular	1.8	2.7	3.6
Hamming	2.9	4.2	6.1
Blackm.-Harris	4.1	6.1	9.5

The same results for Δf are realized for parametrical MSE. The inequality (8) defines concordance region $C(t_k,h)=\{\bar{a} : W(t_k,\bar{a},h) < c/h\}$, and those frequency bands, where (6) and (7) are true.

RC expresses frequency resolution requirement of spectral peaks, corresponding to different components (1) on I_k. It is natural consider only those components, for which (8) is implemented. It permits to use resolution criteria for harmonic signals. Let us introduce the resolution function

$$F(t_k,\bar{a}_j)=\min_{\substack{1\leq l\leq N\\ l\neq j}} \min_{\bar{a}_l}|f(t_k,\bar{a}_l)-f(t_k,\bar{a}_j)|.$$

RC for $\bar{a}_j$ will be implemented if

$$F(t_k, \bar{a}_j) > k/h \tag{9}$$

where k - resolution coefficient, depending on MSE only. (9) allows to define the resolution region as $R(t_k,h)=\{\bar{a}$ is so that: $F(t_k,\bar{a}) > k/h \}$, and those frequency bands, where RC is true. Intersection $R(t_k,h) \cap C(t_k,h)=Q(t_k,h)$ gives the interpretaion region, i.e. those values of parameters $\bar{a}_j$, for which these estimates may be obtained on I_k by use of chosen MSE. When h is increasing, $R(t_k,h)$ is also increasing and $C(t_k,h)$ is decreasing, so it is not possible to estabilish the simple dependence $Q(t_k,h)$ on h. From this point of view it is useful to utilize few values of h to receive maximum wide resulting region of interpretation. The diapason of possible values of h is defined by [$k/\max F(t_k,\bar{a}_j)$; $k/\min F(t_k,\bar{a}_j)$], then it is separated by uniform grid h_m, and

$$Q(t_k)=\bigcup_{h_m} Q(t_k, h_m),$$

which shows maximum acceptable region for estimation of parameters and corresponding frequency diapasons.

Let us note, that it is easy to spread proposed procedures to different phase functions, and also, when analyżed signal has the form

$$u(t)=\sum_{j=1}^{N} b_j(t)\exp\{ig_j(t)\}+n(t),$$

where only bounds $g_{1j}(t) \leq g_j(t) \leq g_{2j}(t)$ for $g_j(t)$ are known. Besides our approach permits preliminary to define the optimum conditions of reception signal.

All those procedure described above, are realized in the form of program complex, containing spectral processing programs (parametric and nonparametric MSE), programs of interpretation region obtaining and of stationarity segments definition, also there are programs of spectral information processing, corresponding to the different types of models (1) and a priori information about estimated parameters. This program complex can be unilized to the solution of applied problems, arising when wave processes are explorated.

CONCLUSIONS

In our work interpretation conditions of spectral processing results for wide class of nonstationary signals are proposed. Those conditions establish the connection between spectral processing parameters and a priori information about signal parameters.

ACKNOWLEDGMENTS

We would like to express our gratitude to prof. I.G.Zurbenko for his valuable recomendations and useful discusson.

REFERENCE

1. M.B.Priestley, Śpectral Analysis and Time Series. New York: Academic, 1982, 890 pp.

2. V.J.Galkin, V.A.Kovrigin, S.J.Pliskin, Numerical Methods of Signal Processing in Identification Problems. Moscow: Edition of Moscow University, 1987, 89 pp. (in Russian).

3. F.D.Harris, "The use of windows under harmonic analysis by method of Descrete Fourier Transform", Proc. IEEE, vol.66, No.1, pp.60-96, 1978.

An Automated Method for Trend Analysis

T. Atilgan, Murray Hill

ABSTRACT

An automated method designed to handle outliers, missing observations and structural changes in smoothing time series is presented. Smoothing of trend, adjusted for seasonality is done using Akaike's Information Criterion (AIC) in the L_1 (least absolute deviation) context. AIC selects approximations to trend among polynominals of up to degree three or cubic cardinal B-splines with a varying number of knots. A short time series is analyzed to demonstrate the method.

1. INTRODUCTION

In analyzing a very large number of short time series (30 to 60 measurements per time series), we are often faced with *outliers, missing observations*, and *structural changes*, requiring us to develop a time series analysis approach which should be *robust* to outliers, have an *automatic* decision making mechanism on how much smoothing needs to be done in the *trend* component of the time series, handle *missing* observations, and be flexible enough, when needed, to handle *structural changes*.

The overall approach of the method introduced here is similar to SABL (see Cleveland, et. al. 1982a, 1982b) in that it uses smoothing and regression to carry out the decomposition of time series. However, it differs from SABL in the method of smoothing and the length of time series may be less than 36.

Three original contributions of this paper are: 1) the introduction of a robust form of Akaike's Information Criterion (AIC) (see Akaike, 1973), and 2) the use of it in estimating seasonally adjusted and smoothed *trend* component of the time series in an automated way (no human interaction is needed). 3) use of cardinal B-splines in trend analysis. The method searches among polynominals of up to degree 3 and cubic cardinal B-splines of differing number of knots, and obtains the AIC optimal *trend* component. This approach provides the flexibility of using simple polynominal fits when they sufficiently explain the data or switching to cubic

Compstat 1990

cardinal B-splines when there are local changes in the time series. The method of smoothing is discussed in detail by Atilgan (1988).

Section 2 of this paper develops the methodology needed to achieve the above objectives. Section 3 demonstrates the method on a time series with missing observations and an outlier.

2. METHOD

Suppose that y_t, t = 1, ..., T is a (possibly transformed) time series made up of three *additive* components,

$$y_t = T_t + S_t + e_t, \quad t = 1,...,T \tag{2.1}$$

where T_t is the *trend* component, which gives the long term variation in the series; S_t is the *seasonal* component describing the variation that repeats itself each year; and e_t is the *irregular* or *noise* component which gives the remaining variation in the series.

In the present paper we will use simple transformations, like log-transforms of the series. More sophisticated transformations to make the series additive is described in Cleveland, et. al. (1982b). In future work, power transformations will be incorporated into the method.

The remainder of this section will discuss the estimation, partitioning and smoothing methods used in this paper.

2.1 Estimation

Suppose that $w_t = y_t - S_t$ denotes deseasonalized time series (trend + noise). Usually the least squares method is used to estimate the parameters in *trend* component of the model; that is, the parameters are estimated by conducting the minimization:

$$\min_{\underset{\sim}{\theta}} \sum_{t=1}^{T} (w_t - g_t(\underset{\sim}{\theta}))^2 ,$$

where g_t represents the trend component. It is well known that least squares is notoriously *non-robust* to *outliers*. Even a single outlier may have a large effect on least squares estimates of the trend component, since we are trying to minimize sums of squares of residuals.

One approach to minimize the influence of outliers is to use

$$\min_{\underset{\sim}{\theta}} \sum_{t=1}^{T} \mid w_t - g_t(\underset{\sim}{\theta}) \mid . \qquad (2.2)$$

This approach is more robust to outliers in y than the least squares method, but it is as bad as least squares for outliers in the x (time) direction. However, almost all of the time, outliers in the time series analyzed here are in the y direction, since the x direction corresponds to months of the years.

Equation 2.2 is known as least absolute deviation or L_1 approach. Note that, estimation by the method of maximum likelihood while assuming the Gaussian distribution for the noise component leads to the least squares approach, and in contrast, the Laplacian, or double exponential, leads to the least absolute deviation method. We utilize this fact to derive and automate smoothing of trend component.

2.2 Partitioning y_t into its Trend, Seasonality and Noise Components

The sequence of operations for partitioning a time series to its components is similar to that of the SABL method of Cleveland, et. al. (1982b), and can be summarized as follows:

1. Create a calendar matrix from the time series; for monthly time series we would have rows corresponding to years and columns corresponding to months.
2. Obtain an initial estimate of the trend component by fitting a cubic polynomial to the time series data.
3. Subtract the estimated trend from the corresponding columns of the calendar matrix. What remains are the seasonal component plus the noise (calendar matrix for the seasonal component). Now, the median of each column of the matrix will give a robust estimate of the seasonal adjustment for the corresponding month of the year (or the day of the week, etc.).
4. Subtract the medians from y_t to obtain the estimated trend plus noise components; w_t.
5. Use the *automatic smoothing* procedure on w_t to obtain a *smooth* estimated *trend* component, say T_{smooth}.
6. Calculate residuals and use boxplots to report outliers.

7. Use T_{smooth} + Medians to estimate missing values and replace outliers.

Note that, as in Cleveland, et. al. (1982b) we could iterate a few times from step (3) to (5), however our experience indicates that usually one iteration is sufficient.

2.3 Automatic Smoothing

Suppose that we have observations w_1 ...., w_T such that

$$w_t = g_t + e_t \ ,$$

where g_t is the unknown function and e_t is the noise. Suppose also that e_t's have zero mean and variance σ^2 , and that they are independently distributed with Laplacian density.

To approximate g_t, we will use linear approximations of the form:

$$g_{t,m} = \sum_{i=1}^{m} \theta_i \psi_i,$$

where θ_i's are unknown parameters and ψ_i's are known functions (e.g., polynominals, splines, etc.). θ_i's are estimated by the L_1 method (2.2), or equivalently by

$$\max_{\underline{\theta}} L_m(\theta) \qquad \text{(log-likelihood method)}$$

where

$$L_m(\theta) = \sum_{t=1}^{T} \log f_m(e_t) \cdot$$

Here, *m* enters the likelihood through g_t and it controls the amount of smoothing; large *m* gives a "rough" function whereas small *m* gives a "smooth" function. How do we choose *m*? Atilgan (1988) demonstrates that Akaike's Information Criterion (AIC) provides a reasonable answer.

AIC is simply defined by

$$\min_{m} \left\{ \hat{L}_m - m \right\} .$$

It gives a *trade-off between bias and variance in estimation.*

The method is easily implemented in "S", Becker et. al.(1988). It selects AIC optimal *trend* components among polynominals of up to degree 3 or cubic cardinal

B-splines with varying number of knots. One provides the time series and its beginning, end and periodicity, and the analysis is done automatically.

Hampel et. al.(1986) give a robust version of AIC (AICR) using M-estimators, defined as:

$$\min_{m} \left\{ 2 \sum_{t=1}^{T} \rho(r_{t,m}) + 2m \right\},$$

where $r_{t,m} = (w_t - g_{t,m}(\theta))/\hat{\sigma}$, $\hat{\sigma}$ is some robust estimate of σ, and $g_{t,m}$ is the M-estimator defined by

$$\sum_{t=1}^{T} \psi(r_{t,m})t = 0.$$

with $\psi(r) = \partial\rho/\partial r$. Note that L_1 is a special case of M-estimators with $\rho = | w_t - g_{t,m}(\theta) | /\hat{\sigma}$ and $\hat{\sigma} = \sum_{t=1}^{T} | w_t - g_{t,m}(\theta) | /n\sqrt{2}$.

Some of the other distributions which can be utilized to give M-estimators leading to robust AICs are: logistic, Cauchy and Huber's least favorable errors distribution. The above ideas on robust AIC can be extended to generalized M-estimators. Generalized M-estimators bound the influence of outlying x's by means of some weight function (see Rousseeuw and Leroy, 1987 for details and further references), while M-estimators are effected by outlying x's.

3. AN EXAMPLE

In this section, we illustrate the method on a monthly (10/83 to 8/85) long distance telephone usage time series at a central switching office. The time series has two missing observations (July 1984 and 1985). Figure 1 shows the time series together with the least squares and least absulute deviation fit. For both fits, the automatic procedure finds that a quadratic fit is sufficient for the trend component. The influence of the outlier on the least squares fit is evident.

Figure 1. Least Squares (--) and Least Absulute Deviation (-) Fits to a Central Switching Office Usage Time Series

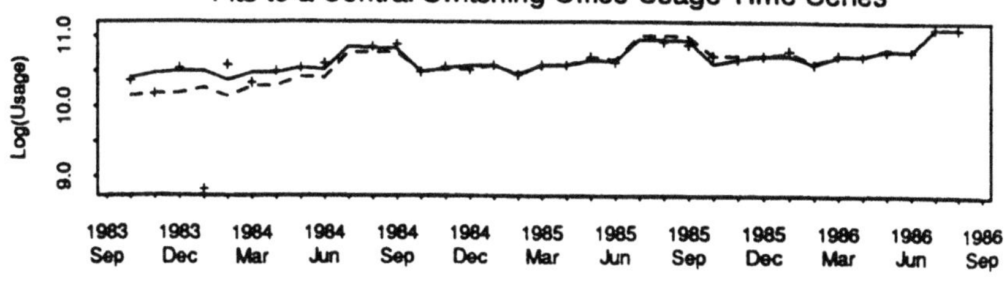

4. DISCUSSION

The preliminary program is written in "S". It takes about 2 minutes to analyze a time series with 35 observations on a VAX/8650 when set up to look for AIC optimal trend using polynomials of up to degree 3 and/or cubic cardinal B-splines of up to 12 knots. We can gain considerable speed by using a cubic polynomial for the trend component as a default model. The current "S" program is a working prototype. We hope to implement the algorithm in FORTRAN, which will improve its performance considerably. As it stands, the procedure offers an automated approach for trend analysis of (especially short; less than 40 measurements) time series with missing observations, outliers and structural changes.

REFERENCES

Akaike, H. (1973), "Information theory and an extension of maximum likelihood principle," In 2nd International Symposium on Information Theory (B. N. Petrov and F. Csáki, eds.), 267-281. Budapest: Akademia Kiadó.

Atilgan, T. (1988), "Basis Selection for Density Estimation and Regression," submitted for publication.

Becker, R. A., Chambers, J. M., and Wilks, A. R. (1988), "The New S Language A Programming Environment for Data Analysis and Graphics," Wadworth and Brook/Cole Computer Science Series: Pacific Grove, California.

Cleveland, W. S. and Terpenning, I. J. (1982a), "Graphical Methods for Seasonal Adjustment," Journal of the American Statistical Association 77, 52-62.

Cleveland, W. S., Devlin, S. J., and Terpenning, I. J. (1982b), "The SABL Seasonal and Calendar Adjustment Procedures," Time Series Analysis: Theory and Practice 1, edited by D. D. Anderson, North-Holland, Amsterdam, 539-564.

Hampel, F. R., Ronchetti, E. M., Rousseeuw, P. J.,and Stahel, W. A. (1986), "Robust Statistics: The Approach Based on Influence Functions," John Wiley and Sons: New York.

Rousseeuw, P. J.,and Leroy, A. M. (1987), "Robust Regression and Outlier Detection," John Wiley and Sons: New York.

Analysis of Spatial Data

A Test of Spatial Isotropy

A. J. Baczkowski, Leeds

KEYWORDS: Geostatistics, isotropy, semi-variogram.

INTRODUCTION

In many practical situations the semi-variogram function is a powerful tool in geostatistical analyses of spatial data. Krige (1976) states that an important requirement of such analyses is the ability to detect the presence or absence of isotropy, in which the semi-variogram exhibits the same behaviour in all directions. Geostatisticians have tended to use graphical methods, essentially plotting the sample semi-variogram for different directions. In an early attempt to use statistical methods to detect anisotropy Matheron (1961) used a χ^2-test of homogeneity of the sample semi-variogram in perpendicular directions, though the dependence of the semi-variogram at different lags and in different directions was neglected in that study. More recently Baczkowski and Mardia (1990) presented tests of symmetry using the sample semi-variogram under Gaussian assumptions. A test is here presented which does not require distributional assumptions for the regionalized variable under study.

Let $\{Z(\mathbf{x})\}$ be a stochastic process defined at all points $\mathbf{x}$ in a region $V \in \mathbb{R}^d$. The population semi-variogram $\gamma(\mathbf{h})$ at a lag $\mathbf{h}$ is defined by

$$\gamma(\mathbf{h}) = E[\{Z(\mathbf{x}+\mathbf{h}) - Z(\mathbf{x})\}^2]/2 .$$

In practice N sampled values $Z(\mathbf{x}_1),\ldots,Z(\mathbf{x}_N)$ will be available and the semi-variogram $\gamma(\mathbf{h})$ is estimated using the sample semi-variogram,

$$g(\mathbf{h}) = \sum \{Z(\mathbf{x}_i) - Z(\mathbf{x}_j)\}^2/(2N_h) ,$$

where the summation is over all N_h pairs $(\mathbf{x}_i,\mathbf{x}_j)$ satisfying $\mathbf{x}_j = \mathbf{x}_i + \mathbf{h}$.

Compstat 1990

Suppose now d = 2 and N = mn values $Z(i,j)$, for $i = 1,\dots,m$ and $j = 1,\dots,n$, are available on a regular (mxn) lattice. The sample semi-variogram at lag (h,k) becomes

$$g(h,k) = \sum_{i=1}^{m-h} \sum_{j=1}^{n-k} \{Z(i+h,j+k) - Z(i,j)\}^2 \,/\, \{2(m-h)(n-k)\}.$$

When the $Z(i,j)$ are Gaussian it can be shown that the variance stabilizing transformation $U_h = \log\{g(h,0)/g(0,h)\}$ is approximately normally distributed; see Baczkowski and Mardia (1987). Under the hypothesis of isotropy, U_h has zero mean. Indeed, Mardia and Gill (1982) used departures of U_h from zero to graphically indicate the presence of an anisotropic covariance function.

In the case m = n, by considering the doubly geometric semi-variogram, see Martin (1979), given by

$$\gamma(h,k) = \sigma^2\{1 - \lambda^{|h|}\, \upsilon^{|k|}\} ,$$

Baczkowski and Mardia (1990) proposed a test of symmetry $\lambda = \nu$ ($= \gamma$) using a bi-directional statistic $BS = \{2(n-1)U_1 - (n-2)U_2\}^2 \,/\, \{10 + 12\hat{\gamma}^2\}$, where $\hat{\gamma}$ is the maximum likelihood estimate for γ. Unfortunately BS only had limited use as a test of isotropy for more general covariance functions.

Other tests of isotropy could be constructed using U_h since, if $\mathbf{U}$ is a vector of U_h's satisfying $\mathbf{U} \simeq N(\boldsymbol{\mu}_u, \boldsymbol{\Sigma}_u)$ under some null hypothesis, then the statistic $(\mathbf{U} - \boldsymbol{\mu}_u)'\boldsymbol{\Sigma}_u^{-1}(\mathbf{U} - \boldsymbol{\mu}_u)$ has an approximate χ^2 distribution. A similar remark applies to tests based directly upon $g(\mathbf{h})$ though U_h is to be preferred as it achieves normality faster even for small lattices. A key problem is that the $g(h,k)$ and hence the U_h have complicated covariance structure even in simple cases. For example, if the $Z(i,j)$ are independent random variables with variance σ^2 then it can be shown for large lattices that $\mathrm{corr}\{g(h,k);g(i,j)\} = 2/3$ for all $(h,k) \neq (i,j)$. To set up tests based upon $\mathbf{U}$ it is necessary to know the form of the underlying population semi-variogram. A further complication is that the model parameters have to be estimated.

RANDOMISATION TEST

Rather than consider the covariance structure of the U_h a randomisation distribution for U_h can be constructed as follows:

a) Let (i,j) be an initial random location on the lattice.
b) Move to one of the four locations $(i',j') = \{(i+h,j),(i-h,j),(i,j+h),(i,j-h)\}$ with equal probability for each location. If the chosen location lies outside the (mxn) lattice then move to one of the other locations.
c) Let $(i,j) = (i',j')$ and repeat step (b) a total $\{n(m-h)\}$ times. If step (b) results in (i',j') being equal to (i,j) on the previous step then move to one of the other available locations.
d) Form the sum $g^*(h,0) = \sum \left\{Z(i',j') - Z(i,j)\right\}^2 / \left\{2n(m-h)\right\}$.
e) Repeat steps (a) to (c) $\{m(n-h)\}$ times and form $g^*(0,h)$ in a similar manner to step (d).
f) Take the ratio $U_h^* = \log\{g^*(h,0)/g^*(0,h)\}$.

The pattern of moves forms a random walk akin to a knight's move in chess. Steps (a) to (f) can be repeated, say a hundred times, and the simulated distribution for U_h^* can be used as an approximation to the distribution of U_h. In practice the observed value for U_h at different lags h can be compared with the simulated distribution for U_h^* although it must be remembered that the results at different lags are not independent.

The justification for this procedure is that if the data is isotropic the differences $\{Z(i',j') - Z(i,j)\}$ have the same properties along the principal axes of the lattice. Hence the sample semi-variograms at lag h in perpendicular directions will be approximately equal and U_h will be close to zero at the centre of the distribution for U_h^*. Under anisotropy the simulated U_h^* values will approximately lie between the extremal values $\pm\log\{g(h,0)/g(0,h)\}$ with U_h in one tail of the distribution.

EXAMPLES

(I) Consider the (17x11) gravimetric readings of Grant (1957) from the Val d'Or in Quebec. Since some values are missing the restriction in step (c) of not being allowed to move to a location which has just been visited can result in no valid available move and so the algorithm is amended by jumping to a new random location as in step (a) and continuing the random walk from there. Without the restriction in step (c), in extreme cases of missing data, the algorithm could oscillate between two locations.

After fitting a quadratic trend surface the values Z(i,j) represent the residuals from the trend surface. Figure (1) shows the simulated distribution for U_h in the case h = 1. As can be seen the observed value $U_1 = -0.309$ lies near the centre of the distribution suggesting there is no evidence to reject the isotropy hypothesis. The probability value p = 0.35 gives the proportion of cases for which $|U_1^*| > |U_1|$.

Figure (1). Histogram of simulated values U_1^* together with U_1

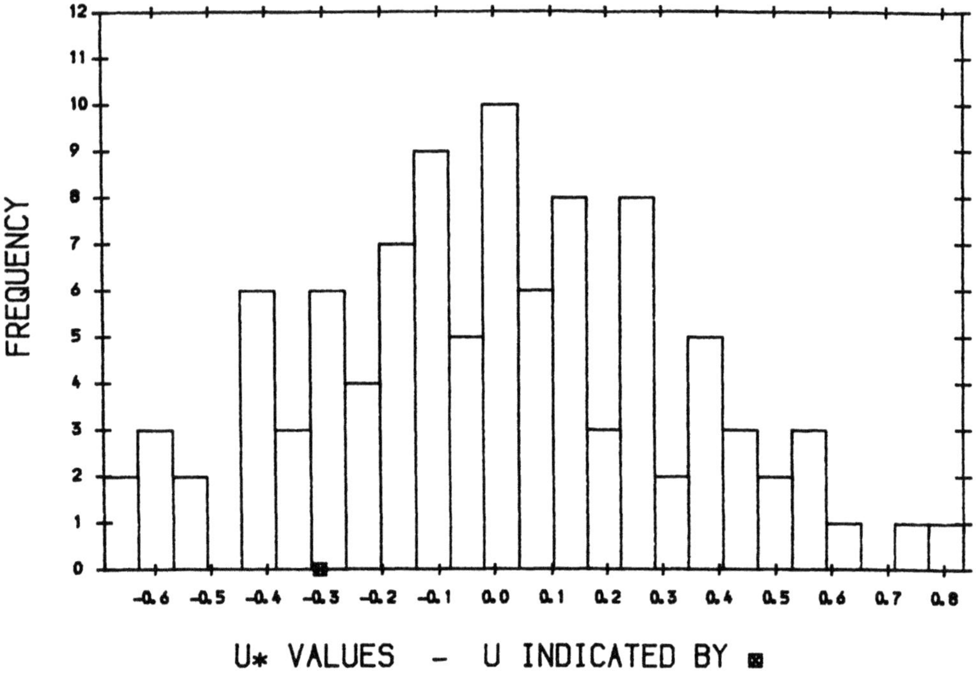

(II) Table (1) gives a (10x10) set of simulated data having a doubly geometric semi-variogram with parameters $\sigma^2 = 2.5$, $\lambda = 0.1$, and $\nu = 0.6$. $U_1 = 0.547$ with probability value p = 0.04. With $\hat{\gamma} = 0.6033$ the statistic BS = 5.042, which is highly significant (p = 0.025) when compared with a χ_1^2 distribution under the null hypothesis $\lambda = \nu = \gamma$.

(III) Mercer and Hall's (1911) set of (25x20) wheat yields gave $U_1 = 0.409$ (p = 0.07) and $U_2 = 0.385$ (p = 0.06) for the residuals from a linear trend surface. Isotropy is rejected at lags one and two.

Table 1. 10x10 grid of simulated doubly geometric values

4.96	3.59	4.97	4.56	2.80	2.26	2.61	2.09	3.97	1.89
2.81	3.71	4.96	4.72	2.78	2.99	1.65	0.84	4.09	3.37
2.61	2.32	7.41	4.19	3.03	2.57	4.60	1.47	4.83	4.82
2.10	1.51	5.02	0.99	4.66	3.88	4.06	1.57	1.79	3.56
1.71	2.88	1.88	1.95	3.83	2.84	3.90	2.83	0.69	1.74
2.26	0.48	3.82	3.34	1.88	2.93	4.62	0.90	0.44	1.75
2.02	2.54	4.59	1.95	3.57	5.14	1.95	2.06	0.14	2.74
5.53	1.96	4.99	2.36	3.76	3.08	2.69	0.68	2.56	2.54
4.87	2.92	3.92	1.95	4.45	4.05	2.75	1.77	3.51	4.28
2.20	3.47	5.02	2.18	1.85	3.22	4.06	3.28	3.37	3.79

DISCUSSION

An alternative way to set construct $g^*(h,0)$ for an (mxn) lattice might be to use n independent random starts, each followed by (m-h) moves, and similarly for $g^*(0,h)$. The algorithm as presented above only allows a shorter run length than n(m-h) if the presence of missing values would force the algorithm to move back upon itself. Another alternative procedure to generate $g^*(h,0)$ would be to choose n(m-h) random locations for (i,j) and from each only make one move of length h.

Extending the method to lattice data in three dimensions presents no obvious difficulty. Extension of the method to directions other than the principal x and y axes is again not difficult.

For data not given on a regular lattice but at a set of N irregularly placed locations $\mathbf{x}_1,\ldots,\mathbf{x}_N$ the sample semi-variogram is calculated using the method of angular regularisation as described by David (1977, pp. 146-7). The algorithm in step (b) can be amended by moving to a location (i',j') where (i',j') is an available sampled location a distance $h \pm \Delta h/2$ from (i,j) so that (i',j') lies in an annulus of width Δh about the point (i,j). The quantity Δh represents a tolerance distance in the angular regularisation method. For a choice of k possible locations (i',j') in the annulus one would be chosen at random with probability 1/k. Care must be taken to avoid situations where, with a sparse data set, cases arise for which the algorithm oscillates between two locations due to a poor initial choice of (i,j) in step (a). The restarting procedure described in Example

(II) can be used.

Although the results at different lags are not independent, in practice several lags h need to be examined. If |**h**| is small then g(**h**) is typically based on a large number N_h of differences and so has small associated standard error. However under mild anisotropy γ(**h**) will not vary much in different directions making detection of the anisotropy difficult. Conversely, for |**h**| large, γ(**h**) in different directions will differ markedly, but detection using g(**h**), and hence U_h, will be difficult as N_h is small and the associated standard error is large.

REFERENCES

Baczkowski, A.J. and Mardia, K.V. (1987) Approximate lognormality of the sample semi-variogram under a Gaussian process. *Communications in Statistics: Simulation and Computation*, **16**, 571-85.

Baczkowski, A.J. and Mardia, K.V. (1990) A test of spatial symmetry with general application. *Communications in Statistics*, to appear.

David, M. (1977) *Geostatistical Ore Reserve Estimation*, Amsterdam, Elsevier.

Grant,F. (1957) A problem in the analysis of geophysical data. *Geophysics*, **22**, 309-44.

Krige, D.G. (1976) Some basic considerations in the application of geostatistics to the valuation of ore in South African gold mines. *JSAIMM*, **76**, 383-91.

Mardia, K.V. and Gill, C.A. (1982) Some statistical inference problems in kriging, I : Numerical applications. *Journal of the Geological Society* (Complete Applications in Geology I and II), 56-72.

Martin, R.J. (1979) A subclass of lattice processes applied to a problem in planar sampling. *Biometrika*, **66**, 209-17.

Matheron, G. (1961) Precision of exploring a stratified formation by boreholes with rigid spacing - application to a bauxite deposit. *International Symposium of Mining Research, University of Missouri*, **1**, Ed. G.B. Clark, Pergamon Press, Oxford (1962), 407-22.

Mercer, W.B. and Hall, A.D. (1911) The experimental error of field trials. *Journal of Agricultural Science*, **4**, 107-32.

Computational Inference

Characteristics of Sequential Sampling Plans for Attributes: Algorithms for Exact Computing

R. Würländer, Munich

Summary

This paper deals with sequential sampling plans for attributive inspection. With a more general definition any kind of sequential sampling plans are included, especially all curtailed single, double and multiple sampling plans as well as all truncated and non-truncated SPRT's (sequential probability ratio tests). In a natural way (by direct methods) we find a recursive formula to compute the termination probabilites for acceptance and rejection of the whole lot. Exact computer algorithms are given in a pseudo-programming language to calculate these probabilities in both cases, sampling with and without replacement. Then there is no difficulty to exactly compute the characteristics of any sequential sampling plan like operating characteristic, power function, average sample number or costs, average outgoing quality, average total inspection and so on.

Keywords: statistical quality control, acceptance sampling by attributes, sampling without replacement, sequential sampling plans, direct methods, numerical algorithms.

1. Introduction

A sampling inspection for attributes can be done by various types of sampling plans: *single, double* or *multiple sampling plans.* For these plans there exist formulas to calculate the characteristics like OC- *(operating characteristic)*, ASN- *(average sample number)*, AOQ- *(average outgoing quality)* and ATI-functions *(average total inspection)* in both cases, sampling with and without replacement (see SCHILLING(1982)). For example the OC-curves of single or curtailed sampling plans are given by the *binomial* respectively the *hypergeometric* distribution functions.

Assuming that inspection can be done item by item, the use of *sequential* sampling plans will reduce the average sample number. Sequential sampling plans are usually designed by the principle of the SPRT *(sequential probability ratio test* by WALD(1947)), but these tests can cause some problems to a user:

— in the binomial case (sampling with replacement) the sequential sampling plan is not a truncated sampling plan, although it terminates with probability one,

— the exact risk probabilities (producer's risk α and consumer's risk β) may differ from the specified risks, $\tilde{\alpha}$ and $\tilde{\beta}$,

— in general the OC- and ASN-curves can be calculated only by approximate formulas, which are no longer valid in the case of sampling without replacement, where the random variables are dependent.

Compstat 1990

Example: We specify two points $(AQL, 1-\tilde{\alpha})$ and $(RQL, \tilde{\beta})$ of the OC-curve with $AQL = 0.1$ *(acceptable quality level)*, $RQL = 0.2$ *(rejectable quality level)*, $\tilde{\alpha} = 10\%$ and $\tilde{\beta} = 5\%$. Designing a SPRT with WALD's boundaries $A_0 = \frac{\tilde{\beta}}{1-\tilde{\alpha}}$ and $A_1 = \frac{1-\tilde{\beta}}{\tilde{\alpha}}$ we expect, that the actual risks α and β coincide with the specified risks $\tilde{\alpha}$ and $\tilde{\beta}$, when using the approximate formula by WALD(1947). However the exact values are $\alpha = 7.59\%$ and $\beta = 4.81\%$. With the approximate formulas for the ASN-curve we calculate $ASN(AQL) = 64.76$ (exact value: 68.83) and $ASN(RQL) = 44.31$ (exact value: 50.01).
In the practical situation of sampling without replacement we also can construct a SPRT in a similar way, but there exist neither approximate nor exact formulas to compute the characteristics like OC- or ASN-curves.

In the binomial case the problems are solved by using *direct methods* (see AROIAN(1976)): Exact recursive formulas can be derived to compute all characteristics. For this purpose numerical algorithms have been developed by HILL(1970) in the case of SPRT's and by MCPHERSON(1974) in a more general case.

However in practical situations sampling plans *without replacement* are of interest. But how can we exactly compute the performance measures of any sequential sampling plan in this case?

2. General definition of sequential sampling plans

Taking over the ideas of direct methods to sampling without replacement, a more general definition of sequential sampling plans will support our aims.

Definition:

Let

$S_n =$ the number of defective items after n inspected items (random variable),
$a_n =$ the acceptance number at stage n,
$r_n =$ the rejection number at stage n,

for $n = 1, \ldots, n_{max}$, where $n_{max} \leq N$ is the truncation sampling number and N is the size of the lot to be inspected.

Then a *sequential sampling plan for attributes* is defined by the

acceptance regions	$\{S_n \leq a_n\}$
rejection regions	$\{S_n \geq r_n\}$
continuation regions	$\{a_n < S_n < r_n\}$

with the (not really restrictive) conditions for all $n \leq n_{max}$:

(C$_1$) a_n integer, r_n positive integer,
(C$_2$) $a_n < r_n$,
(C$_3$) $a_{n-1} \leq a_n \leq a_{n-1} + 1$ with $a_0 = -1$ and
$r_{n-1} \leq r_n \leq r_{n-1} + 1$ with $r_0 = +1$,
(C$_4$) $a_n \geq \max\{-1, m - (n_{max} - n)\}$ and
$r_n \leq \min\{n+1, m+1\}$,

where $m =$ acceptance number at truncation stage n_{max}.

Remarks:

- With the conditions (C_1) – (C_4) the acceptance and rejection regions at each stage n reduce to $\{S_n = a_n\}$ and $\{S_n = r_n\}$. An acceptance point (n, a_n) or a rejection point (n, r_n) is *efficient* only if $a_n = a_{n-1} + 1$ or $r_n = r_{n-1}$.
- The sample number $\mathcal{N}$ is a bounded random variable with $\mathcal{N} \leq n_{max}$.
- Special cases of this definition are all curtailed single, double and multiple sampling plans as well as all truncated and non-truncated SPRT's (with $n_{max} = N$). For example the acceptance and rejection numbers of a curtailed single sampling plan $(N; n_{max}, m)$ are given, if the inequalities in (C_4) become equations.
- It can be proved, that each sequential sampling plan with (C_1) – (C_4) is a *generalized* SPRT with boundaries $A_0(n) \leq A_1(n)$ that can vary with n (see WÜRLÄNDER(1990)). This fact implies a monotonically non-increasing OC-function for each sequential sampling plan.

3. An algorithm for the termination probabilities

The idea of the direct methods is to compute the probabilities for the acceptance, rejection and continuation regions by recursive formulas from one stage to another.

We define the *point probabilities*

$$p_n(s) \stackrel{\text{def}}{=} \mathbb{P}(S_n = s, \mathcal{N} \geq n) = \mathbb{P}(S_n = s, a_i < S_i < r_i \text{ for } 1 \leq i < n)$$

to reach the points (n, s) in a inspection diagram, and the *transition probabilities*

$$q_n(k, s) \stackrel{\text{def}}{=} \mathbb{P}(S_n = s \,|\, S_{n-1} = k)$$

to get s defective items at stage n under the condition that k defective items are found under $n-1$ inspected items.

Regardless of the acceptance and rejection numbers a_n and r_n the recursive formula

$$p_n(s) = p_{n-1}(s) \cdot q_n(s, s) + p_{n-1}(s-1) \cdot q_n(s-1, s)$$

would compute the point probabilities, where

$$q_n(k, s) = \begin{cases} \frac{M-k}{N-(n-1)} & \text{for } k = s-1 \\ \frac{N-M-(n-1-k)}{N-(n-1)} & \text{for } k = s \end{cases}$$

and M denotes the (unknown) number of defective items in the lot of size N.

In the independent case of sampling with replacement we obtain

$$q_n(s-1, s) = \theta \quad \text{and} \quad q_n(s, s) = 1 - \theta \quad \text{for all } n,$$

where $\theta = \frac{M}{N}$ denotes the (unknown) quality level.

Considering the acceptance and rejection numbers of a sequential sampling plan at earlier stages, we have to change the recursive formula to

$$p_n(s) = \begin{cases} p_{n-1}(s)\cdot q_n(s,s) & \text{for } s = a_{n-1}+1 \\ p_{n-1}(s)\cdot q_n(s,s) + p_{n-1}(s-1)\cdot q_n(s-1,s) & \text{for } a_{n-1}+1 < s < r_{n-1} \\ p_{n-1}(s-1)\cdot q_n(s-1,s) & \text{for } s = r_{n-1}\ , \end{cases}$$

starting with $p_0(0) = 1$ and $p_0(s) = 0$ for all $s > 0$. (For a proof see WÜRLÄNDER(1990)). To derive exact formulas for the characteristics we only need the *termination probabilities*

$$p_{acc}(n) \stackrel{\text{def}}{=} p_n(a_n) \qquad \text{and} \qquad p_{rej}(n) \stackrel{\text{def}}{=} p_n(r_n)$$

at each stage n of the sequential sampling plan (see chapter 4).

The following computer algorithm — written in a pseudo-programming language — shows the computation of the termination probabilities for one quality level $\theta = \frac{M}{N}$:

Algorithm:

INPUT: N, M, n_{max}, a_n and r_n for all $n \in \{1,\dots n_{max}\}$

$p(0) \Leftarrow 1$; **for** s **from** 1 **step** 1 **to** $r_{n_{max}}$ **do** $p(s) \Leftarrow 0$ **enddo**
for n **from** 1 **step** 1 **to** n_{max} **do**
  for s **from** r_{n-1} **step** -1 **to** $a_{n-1}+1$ **do**
    $q_1 \Leftarrow \frac{M-s-1}{N-n+1}$; $q_0 \Leftarrow \frac{N-M-n+s+1}{N-n+1}$
    if $(s = 0)$ **then** $p(s) \Leftarrow q_0 \cdot p(s)$
      else $p(s) \Leftarrow q_0 \cdot p(s) + q_1 \cdot p(s-1)$ **endif**
  enddo
  $p_{acc}(n) \Leftarrow p(a_n)$; $p(a_n) \Leftarrow 0$
  $p_{rej}(n) \Leftarrow p(r_n)$; $p(r_n) \Leftarrow 0$
enddo

OUTPUT: $p_{acc}(n)$ probability for acceptance at each stage $n \in \{1,\dots n_{max}\}$
$p_{rej}(n)$ probability for rejection at each stage $n \in \{1,\dots n_{max}\}$

Remark: In the binomial case the algorithm can be simplified by replacing q_0 by the term $(1-\theta)$ and q_1 by θ.

4. Exact formulas for some characteristics

Now we are able to derive *exact formulas* for all imaginable characteristics of any sequential sampling plan. In the following some of these measures are listed and there is no difficulty to program algorithms for computing. (For proofs and details see WÜRLÄNDER(1990)).

OC- and power-function:

The OC-curve shows the probability of acceptance as a function of the quality level $\theta = \frac{M}{N}$, whereas the power-function $G(\theta)$ gives the probability of rejection.

$$OC(\theta) = \sum_{n=1}^{n_{max}} p_{acc}(n) \qquad \text{and} \qquad G(\theta) = \sum_{n=1}^{n_{max}} p_{rej}(n)$$

Distribution of the sample number:

The distribution of the random variable $\mathcal{N}$ allows to compute moments or other statistical measures like median or quantiles.

$$\mathbb{P}_\theta(\mathcal{N} = n) = p_{acc}(n) + p_{rej}(n)$$

Average sample number and costs:

In addition to the OC-curve the ASN-curve is a well-known performance measure of sequential sampling plans. If sampling costs are specified by a non-decreasing function $C_S(n)$ (with $C_S(1) > 0$), also the *average sampling costs* would be of interest.

$$ASN(\theta) \stackrel{\text{def}}{=} \mathbb{E}_\theta(\mathcal{N}) = \sum_{n=1}^{n_{max}} n \cdot \left[p_{acc}(n) + p_{rej}(n)\right] \qquad \text{and}$$

$$ASC(\theta) \stackrel{\text{def}}{=} \mathbb{E}_\theta\Big(C_S(\mathcal{N})\Big) = \sum_{n=1}^{n_{max}} C_S(n) \cdot \left[p_{acc}(n) + p_{rej}(n)\right]$$

AOQ- and ATI-functions:

These characteristics consider the outgoing quality after inspection.
RINNE/MITTAG(1989) distinguish between the following dispositions S_i/L_j:

(S_1) discard the entire sample,

(S_2) sort out the defective items of the sample,

(S_3) sort out and replace all defective items by good ones.

Then we get for the *average outgoing quality*

$$AOQ(\theta) = \sum_{n=1}^{n_{max}} OQ_n\left(\frac{M}{N}\right) \cdot p_{acc}(n),$$

$$\text{where} \quad OQ_n\left(\frac{M}{N}\right) = \begin{cases} \frac{M-a_n}{N-n} & \text{for } S_1 \\ \frac{M-a_n}{N-a_n} & \text{for } S_2 \\ \frac{M-a_n}{N} & \text{for } S_3. \end{cases}$$

If the sampling inspection decides to reject the lot, there are similar dispositions L_j for the rest of the lot: discarding (L_1), sorting out (L_2) and replacing (L_3). So we have to differ nine possible combinations S_i/L_j. Computing the *average total inspection* for a sequential sampling plan, we have to consider the disposition of the sample as well as the rest of the lot:

$$ATI(\theta) = \sum_{n=1}^{n_{max}} \left[TI_n^a\left(\frac{M}{N}\right) \cdot p_{acc}(n) + TI_n^r\left(\frac{M}{N}\right) \cdot p_{rej}(n) \right],$$

$$\text{where} \qquad TI_n^a\left(\frac{M}{N}\right) = \begin{cases} n & \text{for } S_1 \text{ and } S_2 \\ n + a_n & \text{for } S_3, \end{cases}$$

and $TI_n^r(\frac{M}{N})$ is given by the following table:

$TI_n^r(\frac{M}{N})$	L_1	L_2	L_3
S_1/S_2	n	N	$N + M - r_n$
S_3	$n + r_n$	$N + r_n$	$N + M$

5. Conclusions and final remarks

In chapter 4 some exact formulas are given, which show how to compute the characteristics of any sequential sampling plan. Remember, that the definition in chapter 2 also includes all truncated and non-truncated SPRT's as well as all curtailed single or multiple sampling plans. Additional formulas for other characteristics based on cost or Bayesian models will be derived in a similar way.

Based on the fundamental algorithm in chapter 3, the author has developed a computer program called *SeSamPlA* to calculate all interesting measures of one given sequential sampling plan. This can be done in a few seconds with the help of a personal computer. Persons interested in *SeSamPlA* should contact the author, please.

Also some algorithms would be of interest, which search for optimal sampling plans in a sense. The author has found, that a SPRT for the hypergeometric distribution, which is one rival for a sequential sampling plan without replacement, neither has WALD's optimality like the SPRT in the binomial case nor minimizes the maximum ASN.

Acknowledgement: The author is very grateful to Prof. L. Knüsel for inspiration and many useful comments to this publication.

References:

AROIAN,L.A.(1976): Applications of the direct method in sequential analysis, in: *Technometrics* **18**, 301-306.

HILL,I.D.(1970): Algorithm AS 31: Operating characteristic and average sample size for binomial sequential sampling, in: *Applied Statistics* **19**, 197-203.

MCPHERSON,K.(1974): Algorithm AS 67: The evaluation of absorption probabilities in sequential binomial sampling, in: *Applied Statistics* **23**, 83-86.

RINNE,H./MITTAG,H.-J.(1989): *Statistische Methoden der Qualitätssicherung*, Hanser: München.

SCHILLING,E.G.(1982): *Acceptance Sampling in Quality Control*, Marcel Dekker: New York.

WALD,A.(1947): *Sequential Analysis*, Wiley: New York.

WÜRLÄNDER,R.(1990): *Sequentielle Prüfpläne für die Attributenkontrolle: Exakte Berechnung ihrer Charakteristika*, Research report: Department of Statistics, University of Munich (to appear in German).

Exact Experimental Designs Via Stochastic Optimization for Nonlinear Regression Models

J. P. Vila, Versailles

Summary: The aim of this paper is twofold

1. to present a new family of optimal design criteria, based on minimization of the expected volumes of exact confidence regions for nonlinear regression model parameters.
2. to compare the efficiencies of stochastic optimization algorithms of the Kiefer-Wolfowitz type, with those of deterministic algorithms of the Quasi-Newton type, for the minimization of the criteria.

Keywords : Nonlinear Regression ; Optimal Designs ; Stochastic Optimization ; Kiefer-Wolfowitz process ; Fabian process.

1 Introduction

The design of experiments for parameter estimation in nonlinear regression has received a growing attention since the pioneering papers of Box-Lucas (1954), Atkinson & Hunter (1968) and other authors. These works are based on an adaptation of the D-optimality criterion (Kiefer 1961) to the nonlinear situation, in the sense of an implicit model linearization about a priori values for the unknown parameters. However, this first-order approximation can impair the quality of the corresponding "optimal" designs, in case of high nonlinearity. Another drawback of D-optimality is its constitutional inability to take the true limits of the parameter region into account.

More recently, various improvements have been proposed, which are generally based on quadratic approximations (Bates & Watts 1981, Hamilton 1985).

In this paper, we reconsider the design problem from a point of view which is *free of any model approximation*, by proposing as optimal design criterion the minimization of the expected volume of an exact confidence region for the parameters. The statistical and numerical properties of this criterion, X-optimality, were reported elsewhere (Vila 1986). The use of classical descent methods for the minimization of this criterion is possible and most often successful, even if time-consuming. However, the structural form of the X-optimality criterion suggests the alternative use of *Stochastic Optimization algorithms* which allow iterative minimization of the criterion with respect to the design points, *without requiring its intermediate computation.* Among these algorithms, we investigated the class of the multidimensional Kiefer-Wolfowitz stochastic processes (Blum 1954), and found the accelerated process of Fabian (1967) suitable in most situations. We shall address this point more particularly in this paper.

Compstat 1990

2 Notations

let the general regression model be

$$y_i = f(x_i, \theta^\star) + e_i \qquad i = 1, \ldots, n \tag{1}$$

with:
x_i, ith experimental point (ℓ-vector), $x_i \in \chi \subset R^\ell$
$\theta^\star \in \Theta \subset R^p$, unknown p-vector of parameters
$\xi = (x_1, \ldots, x_n) \in \Xi \subset R^m$ (with $m = \ell \times n$), the n-experimental design
$\{e_i\}$ i.i. normally distributed errors with $E(e_i) = 0$ $var(e_i) = \sigma^2$
$Y = (y_1, \ldots, y_n)^t$
$f : \chi \times \Theta \longrightarrow R$ twice continuously differentiable
$F(\xi, \theta) = (f(x_1, \theta), \ldots, f(x_n, \theta))^t$
$X = X(\xi, \theta) = [\{\partial f(x_i, \theta)/\partial \theta_j\}] \quad i = 1, \ldots, n \quad j = 1, \ldots, p$
$P(\xi, \theta) = X(X^t X)^{-1} X^t$
$\Delta(\xi, \theta) = |X^t X|$
s_ν^2 : estimator of σ^2 independent of (1), such that $s_\nu^2 \sim \sigma^2 \chi_\nu^2 / \nu$

Let us recall briefly that, for the a priori parameter values θ^0, the n-design ξ^D will be said D-optimal, if

$$\xi^D = Arg[Max_{\xi \in \Xi} \ \Delta(\xi, \theta^0)],$$

or equivalently,

$$\xi^D = Arg[Min_{\xi \in \Xi} \ \mathcal{V}(\theta^0, \alpha)] \qquad \forall 0 < \alpha < 1,$$

where $\mathcal{V}(\theta^0, \alpha)$ is the anticipated volume of the α-level ellipsoidal confidence region in R^p, of the parameters of the linearized model about θ^0. For model (1), this region is only an asymptotic anticipated confidence region and its level α is only nominal.

3 X-optimality

Very often the approximations on which the application of D-optimality to nonlinear regression is based, are questionable, as in the frequent case of high nonlinearity and of small bounded parameter domain: ellipsoidal confidence regions are not relevant anymore. *Exact confidence regions* used in that case for parameter inference can also be used to define optimal design criteria. Vila (1986) defined and studied a family of criteria based on the following confidence region :

$$\Re(\xi, \theta, Y) \ = \ \{\theta \in \Theta : \quad R(\xi, \theta, Y) < \mathcal{F}_{p,\nu;\alpha}\} \tag{2}$$

where

$$R(\xi, \theta, Y) \ = \ [Y - F(\xi, \theta)]^t \, P(\xi, \theta) \, [Y - F(\xi, \theta)] \, / p s_\nu^2 \ \sim \ \mathcal{F}\prime(p, \nu, \lambda)$$

$\mathcal{F}\prime(p, \nu, \lambda)$ is a non central $\mathcal{F}$ variable with non-centrality parameter λ such that

$$\lambda = \lambda(\xi, \theta) = [F(\xi, \theta^\star) - F(\xi, \theta)]^t \, P(\xi, \theta) \, [F(\xi, \theta^\star) - F(\xi, \theta)] \, / 2\sigma^2$$

Obviously, $Pr\,[\Re \ni \theta^\star] \ = \ 1 - \alpha$. Hence, region (2) is an *exact confidence region* for the p parameters of the model.
It can be shown that its expected volume is given by

$$V = \int_\Theta Pr\,[\mathcal{F}\prime(p, \nu, \lambda\,(\xi, \theta)) < \mathcal{F}_{p,\nu;\alpha}]\, d\theta \tag{3}$$

Minimizing (3) with respect to the design $\xi \in \Xi$, and for a priori values θ^0 for $\theta^\star$, corresponds to the simplest criterion of the X-optimality family.

3.1 Practical considerations

Several expansions for the non central $\mathcal{F}$ cumulative distribution function are available (Abramovitz & Stegun 1972). A more tractable solution for the computation of this c.d.f. is the Tiku approximation (Johnson & Kotz 1970) which affords both good accuracy and rapid computing. It can be profitably used for the calculation of the integrand of (3).

3.2 Properties of X-optimal designs

We shall just recall the most remarkable ones:

1. most frequently, X-optimal designs are made of replications of p support points whatever n. These p support points change with the region Θ and the size n (and converge to a limit support as n tends to infinity).

2. these p support points constitute the X-optimal design of size p which corresponds to the case of heterogeneous variance errors of the form σ^2/r_i for the ith point $(i = 1, \ldots, p)$, if this point is replicated r_i times in the X-optimal design of size n.
   This interesting property has been characterized by a couple of very close necessary condition and sufficient condition which allow the numerical complexity of the minimization of (3) to be reduced.

3. equitability of these replications (i.e. $|r_i - r_j| \leq 1$) is satisfied only when the parameter region Θ is "large enough" relatively to σ^2/n.

4. Parametric nonlinearity, as measured by Beale's criterion (1960) or Bates and Watts parametric curvature (1980), is lower for X-optimal designs than it is for the corresponding D-optimal designs. Moreover, this parametric nonlinearity decreases as the parameter region Θ increases.

5. By definition, X-optimal designs are more efficient than D-optimal designs with respect to expected volumes of exact confidence regions (2). Simulations show that they are also more robust, with respect to inedequacy of θ^0 values.

3.3 Illustration

The previous characteristics can be illustrated by the following model

$$f(x,\theta) = [\theta_1/(\theta_1 - \theta_2)][exp(-\theta_2 x) - exp(-\theta_1 x)] \qquad \theta_1, \theta_2, x > 0 \tag{4}$$

For $\theta^0 = (0.7, 0.2)^t$, Box & Lucas (1959) found the D-optimal design ξ^D made of equitable replications of the two points $x_1 = 1.229, x_2 = 6.858$, whatever n.
Corresponding X-optimal designs were obtained for different rectangular parameter regions Θ and for $\sigma^2 = 0.01, \alpha = 0.05, \nu = 500$ and $n = 4$ (Vila 1986):

i	Θ_i	Vol(Θ_i)	ξ_i^X	$V(\xi_i^X)$	$V(\xi^D)$	$V(\xi_i^X)/V(\xi^D)$
6	$[0., 15.] \times [0., 5.]$	75.	$2(0.939), 2(5.440)$	0.153	0.306	0.499
5	$[0., 6.] \times [0., 3.]$	18.	$2(0.971), 2(5.540)$	0.152	0.208	0.728
4	$[0., 4.] \times [0., 2.]$	8.	$2(0.980), 2(5.577)$	0.151	0.191	0.789
3	$[0., 2.] \times [0., 1.]$	2.	$2(1.014), 2(5.656)$	0.150	0.168	0.890
2	$[0., 1.15] \times [0., 0.5]$	0.575	$1(1.557), 3(6.133)$	0.121	0.127	0.947
1	$[0.45, 0.8] \times [0.1, 0.3]$	0.07	$1(1.649), 3(7.182)$	$0.449e-1$	$0.477e-1$	0.943

The last three columns display the respective values of criterion (3) for the X-optimal designs ξ_i^X, the D-optimal design ξ^D, and the ratios of these values, i.e. the relative efficiencies of the D-optimal design with respect to the X-optimal ones for the six parameter regions Θ_i successively considered.

4 Stochastic minimization

The minimization of (3) can be tackled following a classical Quasi-Newton algorithm, combined with a multiple integration procedure at each step. This was done in a first stage, with rather good results. However, the expectation-like form of (3) suggests the use of a less heavy and more suitable approach, such as *Stochastic Optimization*, which *does not involve* this multiple integral computation.
We investigated the class of the multidimensional Kiefer-Wolfowitz stochastic procedures (Blum 1954), and found the Fabian accelerated process to be significantly the most efficient in most usual cases and the most competitive with the deterministic procedures.
By comparison with the well-known Kiefer-Wolfowitz procedure, the modification of Fabian (1967) consists in taking more observations at every step and utilizing these so as to eliminate the effect of the first odd-orders derivatives of the objective function: the speed of convergence will be increased if the function is approximately even in a neighborhood of the minimum (Burkholder 1956). We shall restrict ourselves here to the main outlines of the application of this stochastic procedure to the minimization of (3).

4.1 The stochastic minimizing process

Following a classical Monte-Carlo approach, $V(\xi)$ can be written with obvious notations

$$V(\xi) \propto E_\theta\left[Z(\theta \mid \xi)\right] \tag{5}$$

the integrand of (3) being considered as a random variable $Z(\theta \mid \xi)$, through the multiuniform variate $\theta \in \Theta$.
For simplicity's sake we shall suppose that $V(\cdot)$ has a unique minimum, located in Ξ (in case of other minima outside Ξ, the following algorithm could be adapted if Ξ is a closed convex set, by the introduction of a projection operator onto Ξ and a few additional hypotheses).
We consider the stochastic iterative process in R^m defined as

$$\xi_{k+1} = \xi_k - a_k W_k \tag{6}$$

for a given ξ_0, and where

- $\{a_k\}$ is a suitable positive decreasing sequence, of the form : $a_k = a/k^\alpha$, $a > 0$ and $0 < \alpha \leq 1$. In the following we shall consider $\alpha = 1$.
- $W_k = \sum_{i=1}^h v_i W_{ki}$ with
  - h, such that $V(\cdot)$ has bounded partial derivatives up to order $(2h+1)$
  - W_{ki} : random m-vector with conditionally (ξ_k) independent components W_{ki}^j $j = 1, \ldots, m$. W_{ki}^j is the mean of n_i conditionally (ξ_k) independent variables, identically distributed as

    $$\left[Z(\theta \mid \xi_k + c_k u_i e_j) - Z(\theta \mid \xi_k - c_k u_i e_j)\right]/2c_k$$

    $c_k = c/k^{1/(4h+2)}$ with $c > 0$, $\{u_i\} : 0 < u_1 < \ldots < u_h \leq 1$,
    e_j : j^{th} canonical unit vector of R^m
  - $v = (v_1, \ldots, v_h)^t = U^{-1}\varepsilon$, with $U = [\{u_i^{2j-1}\}]$ $i, j = 1, \ldots, h$,
    ε : 1st canonical unit vector of R^h.

Under a limited number of conditions generally satisfied by criterion (3), it holds (Fabian 1967)

$$E\left\{\|\xi_k - \xi^\star\|^2\right\} = O\left(k^{-2h/(2h+1)}\right) \tag{7}$$

where $\xi^\star$ is the design minimizing (3) in Ξ. Under weaker conditions it is proved that

$$\|\xi_k - \xi^\star\|^2 k^{2h/(2h+1)-\epsilon} \xrightarrow{a.s.} 0 \quad \forall \epsilon > 0. \tag{8}$$

Let $N = \sum_{i=1}^{h} n_i$, $\mu = (\mu_1, \ldots, \mu_h)$ with $\mu_i = n_i/N$, and $u = (u_1, \ldots, u_h)$.
The couple $d = (u, \mu)$ can be regarded as a "design" for the simulation. There are many such designs leading to the same speed as (7) for process (6). Is it possible to find the "best one"? i.e. the one which minimizes

$$lim_{k\to\infty} k^{2h/(2h+1)} E\left\{\|\xi_k - \xi^\star\|^2\right\}$$

An approximate solution is reachable (Fabian 1968), which we used in our problem :

$$u_i = cos[(h-i)\pi/(2h-1)], \qquad i = 1, \ldots, h, \tag{9}$$

$$\mu_i = [2h(h-1) + 1/2]^{-1} u_i^{-2} (1 - \delta_{hi}/2), \qquad i = 1, \ldots, h, \tag{10}$$

with δ_{hi}, the Kronecker symbol.

4.2 Additions

- *Improvements:*
  the asymptotic variances and the speed of convergence of the process depend on the parameter a of the sequence $\{a_k\}$, and also on the parameter c of the sequence $\{c_k\}$. An adaptive optimal procedure, maintaining the process convergence, could be designed for the best parameter choices, following an approach similar to that of Kadar & Kurz (1980) for the multidimensional Kiefer-Wolfowitz process. Much simpler adaptive strategies, allowing dynamic evolution of a, can be based on the behavior of the inner product $< W_{k-1}, W_k >$ of subsequent step directions: the sign of this stochastic quantity is, to some extent, informative on the regular or irregular behavior of the minimizing process.

- *Termination criteria:*
  this is still also an open question in stochastic optimization, which can be related to the previous issue of the stepsize rule (Pflug 1988). Presently, criteria which are based on the step size and on frequencies of sign changes of first-order differences of successive estimates, are used. We are investigating tentative criteria based on asymptotic confidence region built from the asymptotic normality of the process (Fabian 1968).

4.3 An illustration

Let us come back to model (4). The Fabian process confirmed the results obtained with the previous Quasi-Newton procedure. As expected, the speed of convergence proved very sensitive to parameter a. Provokingly, the best choice for a was not the same for all parameter regions Θ considered. Moreover, this best choice was not relevant for the whole process. This stresses the interest of the availability of an appropriate adaptive procedure, as pointed out previously. However, using an interactive "by hand" adaptive procedure led us to performances at least comparable with those obtained with the Quasi-Newton procedure, from the point of view of the computing time.
The following array compares realizations of three Fabian processes for the minimization of criterion (3) for model (4), with $\Theta_4 = [0., 4.] \times [0., 2.]$ as parameter region, and $n = 4$ (the corresponding X-optimal design solution is displayed in the previous table). The three processes

differ from one another only in the value of h. The common procedure is non adaptive, with $a = 50$. $N = 40$. The common starting point ($k = 0$) is the D-optimal design ξ^D.

k	0	1	5	10	50	100	500	1000	5000	10000
$h = 1$	.1914	.1699	.1590	.1559	.1532	.1527	.1519	.1516	.1513	.1512
$h = 2$	.1914	.1727	.1579	.1549	.1529	.1522	.1516	.1514	.1511	.1510
$h = 6$	.1914	.1703	.2552	.1552	.1523	.1524	.1514	.1512	.1510	.1509

Because of the good starting point and the flatness of the criterion in the vicinity of the solution, the processes slow down regularly after a rapid decrease with few oscillations. The accelerating effect of a higher value of the parameter h, rather moderate here, is demonstrated only slightly as the number of iterations k increases.
This example illustrates the general efficiency of the process in rapidly reaching a reasonable neighborhood of the minimum of the criterion, providing designs with quite good properties in practice.

5 Software

We developed a general interactive Fortran software, for optimal design computation according to X-optimality (and other related criteria), in which this stochastic minimization procedure is available, in addition to a classical Quasi-Newton algorithm.

6 References

Abromowitz, M., Stegun, I.A. (1972). *Handbook of Mathematical Functions.* Dover Publications.

Atkinson, A.C. and Hunter,W.G. (1968). The Design of Experiments for Parameter Estimation. *Technometrics, 10, 271-289.*

Bates, D.M., Watts, D.G. (1980). Relative Curvature Measures of Nonlinearity. *J. R. Statist. Soc., B, 42, 1-25.*

Beale, E.M.L. (1960). Confidence Regions in Nonlinear Estimation (with discussion). *J. R. Statist. Soc., B, 22, 41-88.*

Blum, J.R. (1954). Multidimensional Stochastic Approximation Methods. *Ann. Math. Statist. 25, 737-744.*

Box, G.E.P., Lucas, H.L. (1959). Design of Experiments in Non-linear situations. *Biometrika, 46, 77-90.*

Burkholder, D.L. (1956). On a Class of Stochastic Approximation Procedures. *Ann. Math. Statist. 27, 1044-1059.*

Fabian, V. (1967). Stochastic Approximation of Minima with Improved Asymptotic Speed. *Ann. Math. Statist. 38, 191-200.*

Fabian, V. (1968). On the Choice of Design in Stochastic Approximation Methods. *Ann. Math. Statist. 39, 457-465.*

Fabian, V. (1968). On Asymptotic Normality in Stochastic Approximation. *Ann. Math. Statist., 1327-1332.*

Hamilton, D.C. (1985). A Quadratic Design Criterion for Precise Estimation in Nonlinear Regression Models. *Technometrics, 27, 241-250*

Kadar, I. and Kurz, L. (1980). A Robustized Vector Recursive Stabilizer Algorithm for Image Restoration. *Information and Control 44, 320-338.*

Pflug, G.Ch. (1988). Stepsize Rules, Stopping Times and their Implementation in Stochastic Quasi-gradient Algorithms. in *Numerical Techniques for Stochastic Optimization,* Yu. Ermoliev & R.J-B Wets Eds. Springer-Verlag

Vila, J.P. (1986). Optimal Designs for Parameter Estimation in Nonlinear Regression Models : Exact criteria. *Invited lecture, XIIIth I.B.C, Seattle.*

A Comparison of Algorithms for Combination of Information in Generally Balanced Designs

R. W. Payne and S. J. Welham, Harpenden

Summary: Methods are discussed for producing combined estimates of treatments whose effects can be estimated in more than one stratum of a generally balanced design. The algorithms of Nelder (1968), Wilkinson (1970) and Payne & Wilkinson (1977) provide a method that is specific to generally balanced designs. This method is compared with others that do not assume general balance, namely maximum likelihood and residual maximum likelihood (REML), and its computing requirements are compared with the REML implementation of Robinson, Thompson & Digby (1982).

Keywords: analysis of variance; combination of information; experimental design; generally balanced designs; variance component estimation.

1. Introduction

The concept of general balance, defined by Nelder (1965) is relevant particularly to experimental designs with several *block* (or *error*) terms. The total sum of squares can then be partitioned up into components known as *strata*, one for each block term. Each stratum contains the sum of squares for the treatment terms estimated between the units of that stratum, and a residual representing the random variability of those units. The properties of a generally balanced design are that (1) the block (or error) terms are mutually orthogonal, (2) the treatment terms are also mutually orthogonal, and (3) the contrasts of each treatment term all have equal efficiency factors in each of the strata where they are estimated. Implicit in the definition of general balance is the idea that treatment effects may be estimated in more than one stratum. The simplest example is the balanced incomplete block design, where there are treatment estimates both between blocks, and between the plots within blocks. Under these circumstances, it is advantageous to present treatment estimates that combine the information from each of

Compstat 1990

the strata where the treatment is estimated.

The combined estimates can be defined as those that solve the generalized least squares problem with weight matrix V given, under general balance, by

$$V = \sum_{\alpha=0}^{b} \xi_\alpha S_\alpha ,$$

where S_α are the projection matrices for the strata of the design, and ξ_α are the stratum variances. The orthogonal block structure (condition 1) implies also that

$$S_\alpha S_\beta = \delta_{\alpha\beta} S_\alpha , \qquad \sum_{\alpha=0}^{b} S_\alpha = I .$$

The expectation of the data vector y is given by

$$E(y) = T\tau ,$$

where (under condition 2 of general balance)

$$T = T_0 + \ldots + T_t , \qquad T_i T_j = \delta_{ij} T_i$$

and T_i is the projector for term i of the treatment model. The normal equations then become

$$\sum_{\alpha=0}^{b} (1/\xi_\alpha) T S_\alpha T \hat{\tau} = \sum_{\alpha=0}^{b} (1/\xi_\alpha) T S_\alpha y$$

and condition (3) of general balance allows the estimated vector of fitted values for treatment term i to be calculated as

$$\hat{\tau}_i = \sum_{\alpha=0}^{b} (1/\xi_\alpha) T_i S_\alpha y / \sum_{\beta=0}^{b} (\lambda_{\beta i}/\xi_\beta) = \sum_{\alpha=0}^{b} w_{\alpha i} \hat{\tau}_{i(\alpha)}, \text{ where } w_{\alpha i} = (\lambda_{\alpha i}/\xi_\alpha) / \sum_{\beta=0}^{b} (\lambda_{\beta i}/\xi_\beta),$$

$\lambda_{\beta i}$ is the efficiency factor of term i in stratum β and $\tau_{i(\alpha)}$ is the estimated fitted value vector for term i from stratum α. The treatment estimates in the various strata can readily be calculated, with no need for inversion of matrices, using the algorithm of Wilkinson (1970), Payne & Wilkinson (1977). However, there remains the problem of estimating the variance components, and thus the stratum variances.

2. Estimation of the variance components

One common method of estimation is to equate the stratum residual mean squares to their expectations; but these estimates may be poor if there are few residual degrees of freedom in a stratum, and will not exist when there are no residual degrees of freedom - as for some lattice designs. Nelder (1968) thus suggested an algorithm in which the estimates of the stratum variances are based on the residual from the generalised least squares analysis, projected onto each stratum; this is identical to the usual method when the design is orthogonal. Other, less specialised, methods have also been suggested. Many of these are based on maximum likelihood: apart from an additive constant, the logarithm of the likelihood of the data vector y is given by

$$L = (1/2)\log\{\det(V)\} - (1/2)(Y-T\tau)'V^{-1}(Y-T\tau) \ .$$

Most methods involve first, and possibly also second, differentials of the likelihood. Differentiating L with respect to τ simply leads to the normal equations whose ease of solution for a generally balanced design, with a known (or assumed) matrix V, is illustrated in Section 1, while (again for a generally balanced design) we can differentiate with respect to the stratum variances ξ_α to obtain:

$$\xi_\alpha = (y-T\tau)'S_\alpha(y-T\tau)/\mathrm{trace}(S_\alpha) \ .$$

This is similar to the estimator used in Nelder's method except that the denominator is the total degrees of freedom for stratum α, illustrating the well-known deficiency of maximum likelihood - namely that estimates of variances are biased downwards. Unfortunately, the bias is greater for small degrees of freedom and these usually occur in the higher, more variable, strata - which should not be overweighted. As an alternative, Patterson and Thompson (1971) suggested using only the likelihood of the contrasts with zero expectation (ie. the residual), since the rest of the likelihood (that is the part based on treatment contrasts), can provide no information on error. This *residual*, or *restricted*, maximum likelihood (REML) method gives the same estimates as those that can be obtained by Nelder's method for generally balanced designs, but is applicable also in less specialised situations:

$$RL = -(1/2)\log\{\det(V)\} - (1/2)(y-T\hat{\tau})'V^{-1}(y-T\hat{\tau}) - (1/2)\log\{\det(X^*V^{-1}X^*)\}$$

where, using the notation of Harville (1977), X^* is the design matrix for the treatment effects, reduced to a linearly independent set of contrasts. The equivalence of Nelder's method to REML can be demonstrated by setting $\partial RL/\partial \xi_\alpha$ to zero under the conditions of general balance, to obtain

$$\hat{\xi}_\alpha = (y-T\hat{\tau})' S_\alpha (y-T\hat{\tau}) \; / \; \{\mathrm{trace}(S_\alpha) - \sum_{i=0}^{t} w_{\alpha i} \mathrm{trace}(T_i)\} \; .$$

3. Comparison of algorithms

Algorithmically Nelder's method can be expressed as follows: (1) obtain the treatment estimates within each stratum; (2) assume starting values for the stratum variances; (3) estimate the weights $w_{\alpha i}$ and thus the combined treatment estimates; (4) re-estimate the stratum variances; and (5) test convergence. For comparison, we use the REML implementation of Robinson, Thompson & Digby (1982) which is now available in Genstat 5 (Payne *et al.* 1989); however, we believe that other non-specialised algorithms that involve maximization of likelihoods would have very similar computational requirements. This operates as follows: (1) choose starting values for the variance components; (2) produce the generalized least squares solution for the fixed and random effects, using the current estimates of the variance components; (3) form first differentials and the information matrix, then invert the information matrix (using results of step 2); (4) obtain new estimates of the components (using a modified form of Fisher's scoring technique); (5) test convergence. The obvious difference between the two algorithms is that REML has to recalculate the effects (by calculation and inversion of a large matrix) at every iteration, whereas Nelder's method merely needs to recalculate the weighted combination of the treatment estimates in the various strata, and the resulting stratum residual sums of squares. It needs no matrix inversion, either for estimating the stratum variances, or for the initial analysis of the strata (see Payne & Wilkinson, 1977). Thompson (1977) described a refinement to the REML implementation, in which an *absorbing* factor is used to save space and computing time. The model terms are split into two sets: those terms involving the absorbing factor and those not. The effects for the two parts of the model are estimated separately, and a greater number of rather smaller matrices are required than for the simpler algorithm above. The cost is that standard errors are not available for any effects involving the absorbing factor; but this will be unimportant provided a block factor is used for the absorbing factor, as standard errors will then still be available for treatment effects.

It is easiest to demonstrate how the assumption of general balance saves computing space and time by taking some representative designs. Consider first a balanced incomplete block design with b blocks, k plots per block and t treatments. With no absorbing factor, REML needs to store a symmetric matrix with $b+t+1$ rows, for sums of squares and products, and vectors of size b and t for the block and treatment effects; at each iteration it must invert a matrix with $b+t$ rows. With blocks as absorbing factor, the symmetric matrix to be stored has $t+1$ rows, and the matrix to be inverted has t rows. Conversely, use of the algorithm of Wilkinson (1970), Payne & Wilkinson (1977) for the stratified analysis, and that of Nelder (1968) for the combination of information, requires a working vector of size bt, two vectors of size t for treatment estimates between and within blocks, and a vector of size b if the block residuals are to be stored. For a lattice design with r replicates, k blocks per replicate, k plots per block and k^2 treatments, the situation is as follows: REML with no absorbing factor - storage of a $k^2+rk+r+1$ row symmetric matrix and vectors of size k^2, rk and r for fixed and random effects, inversion of a k^2+rk+r row summetric matrix; REML with replicates as absorbing factor - storage of symmetric matrices with k^2+1 and $k+1$ rows, and a $(k^2+1)\times(k+1)$ rectangular matrix, inversion of one symmetric matrix with k^2 rows and r with $k+1$ rows; Payne & Wilkinson (1977), and Nelder (1968) - storage of vectors of size $r\times k^2$, k^2, k^2, rk and r, no matrix inversion. Finally, a square lattice with r replicates, k rows and k columns per replicate and k^2 treatments: REML with no absorbing factor - storage of a $k^2+2rk+r+1$ row symmetric matrix and vectors of size k^2, rk, rk and r for fixed and random effects, inversion of a $k^2+2rk+r$ row summetric matrix; REML with replicates as absorbing factor - storage of symmetric matrices with k^2+1 and $2k+1$ rows, and a $(k^2+1)\times(2k+1)$ rectangular matrix, inversion of one symmetric matrix with k^2 rows and r with $2k+1$ rows; Payne & Wilkinson (1977), and Nelder (1968) - storage of vectors of size $r\times k^2$, k^2, k^2, k^2, rk, rk and r, again no matrix inversion.

4. Conclusion

The benefits of general balance are firstly that, due to the assumption of orthogonal block structure, the stratum variances can each be calculated from a simple expression, instead of requiring solution of a set of simultaneous equations; secondly that the combined treatment estimates (to solve the generalised least squares problem) are formed merely from a weighted combination of the estimates from the various strata; and thirdly, that these stratified estimates themselves can be calculated by the simple process of sweeping of effects (formed by calculating totals and dividing by effective replication), as described by Wilkinson (1970), Payne & Wilkinson (1977). Such an algorithm is thus very much more efficient computationally than the less specialised REML algorithm. However, it cannot handle covariates (although the necessary extensions are derived in an unpublished paper by R.W. Payne & R.D. Tobias), and REML has the advantage that it also handles unbalanced situations such as arise in variety trials or in the combination of information from several different experiments.

References

Harville, D.A. (1977). Maximum likelihood approaches to variance component estimation and related problems. *J.A.S.A.*, **72**, 320-340.

Nelder, J.A. (1965). The analysis of randomised experiments with orthogonal block structure. I. Block structure and the null analysis of variance. II. Treatment structure and the general analysis of variance. *Proceedings of the Royal Society of London A* **283**, 147-178.

Nelder, J.A. (1968). The combination of information in generally balanced designs. *J. Roy. Statist. Soc, B*, **30**, 303-311.

Patterson. H.D. & Thompson, R. (1971). Recovery of inter-block information when block sizes are unequal. *Biometrika*, **58**, 545-554.

Payne, R.W., Lane, P.W., Ainsley, A.E., Bicknell, K.E., Digby, P.G.N., Harding, S.A., Leech, P.K., Simpson, H.R., Todd, A.D., Welham, S.J., White, R.P., Gower, J.C., Tunnicliffe Wilson, G. & Verrier, P.J. (1989). *Genstat 5 Release 2 Reference Summary*. Oxford: Numerical Algorithms Group.

Payne, R.W. and Wilkinson, G.N. (1977) A general algorithm for analysis of variance. *Applied Statistics*, 26, 251-260.

Robinson, D.L., Thompson, R. & Digby, P.G.N. (1982). REML - a program for the analysis of non-orthogonal data by restricted maximum likelihood. In: *COMPSTAT 1982, part II (supplement)*, pp. 231-232. Wien: Physica-Verlag.

Thompson, R. (1977). The estimation of heritability with unbalanced data. I. Observations available on parents and offspring. II. Data available on more than two generations. *Biometrics*, **33**, 485-504.

Wilkinson, G.N. (1970). A generally recursive algorithm for analysis of variance. *Biometrika*, **57**, 19-46.

A Comparison on Non-Negative Estimators for Ratios of Variance Components

J. T. Lee and K. S. Lee, Seoul

Abstract

The problems of point estimation for ratios of nonnegative variance components in the balanced one way random effects model are considered. Seven estimators are compared with respect to their biases and mean squared error (MSE). A new estimator (New) that dominates ML type estimators in terms of MSE is derived. In conclusion, New and MINQE estimators are recommended that these estimators possess smaller MSE even in the presence of nontrivial bias.

1. Introduction

Consider the balanced one-way classification random effects model

$$y_{ij} = \mu + a_i + e_{ij}, \quad i = 1, \ldots, I, \quad j = 1, \ldots, J,$$

where μ is a fixed effect and the a_i and e_{ij} are independent normal variables with zero means and variance σ_a^2 and σ_e^2, respectively. When this model is analyzed, one of important objective is estimating the ratio $\theta = \sigma_a^2/\sigma_e^2$ and the intraclass correlation $\rho = \sigma_a^2/(\sigma_a^2+\sigma_e^2)$. As is well known, among confidence limits related to variance components, the limits most frequently wanted appear to be those for θ and the intraclass correlation ρ. Statistical inferences concerning the ratio θ and the intraclass correlation ρ are given in Graybill (1976, pp.615 – 621).

Several new techniques for the estimation of variance components in linear models have been introduced. Interested readers may refer to P. S. R. S. Rao and Y. P. Chaubey (1978) for MINQE, Hartung (1981) for MB-MINQE, and to Chaubey (1983) for CMINQUE.

Lately, Loh (1986) considered the problem of estimating a ratio of variance components in the balanced one way random effects model. We extend his result for several modifications of MINQUE estimators. In this paper, the non-negative estimators for ratios of variance components, ML, REML, MINQE, MB-MINQE, and CMINQUE are considered for the balanced one way random effects model. Also, we investigate these estimators with respect to their mean square errors and biases. We derive a new estimator called by New, which dominate ML type (ML, REML, opt) estimators. A numerical comparisons are given among ML, REML, MB-MINQE, CMINQUE, MINQE, opt, and New for selected values of I, J and ρ.

2. Types of Estimators

We shall restrict attention to estimators of θ that are functions of the translation-invariant part of the set of sufficient statistics. We consider three types of estimators of θ:

Type A. $\hat{\theta} = aG$,

Type B. $\hat{\theta} = [bG - c]^+$,

Type C. $\hat{\theta} = f(G)$,

Compstat 1990

where a, b, c represent an arbitrary nonnegative constant and $f(\cdot)$ represents arbitrary nonnegative functions. Here, $[\ \cdot\ ]^+$ denotes the positive part and G is the usual ratio of mean squares for testing $\theta = 0$. Note that G is distributed as $(1 + J\theta)F$, where F is a random variable whose distribution is central F with degrees of freedom $v_1 = I - 1$ and $v_2 = I(J - 1)$.

We refer to the estimator $\hat{\theta}$ as a Type-A, -B or -C, respectively. Among Type-A estimators, the estimators associated with the following values of a are of special interest:

$$a_1 = \frac{1}{J}, \qquad a_2 = \frac{J}{1 + J^2}, \qquad a_3 = \frac{I-1}{I}\frac{J}{(1+J)^2}. \tag{2.1}$$

For convenience, define $\hat{\theta}(a_i) = a_i G$, $i = 1, 2, 3$. We refer to the estimator $\hat{\theta}(a_1)$ as the closest MINQUE. Also the estimators $\hat{\theta}(a_2)$ and $\hat{\theta}(a_3)$ are MB-MINQE and MINQE, respectively. When $I > 1$ and $I(J - 1) > 4$, we have that $a_3 < a_2 < a_1$. We will show that if $v_2 > 4$, the closest MINQUE estimator $\hat{\theta}_C$ is dominated by the MB-MINQE estimator $\hat{\theta}_{MB}$, which in turn is dominated by a MINQE estimator $\hat{\theta}_{MINQE}$ in terms of MSE. The following lemma is necessary for this proof, which is a slightly modified version of a result proved in Loh (1986).

Lemma 1. Suppose $v_2 > 4$. For fixed $\theta \geq 0$, the risk function

$$R(\delta, \theta) = E(\delta G - \theta)^2$$

is strictly increasing in δ for $\delta^* < \delta < \infty$, and it attains its minimum at $\delta = \delta^*$, where

$$\delta^* = K\theta/(1 + J\theta), \quad \text{where} \quad K = v_1(v_2 - 4)[v_2(v_1 + 2)]^{-1} < 1. \tag{2.2}$$

Using Lemma 1, we obtain the following theorem.

Theorem 1. Suppose $v_2 > 4$, and define δ^* as in (2.2). For $\delta^* < \delta_3 < \delta_2 < \delta_1$, let $\hat{\theta}_k = \delta_k G, k = 1, 2, 3$. Then for each $\theta \geq 0$,

$$\text{MSE}(\hat{\theta}_1) > \text{MSE}(\hat{\theta}_2) > \text{MSE}(\hat{\theta}_3).$$

We state the following relationship among Type-A estimators.

Corollary 1. For $I > 1$, $I(J - 1) > 4$, and all $\theta \geq 0$,

$$\text{MSE}(\hat{\theta}_C) > \text{MSE}(\hat{\theta}_{MB}) > \text{MSE}(\hat{\theta}_{MINQE}).$$

Now, we discuss Type-B estimators. Important Type-B estimators are ML and REML estimators. For balanced one way random effects model, the ML and REML estimators of θ are given by the expressions (Loh,1986)

$$\theta_{ML} = \frac{1}{J}[\frac{(I-1)}{I}G - 1]^+ \qquad \text{and} \qquad \theta_{REML} = \frac{1}{J}[G - 1]^+. \tag{2.3}$$

Another Type-B estimators are opt and mode estimators given by

$$\theta_{opt} = \frac{1}{J}[\frac{v_1(v_2-4)}{v_2(v_1+2)}G - 1]^+ \qquad \text{and} \qquad \theta_{mode} = \frac{1}{J}[\frac{v_1(v_2-2)}{v_2(v_1+2)}G - 1]^+. \tag{2.4}$$

Loh (1986) proposed a new estimator θ_{opt} which has minimum MSE among all estimators of θ of the form $J^{-1}[(1-\delta)G - 1]^+$ for $-\infty < \delta < \delta^*$, where $\delta^* = \{v_1(v_2 - 4)\}/\{v_2(v_1 + 2)\}$. The estimator θ_{mode}

means that the Bayes *mode* estimators suggested by Klotz et al. (1969). We can obtain the following relationship for the MSE of the type-B estimators .

$$\mathrm{MSE}(\hat{\theta}_{REML}) > \mathrm{MSE}(\hat{\theta}_{ML}) > \mathrm{MSE}(\hat{\theta}_{mode}) > \mathrm{MSE}(\hat{\theta}_{opt}).$$

One of the Type-C estimators of θ is Bayes model estimator given by the expression

$$\theta_{mean} = \frac{1}{J}[(\frac{v_1}{v_1 - 2}\frac{I_M(v_1/2 - 1, v_2/2 + 1)}{I_M(v_1/2, v_2/2)})G - 1]^+, \tag{2.5}$$

where $M = G(G+v_2/v_1)^{-1}$ and $I_M(p,q)$ is the incomplete beta function ratio. However, this estimator is not recommended because of complex form and larger MSE.

3. New estimator

A new Type-B estimator for θ is proposed as follows. First, fix the value of b as K/J, and then we find the best c that minimizes

$$\mathrm{E}[(KG - c)/J - \theta]^2, \quad \text{where} \quad K = v_1(v_2 - 4)[v_2(v_1 + 2)]^{-1}.$$

Then, the value of c is easily seen to be $c^* = K_1^2/K_2 + J\theta(K_1^2/K_2 - 1)$, where $K_1 = \mathrm{E}(F)$ and $K_2 = \mathrm{E}(F^2)$. Also, it is easy to verify that $c^* < K_1^2/K_2 < 1$. Thus, it is desirable to use the estimator

$$\theta_{New} = \frac{1}{J}[\frac{v_1(v_2 - 4)}{v_2(v_1 + 2)}G - \frac{v_1(v_2 - 4)}{(v_1 + 2)(v_2 - 2)}]^+. \tag{3.1}$$

The opt estimator is dominated by this new estimator. Also, these two estimators are compared analytically with respect to their biases and it follows that

$$(\text{Absolute Bias of } \hat{\theta}_{New}) < (\text{Absolute Bias of } \hat{\theta}_{opt}), \qquad \forall\, \theta \geq 0.$$

4. Numerical results

We computed the MSE and biases of the seven estimators (ML, REML, MB-MINQE, CMINQUE, MINQE, opt, and New) based on several combinations of I, J, and ρ. From Table 1 and Table 2, we have the following conclusions for the one way random effects model.

(1) All the estimators of θ have smaller MSE than $\hat{\theta}_C$.

(2) For all around performance, it appers that $\hat{\theta}_{MINQE}$ or $\hat{\theta}_{New}$ may be recommended.

(3) The $\hat{\theta}_{New}$ has the highest efficiency if $I \leq 3$. The result also holds if $I = 4$ and $J = 3$.

(4) MSE of $\hat{\theta}_{MINQE}$ has poor efficiency rather than that of $\hat{\theta}_{New}$ in the case of $\rho < 0.1$ or $\rho > 0.9$, approximately.

(5) The New and MINQE estimators of θ are competitive and so do REML and MB-MINQE.

(6) ML estimators are always unbiased estimator of θ for $J = 3$.

(7) Although MINQE and New estimators have high efficiency, they have a larger bias rather than other estimators.

5. Some remarks

In general, the good estimators of θ achieve their low MSE's by severely shrinking toward the origin, while other estimators under MSE criterion have a greater tendency to overestimate θ. But this result does not mean that the best estimator of θ has the smallest value of all estimators. As we see, the New estimator is preferable to opt estimator while on the other, the value of New estimator is always greater than that of opt estimator. Also our results indicate that the New and MINQE estimators perform well in spite of nontrivial bias. Bias is a symmetric error. Unbiasedness of estimator says that estimator is correct on the average. When we consider the linear function $f(G)$ as an estimator for θ, unbiasedness condition is not necessarily a desirable characteristic of an estimate. One should keep in mind that the distribution of statistic G is not symmetric.

For the one way classification random model with unbalanced data, our results may not be hold generally. To draw general conclusions for this model, a Monte Carlo comparison of a number of estimators must be considered. The problem of estimating intraclass correlation coefficient is also important. Further investigations in this direction are in progress.

Table 1. Absolute Bias of $\hat{\theta}$ / Absolute Bias of $\hat{\theta}_{REML}$

(I,J,ρ)	$\hat{\theta}_{ML}$	$\hat{\theta}_{MB}$	$\hat{\theta}_{C}$	$\hat{\theta}_{MINQE}$	$\hat{\theta}_{opt}$	$\hat{\theta}_{New}$
(3,3,0.1)	0.00	2.20	2.50	0.63	1.50	0.38
(3,5,0.3)	1.00	2.36	2.59	0.63	3.00	2.05
(3,8,0.7)	2.50	1.32	1.48	3.49	5.25	4.98
(4,3,0.2)	0.00	2.31	2.71	0.40	1.80	0.77
(4,7,0.7)	2.00	1.40	1.64	3.48	5.00	4.71
(4,8,0.9)	2.50	0.96	1.18	4.53	5.80	5.72
(5,5,0.3)	1.00	3.48	3.86	0.58	3.67	2.50
(5,8,0.4)	2.50	3.34	3.61	2.83	6.17	5.19
(5,9,0.9)	3.00	0.99	1.23	5.81	7.00	6.92
(6,4,0.2)	0.50	4.47	5.00	0.80	3.00	1.50
(6,5,0.4)	1.00	3.08	3.54	1.52	3.86	2.97
(6,7,0.7)	2.00	1.62	1.98	4.54	5.57	5.25
(7,3,0.1)	0.00	4.80	5.50	1.88	2.25	0.56
(7,7,0.7)	2.00	1.73	2.15	5.07	5.75	5.42
(7,9,0.9)	3.00	0.99	1.33	7.23	7.50	7.41
(8,3,0.3)	0.00	3.26	4.06	0.00	2.33	1.31
(8,6,0.6)	1.50	2.36	2.90	4.24	5.00	4.50
(8,8,0.2)	2.50	9.57	10.00	1.36	6.78	4.52
(9,3,0.1)	0.00	6.10	7.00	2.50	2.40	0.60
(9,4,0.7)	0.50	1.42	2.21	3.61	3.30	2.98
(9,6,0.5)	1.50	3.46	4.07	3.74	5.10	4.37

Table 2. MSE($\hat{\theta}$) / MSE($\hat{\theta}_{REML}$)

(I,J,ρ)	$\hat{\theta}_{ML}$	$\hat{\theta}_{MB}$	$\hat{\theta}_{C}$	$\hat{\theta}_{MINQE}$	$\hat{\theta}_{opt}$	$\hat{\theta}_{New}$
(3,3,0.1)	0.43	0.95	1.19	0.15	0.11	0.03
(3,4,0.6)	0.44	0.90	1.03	0.21	0.21	0.17
(3,5,0.3)	0.46	1.01	1.10	0.22	0.27	0.19
(3,8,0.7)	0.49	0.98	1.01	0.36	0.36	0.34
(4,3,0.2)	0.54	0.99	1.26	0.18	0.22	0.11
(4,5,0.5)	0.57	0.97	1.06	0.31	0.39	0.33
(4,7,0.7)	0.59	0.97	1.01	0.43	0.46	0.43
(4,8,0.9)	0.60	0.97	1.00	0.48	0.48	0.47
(5,3,0.1)	0.62	1.25	1.58	0.26	0.31	0.16
(5,5,0.3)	0.65	1.09	1.20	0.31	0.47	0.37
(5,8,0.4)	0.67	1.03	1.07	0.44	0.55	0.49
(5,9,0.9)	0.68	0.98	1.00	0.56	0.57	0.56
(6,4,0.2)	0.69	1.27	1.48	0.29	0.48	0.35
(6,5,0.4)	0.70	1.03	1.15	0.36	0.54	0.46
(6,7,0.7)	0.71	0.97	1.02	0.53	0.60	0.57
(6,9,0.8)	0.72	0.98	1.00	0.61	0.63	0.62
(7,3,0.1)	0.71	1.54	1.96	0.34	0.44	0.29
(7,5,0.2)	0.74	1.36	1.51	0.36	0.59	0.47
(7,7,0.7)	0.75	0.97	1.02	0.57	0.64	0.62
(7,9,0.9)	0.76	0.98	1.00	0.65	0.67	0.67
(8,3,0.3)	0.74	1.11	1.47	0.24	0.50	0.38
(8,6,0.6)	0.78	0.98	1.05	0.54	0.66	0.63
(8,7,0.4)	0.78	1.06	1.12	0.49	0.68	0.62
(8,9,0.8)	0.79	0.98	1.01	0.67	0.70	0.69
(9,3,0.1)	0.77	1.83	2.34	0.42	0.54	0.39
(9,6,0.5)	0.80	1.02	1.10	0.51	0.69	0.65
(9,7,0.6)	0.81	0.99	1.04	0.59	0.71	0.68
(9,9,0.8)	0.81	0.98	1.01	0.70	0.73	0.72

Acknowledgement

This reserach was supported by a grant of the Korea Science Foundation.

References

1. Chaubey, Y. P. (1983)."A Non Negative Estimator of Variance Component Closest to MINQUE," *Sankhyā, Ser. A,* Vol. 45, 201 - 211.

2. Chaubey, Y. P. (1984)."On the Comparison of Some Non-Negative Estimators of Variance Component For Two Models," *Commun. Statist.- Simula. Computa.,* Vol. 13(5), 619 - 633.

3. Conerly, M. D., and Webster, J. T. (1987). "MINQE for the One-Way Classification," *Technometrics,* Vol. 29(2), 229 - 236.

4. Graybill, F. A. (1976). *Theory and Applications of the Linear Models,* Duxbury Press, North Scituate, Massachusetts.

5. Hartung, J. (1981). "Nonnegative Minimum Biased Invariant Estimation in Variance Components Models," *Annals of Statistics*, Vol. 9, 278 - 292.

6. Loh, Wei-Yin. (1986). "Improved Estimators for Ratios of Variance Components," *Journal of the American Statistical Association*, Vol. 81, 699 - 702.

7. Klotz, J. H., Milton, R. C., and Zacks, S. (1969). "Mean Square Efficiency of Estimators of Variance Components," *Journal of the American Statistical Association*, Vol. 64, 1383 - 1402.

8. Peixoto, J. L., and Harville, D. A. (1986). "Comparisons of Alternative Predictors Under Balanced One-way Random Model," *Journal of the American Statistical Association*, Vol. 81, 431 - 436.

9. Rao, P. S. R. S., and Chaubey, Y. P. (1978). "Three Modifications of the principle of MINQUE," *Commun. Statist.- Theor. Meth.*, Vol. A7(8), 767 - 778.

10. Scheffe, H. (1959). *The Analysis of Variance*, New York: John Wiley.

11. Searle, S. R. (1971). *Linear Models*, New York: John Wiley.

12. Snedecor, G, W., and Cochran, W. G. (1980). *Statistical Methods* 3rd ed., Ames, IA: Iowa state University Press.

13. Swallow, W. H., and Monahan, J. F. (1984). "Monte Carlo Comparison of ANOVA, MINQUE, REML, and ML Estimators of Variance Components," *Technometrics*, Vol. 26, 47 - 57.

Optimal Fit in Non-Parametric Modelling Via Computationally Intensive Inference

M. Abrahamowicz and A. Ciampi, Montreal

Abstract

We present a criterion for model selection in non-parametric inference based on comparing the confidence intervals of a family of competing estimators (CIC). Application to M-spline density estimation is dicussed: here the family of density estimators is indexed by the meta-parameter "number of knots". Numerical examples show the relationship between CIC model choice and model choice based on the AIC and the BIC.

1. Introduction

Non-parametric models are becomimg increasingly important in data analysis, since they have been shown to perform extremely well in a broad variety of situations. Indeed, when little is known about the nature of the random processes generating the data, a non-parametric model is considered the tool of choice.

One approach to fitting non-parametric models is maximum-likelihood estimation (mle). While this approach enjoys well known optimal properties, the quality of the fit depends, often in a crucial way, on the choice of some "meta-parameters", such as window-width for kernel estimation of densities, penalty parameter for smoothing splines models and number of knots for regression splines models.

The choice of meta-parameters is usually determined from the data by optimizing a specified criterion which compares point-estimates obtained for different the meta-parameter values. Perhaps the most popular criteria for model selection are the AIC (Akaike, 1974) and the BIC (Schwartz, 1978). An alternative is cross-validation applied to the likelihood itself, a computationally intensive approach which the AIC is considered to approximate well for large samples(Stone, 1977). There is no universal agreement on the choice of the criterion and, indeed, different criteria lead often to different model choices, as found, for instance, in (Abrahamowicz & Ciampi, 1990). In (Abrahamowicz & Ciampi, 1990) it is also shown by simulations that both AIC and BIC may fail to pick the correct model.

In this paper we present an approach to the selection of meta-parameters based on a confidence interval (CI) rather than on a point-estimate and illustrate this approach in the context of non-parametric density estimation. The approach is described in section 2. Section 3 is devoted to a summary of the regression spline approach to density estimation (Abrahamowicz, Ciampi, Ramsay, 1989), and section 4 presents examples of criterion applications to the selection of the "optimal" dimensionality of the spline space. We conclude in section 5, with a short summary of a limited simulation study and a brief discussion.

Compstat 1990

2. CIC: Confidence Interval Criterion for Model Selection

Let $\mathcal{D}$ denote a data set, $\phi(t)$ a function of a variable t, $\tilde{\phi}(t;\kappa)$ any estimator of $\phi(t)$ constructed from D and dependent on a meta-parameter κ. We assume that we have chosen a measure of the lack-of-fit of an estimator to D, LFD($\tilde{\phi}:\mathcal{D}$), and we denote by $\hat{\phi}(t;\kappa)$ an estimator of $\phi(t)$ that minimizes LFD.

We assume also that the meta-parameter κ is such that the LFD is expected to decrease with increasing κ, as is the case in regression splines (Ramsay,1988) when κ is the number of knots and in smoothing spline estimation (Silverman,1985), when it is the inverse of the penalty parameter. The meta-parameter is associated also with the complexity of the model. In the case of regression splines complexity is determined by the dimensionality of the subspace of the function space to which the estimator belongs; in that of smoothing splines by the relative lack of restrictions on the estimate.
If the true model is known, as is the case in simulation studies, we also assume that a measure of the lack-of-fit of an estimator to the true ϕ, LFT($\tilde{\phi}:\phi$) has been chosen. Choices for LFD and LFT based on the likelihood of the data are discussed in (Abrahamowicz & Ciampi,1990).

The model selection algorithm proposed here is very simple. Essentially, we construct a sequence of estimators of $\phi(t)$ of increasing κ, each of which minimizes LFD. Each estimator is compared to the following ones: the disagreement of a simpler estimator with respect to a more complex one, is measured by the percentage of the point estimates of $\phi(t)$, based on the former, which fall out of a chosen confidence interval for the latter. We chose the simplest estimator of the sequence whose disagreement with every more complex one, is smaller than a pre-fixed threshold. We call such a selection criterion CIC from Confidence Interval Criterion. A more formal description of the algorithm follows.

<u>ALGORITHM FOR CIC SELECTION</u>

<u>Step 1</u>: 1.1. Input:
data D ; sequence of meta-parameter values $\kappa_1<\ldots<\kappa_r<\ldots<\kappa_v$; a constant Δ which determines the half-width of the pointwise CI for an estimate in terms of its standard error (CI=$\hat{\phi}(t)\pm\Delta$*SE); vector $y_i=y_1,\ldots,y_n$ of n points (selected from the domain of $\phi(t)$) used for comparison of estimates; tolerance threshold τ for the acceptable (in the case of an agreement between two estimates) proportion of points where one estimate falls outside the confidence limits for the other one.

1.2. Initiate the vector κ^* of "acceptable" estimates by including the simplest model (for the lowest meta-parameter κ_1):

$$p = 1; \quad \kappa^*_1=\kappa_1.$$

<u>Step 2:</u> For $\kappa_r=\kappa_1,\ldots,\kappa_v$:

2.1. for $y_i=y_1,\ldots,y_n$ find :
point estimate $\hat{\phi}(y_i,\kappa_r)$; its standard error $SE(y_i,\kappa_r)$ and confidence limits $CI(y_i,\kappa_r)=\hat{\phi}(y_i,\kappa_r)\pm\Delta*SE(y_i,\kappa_r)$.

2.2. compute LFD[$D,\hat{\phi}(t,\kappa_r)$]

2.3. for $j=\kappa_1,\ldots,\kappa_{r-1}$:

2.3.1. if LFD[D,$\hat{\phi}(t,\kappa_r)$] ≥ LFD[D,$\hat{\phi}(t,j)$] go to 2.

2.4. include κ_r-based estimate in the sequence of "acceptable" solutions iff it hasn't been rejected in 2.3 : $p=p+1$, $\kappa_p^* = \kappa_r$.

Step 3: Assess agreement of each "acceptable estimate" with all more complex ones (which fit D better) :

For $\kappa_r^* = \kappa_1^*, \ldots, \kappa_p^*$:

3.1. for $j = \kappa^*_{r+1}, \ldots, \kappa^*_p$:

set pdis(κ^*_r, j)=0.0

3.1.1. for $y_i = y_1, \ldots, y_n$:

iff $\hat{\phi}(y_i,\kappa^*_r) \notin CI(y_i, j)$ then

pdis(κ^*_r, j) = pdis(κ^*_r, j) + 1/n.

3.1.2. iff pdis(κ^*_r, j) > τ go to 3.

3.2. Chose κ^*_r as the optimal meta-parameter value if it has not been rejected in 3.1.2. :

set $\kappa_{opt} = \kappa^*_r$ and STOP.

3. CIC choice of number of knots in M-spline density estimation

One of the most important situations in which non-parametric methods are widely applied is density estimation (Silverman, 1986). According to the particular approach chosen, the meta-parameter may be a penalty parameter, a window width or the number of knots that defines the spline space to which the estimate is constrained. Here we present the CIC application to a method recently proposed by us (Abrahamowicz,Ciampi,Ramsay,1989) in which the mle density is obtained as a linear expansion of cubic M-splines i.e. cubic piecewise normalized polynomials. Subsequent polynomial pieces are constrained to join smoothly with each other at m points called *knots* which are placed at (m+1)-th quantiles of the empirical distribution.

The method can be seen as a reasonable compromise between parametric and fully non-parametric approaches as it combines substantial flexibility with low dimensionality of the function space. As a result it has been shown to produce relatively unbiased and precise estimates for simulated data and to uncover some unusual features of actual data (Abrahamowicz, Ciampi, Ramsay, 1989). On the other hand the quality of the solution changes as a function of the number of knots and the "optimal" value of the latter varies depending on the complexity of the underlying population distribution. This situation emphasizes the importance of an appropriate selection of the meta-parameter (number of interior knots) which determines the dimensionality of the spline space to which the estimator is constrained.

Based on results of previous simulations we can recommend that the sequence of meta-parameters includes values from 0 to 6 or to 10% of the sample size, whichever the lower. Negative loglikelihood of D provides, in this context, a natural measure of LFD and Kullback-Leibler number (KL) that of LFT (see Abrahamowicz & Ciampi,1990 for details). Notice here that the rule for placing knots at quantiles causes that models are not nested and occasionally a fewer-knots estimate may fit D better than a higher dimensionality one (which should be then eliminated at step 2.3.1 of the algorithm).Finally CI's are based on the large-sample mle theory while taking into account the linear equality constraint on the estimated coefficients of the linear expansion as described in (Abrahamowicz,Ciampi,Ramsay,1989).

4. Numerical Examples

We present here three examples of M-spline density estimation from simulated data which illustrate the salient aspects of the relative performances of AIC, BIC and CIC.

In each case, 100 data points were generated from a fixed distribution. The sequence of meta-parameters ran from 0 (simple polynomial) to 6 knots. The tolerance threshold for agreement was set to 5%. The half-width of the (symmetric) CI was 2SE (Δ=2.0). Finally, the agreement was assessed on a series of equally spaced 300 points.

Panel a) of the table below refers to data generated from a mixture of exponential (mixing parameter .4, mean 200) and Weibull (shape parameter 3.0 and mean 1200); this results in a sharply bimodal and skewed distribution. Panel b) refers to another bimodal distribution, this time roughly symmetric, obtained as a mixture of two normals: .4 N(300; 100) + .6 N(1500; 500). Panel c) refers to a moderately skewed unimodal Weibull distribution (shape parameter 1.5 and mean 1000).

AIC, BIC and CIC model selections
in simulated examples of M-spline density estimation

# knots	LFT (KL)	LFD (-LogL)	AIC	BIC	% of disagreement with respect to estimate using : 1	2	3	4	5	6	
					k n o t s						
a)											
0	0.0865	755.9	1515.7	1520.9	42	19	32	30	28	23	
1	0.0413	752.8	1511.6	1519.5		5	33	24	18	10	
2	0.0394	752.4	1512.8	1523.2			20	10	11	7	
3	0.0247	751.1	1512.2	1525.2				0	0	2	CIC
4	0.0237	751.1	1514.2	1529.8					0	0	
5	0.0241	750.9	1515.8	1534.0						0	
6	0.0339	749.5	1515.0	1535.9							
b)											
0	0.2421	787.4	1578.9	1584.1	0	43	*53*	45	50	54	
1	0.2353	786.8	1579.7	1587.5		23	*30*	39	47	49	
2	0.1171	776.3	1560.6	1571.0			*22*	10	38	28	
3	*0.1468*	*777.0*	*1563.9*	*1577.0*				*20*	*29*	*33*	
4	0.1116	773.1	1558.2	1573.9					22	6	
5	0.0571	769.9	1553.8	1572.0						0	CIC
6	0.0659	769.5	1555.1	1571.9							
c)											
0	0.0415	773.6	1551.3	1556.5	52	57	50	50	48	48	
1	0.0052	769.9	1545.8	1554.0		0	8	10	14	22	
2	0.0087	769.3	1546.7	1554.0			0	0	7	10	
3	0.0098	768.9	1547.8	1560.8				0	0	3	CIC
4	0.0141	768.4	1548.8	1564.5					0	1	
5	0.0202	767.9	1549.7	1567.9						0	
6	0.0263	767.3	1550.7	1571.5							

The first five columns of the table represent the value of κ, the measure of lack-of-fit to the true model, lack-of-fitto data, the AIC and the BIC. The next six columns contain the triangular matrix of the proportion of disagreement of the estimate in the respective row with respectto the more complex esimates. The values in italics refer to estimates which have worse fit to the data than some simpler estimates, and are therefore excluded from the set of acceptable solutions. AIC and BIC selections (minimum values) are underlined and so are the rows corresponding to CIC selections in the triangular matrix.

Example a) shows a situation in which AIC, and BIC choices (k=1) agree, whereas the CIC rejects this choice. This can be seen from the percentages in the k=1 rowof the triangular matrix that are all above the 5% threshold. The CIC selection is k = 3. This choice is substantially better in terms of LFT; furthermore, a graphical inspection of the CIC solution shows that it is the simplest model that recovers the bimodality of the true distribution.

In example b), AIC and BIC choices differ, the AIC favoring the more complex model and coinciding with the CIC choice. The latterchoice is "best", in the sense that it has the lowest LFT.

The two examples above are typical of what we have found in our simulations: when the true distribution is complex enough to require more knots, the CIC choice does capture its essential features and is preferable to the AIC and BIC solutions. For a choice of τ close to 5%, the CIC choice tends to resemble more the AIC choice.

Example c) shows a somewhat exceptional situation, in which the CIC choice (3 knots) is more complex than needed, whereas AIC and BIC agree in chosing the simpler and better 1-knot estimator. Notice, however, that the 3-knot model is still a reasonable one, as opposed to the 0-knot or higher knot solutions. The relative weakness of the CIC in this case can be possibly explained by the choice of Δ and τ, as well as by characteristics of the sample. We have calculated a meaure of the *exceptionality of the sample within the Weibull family*, as defined in (Abrahamowicz & Ciampi, 1990), and found that it is quite high (this means that though the sample looks like a Weibull sample, due to random variation its mean is much different from the actual mean).

5. Discussion and Conclusions

We have proposed a new algorithm, for model selection, the CIC, which is especially useful in chosing a meta-parameter in non-parametric modelling. Like other approaches, the CIC is based on a compromise between goodness-of-fit and economy of description (Occam's razor). Unlike other approaches, the CIC choice depends on the *pointwise* performance of the model: a simpler model is preferred to a more complex one with better fit to the data, if the predictions of the former agree with those of the latter. The agreement is defined in terms of a confidence interval CI and a tolerance parameter τ: model 1 agrees with model 2 if the proportion of itspredictions that fall out of the CI of model 2 is acceptable according to a tolerance threshold specified by τ.

Some general conclusions can be tentatively drawn on the basis of limited simulations performed so far, in which we have compared the CIC with the AIC and the BIC. As is well known, the BIC is more conservative than the AIC (for moderate or large sample sizes) in the sense that it gives more importance to economy of description and less importance to goodness of fit than the AIC. The conservativeness of the CIC, on the other hand, can be adjusted by controlling τ. As τ tends to zero, CIC choices tend to coincide with AIC choices; on the other hand, as τ increases, CIC choices tend to coincide with BIC choices.

For the small value of τ used here (.05), we have observed that the CIC is "almost always" the best of the three criteria: a) its choice is never too far from the true model; b) when the true model is complex (for example, mixture of distributions in density estimation), the CIC choice is uniformly as good or better than the choices based on the AIC and the BIC;; c) the only situations in which the CIC choice may be weak is when the true model belongs to a simple parametric family *and* the sample on which the estimate is based is exceptional *within the family* (see Abrahamowicz & Ciampi, 1990); in this case, the more conservative choice given by the BIC may be closer to the true model.More simulations are clearly needed in order to support the above conclusions in the context of a broader spectrum of non-parametric function estimation including e.g. hazard estimation (Etezadi-Amoli & Ciampi,1987) and binary regression (Ramsay & Abrahamowicz,1989).

In view of the computational expenses associated to CIC model selection, we recommend its use when AIC and BIC choices disagree and it is important to obtain a pointwise accurate estimate.

References

Abrahamowicz, M.; Ciampi, A; & Ramsay, J.O. (1989). Survival analysis using regression splines. I: cubic M-spline density algorithm for density estimation. McGill University, Mathematics & Statistics Dept.; Report 89-32.

Abrahamowicz, M. & Ciampi,A. (1990). Informatic theoretic criteria in non-parametric density estimation: bias and variance in the infinite dimensional case. (Computational Statistics and Data Analysis - In Press).

Akaike, H. (1974) A new look at the statistical model identification. IEEE Trans. on Automatic Control, AC-**19**, 716-723

Etezadi-Amoli, J. & Ciampi, A. (1987). Extended hazard regression for censored survival data with covariates: a spline approximation for the baseline hazard function. Biometrics, 43: 181-92.

Ramsay, J.O. (1988). Monotone regression splines in action (with discussion). Statistical Science, 3 : 425-61.

Ramsay, J.O. & Abrahamowicz, M. (1989). Binomial regression with monotone splines: a psychometric application. JASA, 84: 916-25.

Schwartz, G. (1978) Estimating the dimension of a model Ann. of Statist. **6**, 461-464

Silverman, B.W. (1985). Some aspects of the spline smoothing approach to nonparametric regression curve fitting (with discussion). J.Roy Stat.Soc., B, 47, 1-50.

Silverman, B.W. (1986). Density Estimation for Statistics and Data Analysis. London: Chapman and Hall.

Stone, M. (1977)An asymptotic equivalence of choice of model by cross-validation and Akaike's criterion. J. Royal Statist. Soc. **B39, 44-47**

Notes About New Developments in Statistical Software

Statistical Models in S

J. Chambers, T. Hastie and D. Pregibon, Murray Hill

1. Introduction.

The interactive data analysis and graphics language S (Becker, Chambers and Wilks, 1988) has become a popular environment for both data analysts and research statisticians. A common complaint, however, has concerned the lack of statistical modeling tools, such as those provided by GLIM© or GENSTAT©.

We and others are developing an integrated approach to software for statistical models, to be included in a future version of S. As a research project in statistical computing, our work emphasizes structures to represent the model specification and the underlying data, and an organization of the computations that utilizes an object-oriented approach (specifically, methods that allow users to treat different classes of models uniformally and simply). The methodological coverage includes linear models, analysis of variance, generalized linear models, additive models, smooth curves and surfaces, tree-based models, and general nonlinear models. The computational organization allows a pleasing degree of uniformity among the kinds of models. The general environment provided by S permits emphasis on interactive computing, on graphics, and on the easy extensibility of the software by users. Because of time limitations, the present paper concentrates on a subset of the methodological coverage; specifically, on functions for fitting linear models (`lm`), generalized linear models (`glm`) and generalized additive models (`gam`). We also concentrate on the computational issues involved in representing model and in the use of data frames to represent the data.

2. The model formula.

A call to the linear model function has the form `fit1 <- lm(formula)`. The model `formula` is an important feature of the software, and allows the analyst to specify a variety of models in a natural form. An example of a formula is:

```
chd ~ age + sex*race + log(cholesterol) + poly(sbp,4)
```

and reads `chd` is modelled by a linear term in `age` plus a crossed term in `sex` and `race`, and so on. In this instance the call would be

```
fit1 <- lm(chd ~ age + sex*race + log(cholesterol) + poly(sbp,4))
```

The formula is parsed, and results in a linear model as follows:

(i) `chd` is identified as the response

Compstat 1990

(ii) **age** and **log(cholesterol)** are fitted linearly.

(iii) **sex** and **race** are categorical and the term **sex*race** is broken down into the terms **sex + race + sex:race** where **sex** and **race** are main-effects and **sex:race** denotes their interaction. These terms are parametrized using appropriate contrasts (controllable by the user) to represent the main effect and interactions.

(iv) poly(sbp,4) represents a fourth degree polynomial.

As in the GLIM language, **sex/age** means **sex** plus **age** nested within **sex**. Parentheses can be used to group terms, as in a*(b+c) which is equivalent to a*b + a*c. Processing of a formula thus defines expressions for the terms that will be used in fitting a model. These expressions will be evaluated in the context of the relevant data frame (see the next section. The terms can be any of the following:

(i) Numeric vectors or matrices — these are simply placed in the model matrix.

(ii) S categories (categorical variables or factors), optionally ordered. Character and logical vectors are converted to categories. For computations in linear-type models, categories generate a matrix of contrasts. The choice of contrasts can be controlled by the user by giving the category a contrast attribute.

(iii) Calls to S functions, of arbitrary complexity, that evaluate to any of the objects in (i) or (ii).

For example, the function poly in the formula above takes the numeric vector sbp and returns a matrix of orthogonal polynomials. The bases representing interactions are simply formed by taking the tensor-product of the columns of the bases representing the component objects. Other useful functions in a similar vein are

- bs(var, df, knots, degree=3) — which evaluates to a B-spline basis for representing polynomial splines. The default degree is 3, and the knots can be explicitly supplied via the knots argument, or else implicitly by specifying the number of *degrees of freedom* using the df argument. In the latter case, for a cubic spline, df-3 knots are placed at suitable quintiles of var.

- cut(var, breaks) — this is a standard S function which creates a categorical variable or category by cutting var at the given vector of breakpoints.

- lag(var, k) — for creating lagged versions of var, which may be the response as well.

- Any transformations of arbitrary complexity, such as sqrt(var-min(var)) or var>30.

- The identity function allows operators that define crossed and nested terms to be used in their ordinary sense, as in I(1/(1+var)). Without the protection of the identity function, the "/" would be interpreted as the nesting operator.

The only additional constraint on the value of the expressions for terms is that they evaluate to objects on the same set of observations (the same "rows").

Formulae such as these can be interpreted by the `lm`, `aov`, `glm` and `gam` functions, and, with some variation in meaning, by the functions that fit smooth surfaces and tree-based models. Formulas for general nonlinear models are obviously different, in that the implicit shorthand that omits the coefficients in the above is no longer available. Even here, the user's view is in many respects similar, however.

3. Data frames.

We have introduced into S a class of objects called `data.frames`, which can be used if convenient to organize all the variables relevant to a particular analysis using the model-fitting functions. A data frame behaves like a matrix in that each "column" has the same number of rows, but the columns can be numeric vectors, categorical vectors or matrices. The user should be able to think of a data frame as a matrix whenever convenient, but also as a list or frame, containing variables of the sort discussed in the previous section. As a matrix, it can for example be subset by row; this is useful for deleting particular observations, or for resampling methods that resample entire observations. On the other hand, as a list or frame, its variables can be referred to by name. So in our example above, suppose `chd.data` is a data frame with variables named `chd`, `age`, `sex`, `race`, `cholesterol` and `sbp`. The call

```
lm(chd ~ age + sex*race + log(cholesterol) + poly(sbp,4), data=chd.data)
```

would build the model matrix and the response by referencing the named variables in `chd.data`. Alternatively, an extension in the version of S we are using allows the data frame to be attached to the search list for S objects, in which case no second argument to `lm` is needed.

4. GLMs and families.

The set of GLM functions are designed to handle all the exponential family and quasi-likelihood linear models as described in McCullagh and Nelder (1983). Let's focus on the logistic regression model for binary response data. The model has the form

$$\log\left(\frac{P(\boldsymbol{x})}{1-P(\boldsymbol{x})}\right) = \boldsymbol{\beta}^T\boldsymbol{x}$$

where $P(\boldsymbol{x}) = E(y \mid \boldsymbol{x})$ is the conditional probability of y being a one give $\boldsymbol{x}$. The *logit* transformation is called the *link* function and relates the regression function $E(y \mid \boldsymbol{x})$ to the *linear predictor* $\boldsymbol{\beta}^T\boldsymbol{x}$. Using the binomial error model, we estimate the parameters by maximum likelihood. A typical call to `glm` has the form

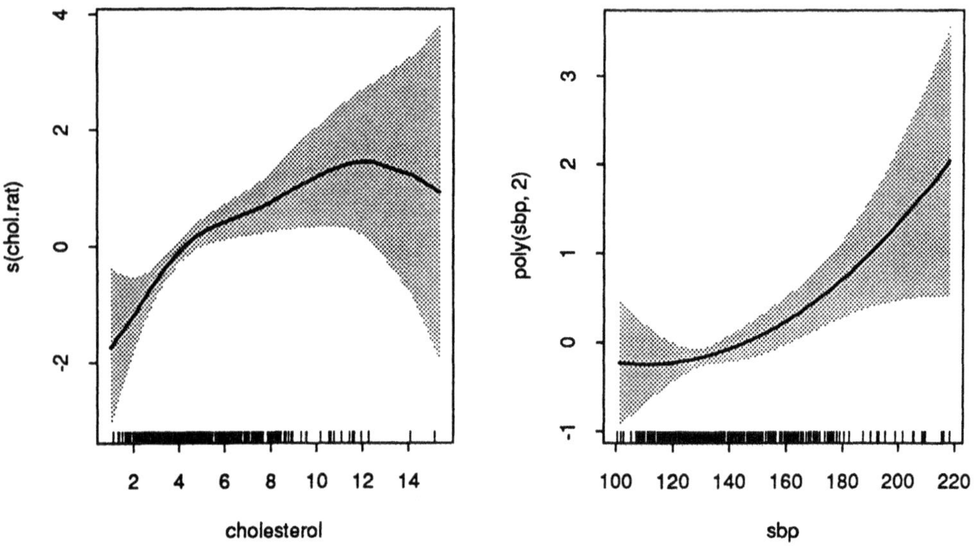

Figure 1. A plot of two of the terms in the fitted generalized additive model of section 5. The spikes at the base of the plots indicate the occurrence of data points. The shaded regions represent an approximate pointwise 2× standard-error band about the fitted curves. The plotting procedure identifies automatically the two "x" variables for plotting, in this case `cholesterol` *and* `sbp`*, while the "y" variables are the fitted functions themselves.*

```
glm(formula, family=binomial, ...)
```

for example, where `binomial` is a function that generates a *family object* for the logistic regression model with binomial errors. A family object is itself a list of functions such as a link function, its inverse and derivative, a variance function, deviance and so on. The function `binomial` is just a convenient way of creating this particular family object. By giving it arguments as in `...family=binomial(link=probit),...` alternative link functions and/or variance functions can be specified. Other family functions are `gaussian`, `gamma`, `poisson`, `inverse.gaussian` and the fully general `quasi` family for arbitrary link and variance functions. In additon, a user can provide a family object constructed by any means, perhaps with the aid of `quasi` and a library of standard `link` and `variance` objects.

5. Generalized additive models.

The `gam` function fits the generalized additive model as described in Hastie and Tibshirani (1990). These models allow nonparametric terms to be combined via an additive predictor. The formula language accommodates these extensions; for example, the call

```
gam(chd ~ age + sex*race + s(cholesterol) + poly(sbp,2), family=binomial)
```

indicates that `cholesterol` is to be fit nonparametrically using a smoothing spline, with the default 5 *effective degrees of freedom*, while `sbp` is to be fit using a quadratic polynomial. Any number of terms can be specified as a "smooth" term. The functions are estimated in an iterative fashion, and minimize a penalized likelihood criterion appropriate for these generalized models. Flexible plotting functions are designed to display the composite terms comprising the additive predictor. Figure 1 shows a plot of the `cholesterol` and `sbp` curves for the above fit, together with pointwise approximate 2× standard-error curves.

References

Becker, R., Chambers, J.M. and Wilks. A. (1988), *The New S language*, Wadsworth, Belmont CA.

McCullagh, P. and Nelder, J. (1983), *Generalized Linear Models*, Chapman and Hall, London.

Hastie, T. and Tibshirani, R. (1990), *Generalized Additive Models*, (in press) Chapman and Hall, London.

GLIM4 - Developments in Model Fitting

B. Francis, M. Green and M. Bradley, Lancaster

Keywords: GLIM, Generalised Linear Models, Statistical Software, Givens algorithm

Introduction

GLIM™ was first developed with the simple aim of providing a software tool that would allow the fitting of Generalised Linear Models, with a small number of utilities necessary for practical data analysis. Today, using GLIM377 (Payne *et al, 1985)* the variety of models that GLIM has been used to fit (Aitkin *et al, 1989)* is far wider than anyone could have imagined in the early days of release 1. For the new release, the temptation to extend the fitting procedure of GLIM to include models well outside those of GLMs has been resisted. Rather, it is the intention of the current enhancement to incorporate some of the facilities that will extend the range of GLMs and their extensions which can be fitted; most effort has been concentrated on extending the model formula syntax, to allow the user greater control and access to system structures, to improve the operation and readability of GLIM macros and to provide simple high-quality graphics to allow graphical techniques for model checking and display to be used.

The Language

The opportunity has been taken to rewrite much of the original GLIM code dealing with data storage and housekeeping. The user will see litle difference in the syntax of GLIM, but the package although still written in FORTRAN 77, is now far more modular, with consequent advantages for development and implementation work. GLIM now allows LIST and NUMBER structure types and associated LIST and NUMBER directives. The length of structure names has been increased to a maximum of eight characters, removing an often complained about restriction.

Model Fitting

(a) Numerical Algorithm: In addition to the Gauss-Jordan algorithm of GLIM377, the user will have the choice of an alternative - the Givens algorithm (Clarke, 1980, 1981), which gives greater numerical stability for ill-conditioned problems. Certain limits, such asthat for the number of terms in a model, have been removed.

GLIM is a trademark of the Royal Statistical Society

(b) Model Formulae: Some minor extensions to the ERROR and LINK directives have been made, for example the inclusion of the Inverse Gaussian distribution. A major change is the extension of the syntax of the model formula to include terms of the following form:

(i) **F<integer>**	orthogonal polynomials of **F** of order *integer.*
(ii) **F<M>**	contrasts defined by **M** of factor **F**.
(iii) **<M>**	design matrix.
(iv) **X1.X2**	cross product "interactions"
(v) **(X1+X2+F)**integer**	exponentiation giving easy specification of n-way interactions

where **X1** and **X2** are variates, **F** is a factor and **M** a matrix.

(c) Indexing of model terms The interpretation of model term indexing is as used in the CALCULATE directive. For example if a model contains the term **X** which has been defined to be indexed by the vector **I**, then the value for the kth unit will be the I_kth value of **X**. This allows model vectors to be stored in a compresed form, often giving substantial saving in the storage of the design matrix.

(d) Access to System Structures: The use of GLIM to fit non-standard models often requires access and manipulation of the system structures beyond that provided routinely in GLIM377. GLIM4 will provide:

(i) Extraction of rows of the design matrix defined by a model formula.

(ii) extraction of subsets of the parameter estimates defined by a term of the model formula.

(iii) greater access to internal model quantities. For example the user now has access to the standardised residuals, design matrix, leverage values, deviance increments and SSP matrix.

(e) Elimination of nuisance model terms

This facility allows the user to specify a model term which will be eliminated from subsequent fits.The term is implicitly still part of the model formula, but its parameters are not explicitly estimated. This results in savings in both time and the size of the working triangle. This facility is useful in many standard GLIM applications such as the multinomial, matched case-control analyses and survival analysis.

(f) OWN models: In GLIM377 the definition of an OWN model required four macros; two dependent on the "error" distribution and two on the link. In GLIM4 these will be defined separately so that an OWN link can be defined with a standard ERROR specification or an OWN error with a standard LINK. The OWN model facility has also been extended; new macros may be specified which may be called

(i) before the first iteration

(ii) after the formation of the model triangle (the SSP matrix in the Gauss-Jordan algorithm) but before the inversion.

(iii) at the end of each iteration

Any structure available to the user can be modified during the iterative fit.

User Facilities

Many user facilities have been enhanced to allow the package to be used more effectively as a statistical programming language. File handling has been improved, with the ability to open named files. All calculation is now carried out in double precision (but stored in named structures in single precision). A large number of new functions have been introduced. Some are statistical, such as functions for the chi-squared, F and t probabilities - others are mathematical, such as the gamma and beta functions and others are internal enquiry fuctions which return for example the length, type and number of levels of a variate.

The sorting facility in GLIM has been generalised, allowing sorting of an arbitrary number of variates by an arbitrary number of key variates into a set of destination variates. The direction of sort (ascending or descending) may also be specified. The tabulation directive has also been improved, with better labelling of the dimensions and greater user control over table printing.

The HISTOGRAM and PLOT directives have both been revised. The PLOT directive now allows multiple plots with weights and user determined scale points, together with titling and x and y axis labels.

Where the operating system allows, an enhanced PAUSE directive allows either single operating system commands to be issued or a "shell-out" to the operating system from within GLIM. An system dependent initialisation file which is read automatically at the start of a GLIM session will allow those using GLIM in a teaching context to set up datasets and a suitable environment for their students.

Finally, a summary of the user manual is available on-line through the MANUAL directive.

Macros and macro editing

We have aimed to make macros more readable, easier to develop and easier to debug and to change when necessary. We have also ensured compatibility with the previous release of GLIM, and with minor exceptions, all macros written in GLIM377 should work in GLIM4.

Keyword arguments provide names to formal arguments within the contents of a macro, and provide an alternative to the usual method of referring to arguments (%1,%2 etc.) They are not mandatory, and formal arguments and keyword arguments may be mixed.

Local structures provide a way to define structure names which have context only within the body of the macro.

The macro editor is a simple line editor which may be used not only to edit existing macros but also for the creation of new macros. Line numbers are displayed at the beginning of each line. Any keyword arguments and local structures are located at the beginning of the macro, where the user has the ability to redefine, delete or create new keyword arguments and local structures.

A simple macro debugger is now provided, giving users the ability to stop execution of the macro after every directive symbol, and to print and if necessary change quantities.

Another useful facility is user-prompting for values: the ability to prompt the user from within the macro for a scalar, macro contents or macro argument.

High-quality graphics

The graphics facilities available in GLIM4 through the GRAPH and GSTYLE directives are a substantial enhancement to the trial facilities available to some users in GLIM377 and described by Reese(1989). Our philosophy has been to provide a set of simple graphics commands which can be used to investigate models and to aid in the display of diagnostics and results, rather than aim for presentation graphics. Lines are defined by vector pairs and may have a symbol type, a line style and a colour. The syntax is similar to the PLOT directive, and allows for multiple vector-pairs, labelling, user-defined scale points and titling. Optional weights for each vector-pair allow points to be weighted out , and also give scaling for symbols if required. The LAYOUT directive provides additional control over the positioning of the graph, and additionally allows multiple graphs on a page.

A standard range of device drivers are provided to all mainframe sites by default; these include Tektronix, HPGL and Postscript.

References

Aitkin, M.A., Anderson, D.A., Francis, B.J. and Hinde J.P. (1989). *Statistical Modelling in GLIM.* Oxford University Press, Oxford.

Baker, R., Clarke, M., Francis, B., Green, M., Nelder, J., Payne, C. and Webb, J. (1985). *GLIM3.77 User Guide.* NAG, Oxford.

Clarke, M.R.B. (1980) *Choice of algorithm for a model-fiting system.* COMPSTAT 80. Proceedings in Computational Statistics. Vienna, Physica-Verlag

Clarke, M.R.B. (1981) *Algorithm AS 163. A Givens algorithm for moving from one linear model to another without going back to the data.* Applied .Statistics., 30, 2, 198-203

Reese, R.A. (1989) *The optional Graphics Command Extension to GLIM377.* In *Statistical Modelling- Proceedings of GLIM89 and the 4th International Workshop on Statistical Modelling,* Lecture Notes in Statistics, 57, Springer-Verlag.

Address List of Authors

dr
ABRAHAMOWICZ MICHAL
THE MONTREAL GENERAL HOSP,DEPT
OF MED,DIV.OF CLINICAL EPIDEM.
1650 CEDAR AVE.
H3G 1A4 MONTREAL, QUEBEC
CANADA

dr
ARTHANARI T.S
SQC & OR UNIT
INDIAN STATISTICAL INSTITUTE
143 NUNGAMBAKKAM HIGH ROAD
600034 MADRAS
INDIA

ph.dr
ATILGAN TASKIN
AT & T BELL LABORATORIES
ROOM 30J-014
600 MOUNTAIN AVE. MURRAY HILL
07974-2070 NEW JERSEY
USA

dr
BACZKOWSKI ANDREW JOSEPH
DEPARTMENT OF STATISTICS,
UNIVERSITY OF LEEDS
LS2 9JT LEEDS
UNITED KINGDOM

BATAGELJ VLADIMIR
VTOZD MATEMATIKA IN MEHANIKA
UNIVERZA V LJUBLJANI
JADRANSKA 19
61000 LJUBLJANA
YU

BOLLA MARIANNA
MATHEMATICAL INST. HUNGARIAN
ACADEMY OF SCIENCES
P.O.BOX 127
1364 BUDAPEST
HUNGARY

dr
BOOMSMA ANNE
VAKGROEP STATIST.& MEETTHEORIE
RYKSUNIVERSITEIT GRONINGEN
OUDE BOTERINGE- STRAAT 23
9712 GC GRONINGEN
THE NETHERLANDS

BRADLEY MALCOLM
CENTRE FOR APPLIED STATISTICS,
LANCASTER UNIVERSITY
LA1 4YF LANCASTER
UK

CASTANO-TOSTADO EDUARDO
DEPARTMENT OF STATISTICS,
OKAYAMA UNIVERSITY
TSUSHIMA-NAKA
700 OKAYAMA
JAPAN

CATARCI TIZIANA
DIP.DI INFORMATICA E SISTEMIST
ICA,UNIV.DI ROMA "LA SAPIENZA"
VIA BUONARROTI 12
00185 ROMA
ITALY

CAUSSINUS HENRI
LAB.DE STATIST.ET PROBABILITES
U.R.A.-C.N.R.S. D 745
UNIV.PAUL SABATIER 118 ROUTE D.NARBONNE
31026 TOULOUSE CEDEX
FRANCE

CELEUX GILLES
I.N.R.I.A.
ROCQUENCOURT P.O.BOX 105
78153 LE CHESNAY CEDEX
FRANCE

dr
CHAMBERS JOHN
AT AND T BELL LABORATORIES
ROOM MH 2C-282
600 MOUNTAIN AVENUE MURRAY HILL
07974 NEW JERSEY
USA

CHURCH KENNETH W.
AT & T BELL LABORATORIES
600 MOUNTAIN AVE MURRAY HILL
NJ 07079 NEW JERSEY
USA

CIAMPI ANTONIO
MCGILL UNIV, MONTREAL CHIL-
DREN'S HOSPITAL, RESEARCH INST
2300 TUPPER ST.
MONTREAL
QUEBEC H3H 1P3
CANADA

CLARK LINDA
AT & T BELL LABORATORIES
ROOM 2C-264
600 MOUNTAIN AVE MURRAY HILL
07974 NEW JERSEY
USA

CLEROUX R.
DEPARTMENT D'I.R.O.,
UNIVERSITE DE MONTREAL
C.P. 6128
H3C 3J7 MONTREAL, QUEBEC
CANADA

COTTEE MICHAELA
FACULTY OF MATHEMATICS
THE OPEN UNIVERSITY
WALTON HALL MILTON KEYNES
MK7 6AA BUCKINGHAMSHIRE
UK

D'ANGIOLINI G.
ISTITUTO NAZIONALE DI STATIS-
TICA (ISTAT)
VIA C. BALBO 16
00185 ROMA
ITALY

DANICIC SEBASTIAN
PNL - DEPT OF COMPUTING, MATHE
MATICS AND STATISTICS
EDEN GROVE
N7 8DB LONDON
UK

dr
DORDA WOLFGANG
DEPT.OF MEDICAL COMPUTER
SCIENCE, UNIV.OF VIENNA
GARNISONGASSE 13 (AKH, 8.HOF)
1090 VIENNA
AUSTRIA

DURAND J.F
ENSAM-INRA-USTL
UNITE DE BIOMETRIE
9 PLACE PIERRE VIALA
34060 MONTPELLIER
FRANCE

dr
EKBLOM HAKAN
DEPT. OF MATHEMATICS
LULEA UNIVERSITY
95187 LULEA
SWEDEN

FERLIGOJ ANUSKA
FSPN, UNIVERSITY OF LJUBLJANA
KARDELJEVA PLOSCAD 5
61000 LJUBLJANA
YU

FRANCIS BRIAN
CENTRE FOR APPLIED STATISTICS,
LANCASTER UNIVERSITY
LA1 4YF LANCASTER
UK

FROESCHL K.A.
DEPT.OF STATISTICS & COMPUTER
SCIENCE, UNIVERSITY OF VIENNA
UNIVERSITAETSSTR. 5
1010 VIENNA
AUSTRIA

prof
GALE WILLIAM
AT & T BELL LABORATORIES,
ROOM 2C 278
600 MOUNTAIN AVE MURRAY HILL
NJ 07974 NEW JERSEY
USA

dr
GEBHARDT FRIEDRICH
GESS.FUER MATHEMATIK UND
DATENVERARBEITUNG MBH
SCHLOSS BIRLINGHOVEN P.O.BOX 1240
5205 SANKT AUGUSTIN 1
FRG

GILCHRIST ROBERT
PNL - DEPT.OF COMPUTING, MATHE
MATICS AND STATISTICS
EDEN GROVE
N7 8DB LONDON
UK

prof.
GOMES MARIA IVETTE
DEPT. OF STATISTICS - DEIOC
FACULTY OF SCIENCE OF LISBON
BLC C/2,CAMPO GRANDE CIDADE UNIVERSITARIA
1700 LISBOA
PORTUGAL

dr
GONZALES PIERRE-LOUIS
INST D'INFORMATIQUE D'ENTERPR.
CONS.NAT.DES ARTS ET METIERS
LES PASSAGES 18, ALLEE J.ROSTAND
91002 EVRY CEDEX
FRANCE

prof.
GOVAERT GERARD
I.N.R.I.A. AND METZ UNIVERSITY
LORRAINE BP 239
54506 VANDAUVRE
FRANCE

prof.
GOWER JOHN
AFRC INST OF ARABLE CORPS RES.
ROTHAMSTED EXPERIMENTAL STAT.
HARPENDEN
AL5 2JQ HERTS
UK

dr
GREEN MICHAEL
CENTRE FOR APPLIED STATISTICS,
LANCASTER UNIVERSITY
LA1 4YF LANCASTER
UK

prof.
GROSSMANN WILFRIED
DEPT.OF STATISTICS & COMPUTER
SCIENCE, UNIV.OF VIENNA
UNIVERSITAETSSTR. 5
1010 VIENNA
AUSTRIA

HARMAN MARC
PNL - DEPT OF COMPUTING, MATHE
MATICS AND STATISTICS
EDEN GROVE
N7 8DB LONDON
UK

DR
HASTIE TREVOR
AT & T BELL LABORATORIES
600 MOUNTAIN AVE MURRAY HILL
07974 NEW YERSEY
USA

dr
HAVRANEK TOMAS
SVT CSAV
POD VODARENSKOU VEŽI 2
18207 PRAHA 2
CZECHOSLOVAKIA

HEBRAIL GEORGES
ELECTRICITE DE FRANCE,
RESEARCH CENTER-DER, G1/011
1, AV.DU GAL DE GAULLE
92141 CLAMART CEDEX
FRANCE

dr
KARNEL GERALD
INSTITUT FUER STATISTIK,
TECHNISCHE UNIVERSITAET WIEN
WIEDNER H.STR. 8-10
1040 WIEN
AUSTRIA

KARPENKO VLADIMIR A.
DEPARTMENT FO CALCULATING,
MATHEMATICS AN CYBERNETICS
MOSCOW STATE UNIV. LENINSKIE GORY
117234 MOSCOW
USSR

dr
KIERS HENK
UNIVERSITY OF GRONINGEN,
DEPT. OF PSYCHOLOGY
GROTE MARKT 31/32
9712 HV GRONINGEN
THE NETHERLANDS

KUZMENKOV VJACHESLAV
MOSCOW STATE TECHNICAL UNIV.
DEPT.OF APPLIED MATHEMATICS
2-AYA BAUMANSKAJA 5
107005 MOSCOW
USSR

prof.
LAURO CARLO NATALE
UNIV.DI NAPOLI, DIPARTIMENTO
DI MATEMATICA E STATISTICA
VIA PARTENOPE 38
80121 NAPOLI
ITALY

LEE JANG TAEK
DEPT.OF COMP. SC. & STAT.,
DANKOOK UNIVERSITY
8 HANNAM-DONG YONGSAN-KU
140-714 SEOUL
KOREA

prof.
LEE KANG SUP
DEPT.OF COMP. SC. AND STAT.,
DANKOOK UNIVERSITY
8 HANNAM-DONG YONGSAN-KU
140-714 SEOUL
KOREA

LENZERINI MAURIZIO
DIP.DI INFORMATICA E SISTEMIST
ICA,UNIV.DI ROMA "LA SAPIENZA"
VIA BUONAROTTI 12
00185 ROMA
ITALY

LESSI O.
DIPARTIMENTO DI MATEMATICA
UNIVERSITA DELLA BASILICATA
POTENZA
ITALY

LIN QIAN
MCGILL UNIV, MONTREAL CHIL-
DREN'S HOSPITAL, RESEARCH INST
2300 TUPPER ST
MONTREAL
QUEBEC H3H 1P3
CANADA

MADSEN KAJ
INST. FOR NUMERICAL ANALYSIS
DTH
2800 LYNGBY
DENMARK

MALVESTUTO FRANCESCO
STUDI-DEPT, E.N.E.A.
V.REG.MARGHERITA 125
00198 ROMA
ITALY

ass.prof.
MEHTA CYRUS R.
DANA FARBER CANCER INSTITUTE
DEPT. OF BIOSTATISTICS
44 BINNEY STREET
BOSTON
MA 02215
USA

MOOIJAART AB
DEPT.OF PSYCHOMETRICS AND
RESEARCH METHODS, UNIV. LEIDEN
WASSENAARSEWEG 52
2300 RB LEIDEN
THE NETHERLANDS

prof.dr
NEUMANN KLAUS
STATE CENTRAL STATISTICAL OFF.
HANS-BEIMLER STRASSE 70/72
1026 BERLIN
GDR

NEYKOV NEYKO
LAB.OF INFORMATICS,
HYDROMETEOROLOGICAL SERVICE
BLVD LENIN 66
1184 SOFIA
BULGARIA

NEYTCHEV P.N.
LAB.OF INFORMATICS,
HYDROMETEOROLOGICAL SERVICE
BLVD LENIN 66
1184 SOFIA
BULGARIA

ODAKA YOSHIMASA
INFORMATION PROCESSING CENTER,
OKAYAMA UNIVERSITY OF SCIENCE
RIDAICHO
700 OKAYAMA
JAPAN

dr
OLIVI LUCIANO
COMMISION OF EUROPEAN COMMUN.
JOINT RESEARCH CENTER
21020 ISPRA
ITALY

PARISI D.
CEC-JOINT RESEARCH CENTRE

21020 ISPRA
ITALY

PATEL NITIN
DANA FARBER CANCER INSTITUTE
DEPT. OF BIOSTATISTICS
44 BINNEY STREET
BOSTON
MA 02215
USA

PAYNE ROGER W.
AFRC INST OF ARABLE CROPS RES.
ROTHAMSTED EXPERIMENTAL STAT.
HARPENDEN
AL5 2JQ HERTS
UK

PLISKIN S.J.
THE CHAIR OF HIGH MATHEMATICS,
MOSCOW STATE TECHNICAL UNIV.
2-D BAUMANSKAYA 5
107005 MOSCOW
USSR

dr
PREGIBON DARYL
AT&T BELL LABORATORIES
ROOM MH 2C-264
600 MOUNTAIN AVE MURRAY HILL
07974 NEW JERSEY
USA

RIOUX B.
DEPARTMENT D'I.R.O.
UNIVERSITE DE MONTREAL
C.P. 6128
H3C 3J7 MONTREAL, QUEBEC
CANADA

dr
ROMANAZZI MARIO
STATISTICAL LABORATORY,
UNIV.OF VENICE "CA'FOSCARI"
3246 DORSODURO
30123 VENICE
ITALY

ROSS GAVIN
STATISTICS DEPARTMENT, IACR
ROTHAMPSTED EXPERIMENTAL STAT.
HARPENDEN
AL5 2JQ HERTS
UK

prof.dr
ROUSSEEUW PETER
U.I.A.
VESALIUSLAAN 24
2520 EDEGEM
BELGIUM

dr
RUDAS TAMAS
EOTVOS UNIV.& SOCIAL SCIENCE
INFORMATICS CENTER -TARKI
FRANKEL L.U. 11
1027 BUDAPEST
HUNGARY

RUIZ ANNE
LAB.DE STATIST.ET PROBABILITES
U.R.A. C.N.R.S. D 745
UNIV.PAUL SABATIER 118 ROUTE D.NARBONNE
31026 TOULOUSE CEDEX
FRANCE

SCOTT WILLIAM ANDREW
CENTRE FOR APPLIED STATISTICS
LANCASTER UNIVERSITY
LA1 4YF LANCASTER
UK

SENCHAUDHURI PRALAY
DANA FARBER CANCER INSTITUTE,
DEPT OF BIOSTATISTICS
44 BINNEY STREET
BOSTON
MA 02215
USA

dr
SICILIANO ROBERTA
DEPT.OF MATHEMATICS AND STATIS
TICS, UNIVERSITY OF NAPLES
VIA PARTENOPE 36
80121 NAPOLI
ITALY

SUCHARD MARIE
ELECTRICITE DE FRANCE,
RESEARCH CENTER - DER
1, AV.DU GAL DE GAULLE
92141 CLAMART - CEDEX
FRANCE

prof
TANAKA YUTAKA
DEPT. OF STATISTICS
OKAYAMA UNIVERSITY
2-1-1 TSUSHIMA-NAKA
700 OKAYAMA
JAPAN

TEN BERGE JOS M.F
UNIVERSITY·OF GRONINGEN,
DEPT OF PSYCHOLOGY
GROTE MARKT 31/32
9712 HV GRONINGEN
THE NETHERLANDS

TERSKIN OLEG
MOSCOW STATE TECHNICAL UNIV.
DEPT. OF APPLIED MATHEMATICS
2-AYA BAUMANSKAJA 5
107005 MOSCOW
USSR

TODOROV VALENTIN V.
LAB. OF INFORMATICS,
HYDROMETEOROLOGICAL SERVICE
BLVD LENIN 66
1184 SOFIA
BULGARIA

drs.
VAN DEN BERG GERDA
DEPT.OF BEHAVIORAL COMPUTER
SCIENCE, LEIDEN UNIVERSITY
PO BOX 9555
2300 RB LEIDEN
THE NETHERLANDS

VAN ZOMEREN BERT C.
DELFT UNIVERSITY OF
TECHNOLOGY
DELFT
THE NETHERLANDS

dr
VILA JEAN-PIERRE
I.N.R.A.
LABORATOIRE DE BIOMETRIE
ROUTE DE SAINT-CYR
78026 VERSAILLES
FRANCE

dr
VISSER RON
DEPT.OF BEHAVIORAL COMPUTER
SCIENCE, LEIDEN UNIVERSITY
PO BOX 9555
2300 RB LEIDEN
THE NETHERLANDS

WELHAM S.J.
AFRC INST OF ARABLE CROPS RES.
ROTHAMSTED EXPERIMENTAL STAT.
HARPENDEN
AL5 2JQ HERTS
UK

dr
WHITTAKER JOSEPH
DEPT. OF MATHEMATICS
LANCASTER UNIVERSITY
LA1 4YF LANCASTER
UK

prof.dr
WITTKOWSKI KNUT
EBERHARD KARL UNIVERSITY
DEPT. OF MED. BIOMETRY
WESTBANHOFSTR. 55
7400 TUEBINGEN
FRG

dipl.stat.
WUERLAENDER RAINER
DEPARTMENT OF STATISTICS
UNIVERSITY OF MUNICH
GESCHWISTER-SCHOLL PLATZ 1
8000 MUNICH 22
FRG

YOUSIF GAFAR
MCGILL UNIV, MONTREAL CHIL-
DREN'S HOSPITAL RESEARCH INST
2300 TUPPER ST
MONTREAL
QUEBEC H3H 1P3
CANADA

prof
ZHURBENKO IGOR GEORGE
STATISTICAL LAB. OF MOSCOW UN.
MATHEMATICAL DEPARTMENT
119899 MOSCOW
USSR

GPSR Compliance

The European Union's (EU) General Product Safety Regulation (GPSR) is a set of rules that requires consumer products to be safe and our obligations to ensure this.

If you have any concerns about our products, you can contact us on ProductSafety@springernature.com

In case Publisher is established outside the EU, the EU authorized representative is:

Springer Nature Customer Service Center GmbH
Europaplatz 3
69115 Heidelberg, Germany

Batch number: 10726298

Printed by Printforce, the Netherlands